Molecular Diseases 5

Prof. Dr. sami A. AL-Mudhaffar

Hormones in Diseases

Chapter one

Hormones in Diseases

Contents

- Part A Hormones
- Part B (hCG) in Gynecologic Cancers
- Part (c) Human Chorionic Gonadotropin in Breast Tumors

Part A

Hormones

First: the general features of the hormones

Permanence of an individual's life require coordination of transmission of information from the outside of the individual to the inside or among organs of the same individual. The control systems in the body of the animal and human being represented by endocrinology and nervous system, and the growth processes in the animal are under the endocrine glands beside that the changes are largely under the control of the glands despite the fact that hormonal control are slow in their work, compared with the rapid neurological control.

The endocrine system control many functions such as growth, reproduction, metabolism as well as regulation of the b internal temperature and others, such as secretion an chemicals from endocrine cells in small quantities up to pico grams, that called hormones and accomplish specialized task in other cells "target cells".

Hormones being known as chemical messengers, that perform important roles in the control of the vital actions that occur in living organisms and excreted in mammals by endocrine glands to move to various parts of the body by the blood. The hormones are divided in mammals to three groups, the first group contains peptide hormones that include the hypothalamus and pituitary as well as insulin and glucagon. The second group is called steroidal hormones, which includes hormones of the adrenal cortex and the sex glands, such as androgens and estrogens and the third group include amino derivative hormones such as adrenaline and hormones of the thyroid hormones.

The hormones are synthesized in endocrine gland, which has no channels "Ductless". It can be defined as chemicals that appear as a result of its composition in the endocrine gland of the human body excreted directly into the blood and transported to other organs of the body by the blood streams directly affecting the effectiveness of the body's life. This definition could be applied to non-hormonal substances such as carbon dioxide, which can be transferred to various organs of the body during breathing and neurological process such as acetyl choline ,synthesized by the nerve tissue and transported to other locations in the body.

It is possible to have a precise definition of the hormone to be represented (the chemicals that make up a very small quantities, which is organizing various biochemical reactions to configure the appropriate quantities of various substances in accordance with the body's need). Accordingly the following are characteristics of hormones, several hormones control

reproduction, which is a complex physiological phenomena, including:
- Steroidal sex hormones "Sex Gonads" which is produced by the gonads.
- Hormones that stimulate the Follicles such as Follicle stimulating hormone (FSH)" that control the biosynthesis steroidal hormones as well as sex cells.
- Stimulating hormone of yellow body "Lutenizing hormone (LH)" and this hormone also controls the
steroidal hormones ,sexual cells and ovulation.

- The "Prolactin"

The pituitary gland produce (LH, FSH, and prolactin). There are other hormones that regulate hormones of the anterior lobe of the pituitary gland which is called Gn- RH and hormone secretion inhibitor for Prolactin wich is secreted (PIH, PRH) by the hypothalamus region .

- The Oxytocin" hormone is excreted from the posterior lobe of the pituitary gland and plays an important role in various reproductive processes, such as child birth and lactation.
Mechanisms of actions of hormones
There are different mechanisms of hormones , illustrated according to the following :-

- The peptide hormones, for example, are in conjunction with receptors located in particular, in the cell membrane and that the number of receptors may range between 2000-100000 per cell or more in some cells.

- Some steroidal and thyroid hormones are connected with the receptors located in the cytoplasm and nucleus, generating certain types of mRNA. It seems that the receptors of the hormones are for each particular hormone and the chemical structures vary according to hormone.

-The hormones Influence a lot of enzymes directly or indirectly through their impact on the cell membrane and the best example to illustrate this is the insulin. The hormones are characterized by general specifications through its influence on a number of organs and "target tissues", and are affected by, for example, the "Adrenal cortex", "Thyroid gland" "Gonads" hormone stimulating the adrenal cortex "Adrenocorticotropin".

The movement of hormones in blood leads to the specific organs to do their functions and the hormones are united with proteins found in plasma to accelerates its influence in the tissues which they operate.

Hormonal Therapy

Recent studies have revealed that hormones, hormone agonists, hormone antagonists provide beneficial and sometime dramatic results in cancer treatment and chemoprevention. In many cases they provide similar therapeutic results and can replace the conventional cancer therapies (chemotherapy, surgical therapy and radiation therapy) thus becoming new adjuvant systemic therapy for cancer .

Progestational agents and estrogen antagonists such as tamoxifen have been used either single or in combination with cytotoxic chemotherapy in patient with advanced disease . Some patients with epithelial ovarian tumors have had response to endocrine therapy although it appears that the overall response rate is 10%.
The trials using flutamide which is antiandrogen drug, these trials suggest that flutamide does not have significant activity against ovarian cancer and it is not recommended in patients pretreated with more than one chemotherapy regimen . Therefore, hormonal therapy for patients with advanced ovarian cancer should primarily be reserved for those who have failed chemotherapeutic regimens and who are not candidates for aggressive salvage drug regimen or investigative approaches .

Radiotherapy has been used in ovarian cancer as an adjuvant where optimal debulking surgery has been performed and as a consolidation therapy when added to surgically and chemotherapeutically complete treatments.

Increase and decrease of hormones

The pathological changes resulting from the increase or lack of hormone can trigger the death after the progress of the condition. The increase and decrease of hormones based on emotional or psychological factors, as the body's need of hormones linked to the case of Endocrine glands .

It may be "the reason for the increase or decrease of hormones" to an imbalance of congenital or acquired. The condition relating to the excessive increase of the natural state of hormones is called "Hyperfunction" but the increase in secretion is "Hypersecretion"but the decrease in the secretion is called "Hyposecretion" and the related disorders in the job "Dysfunction". The disorder may be due to one of the gland such as thyroid gland, or as double glands such as ovary . It seems that the dysfunction could be primary or secondary.

Specificity of hormones

The phenomenon specificity is characterized by hormones, for example, adrenal gland secretes several different hormones with the action of the hormone on specific site.

The specificity is not absolute and does not mean any limitation on the activity of the hormone , for example the hormones of"Estrogens that secrete follicles eggs cause the following
- Stimulate the direct cell division in the skin and the formation of antibody.
- Several changes in the organs of the secondary female system such as uterus .

It is rare to find Hormone acting alone, where it is believed that the hormones in general relationship to each other is characterized by a functional relationship between the nervous system and the endocrine and for examples the relationship between hormones and the menstrual cycle in which many of those changes include a number of hormones that occur.

There is, for example, adrenal gland cortex, the testis, ovary and the placenta, which produces several steroid hormones, for example, the hormone "Somatotropin" in the pituitary gland affect other glands also.

Functional characteristics of the hormones

The hormones have a number of obvious characters such as :
- Small quantities are produced by the endocrine glands, which is estimated to the extent of the number of nanograms to milligrams each day and the concentrations of the hormones in the blood is low but the concentrations in the tissues is lower.
- These hormones do not affect the glands that secrete them, for example, the metabolic reactions which are located in the thyroid gland are not affected by these thyroid hormones. , for example, insulin works on the first type of Diabetes , while the hormone LH affect the second type.
- The effect of the hormone on tissues depends on the viability of tissue response and the amount of the hormone in these tissues, for example, the growth hormone in the small living organisms control the growth, while in the young people this hormone is releasing fatty acids from fat storage centers.
-There are both of hormones and nutrients present in very small amounts in the blood of the organism, for example, vitamins concentration in the blood is influenced by the amount of these substances in food, while the hormones in the blood are not affected except thyroid hormones that

are affected by the amount of iodine in food. The food affect the rate of bio synthesis and secretion of these hormones, for example, blood glucose concentration affect the secretion of both insulin and growth hormone. Also, the lack of protein in the diet influences the hormones making peptides.

The functions of the hormones can be divided into various types such as :
- Regulatory or homeostatic function or Organized Functions such as
- Integrative function
- Morphogenetic function
- Permissive function
- Organized Functions

It can be explained by the mechanism of "Control theory" and include:
- Activation of blood glucose and pancreas by insulin secretion or glucagon.
- Activation of alpha cell of pancrease to secrete glucagon from the movement of glucose into these cells.
- Activate the pancreatic alpha cells of blood glucose by the secretion of glucagon.

There are three ways to regulate blood glucose:
- Participation of insulin to reduce blood glucose concentration, and then generates the case of low blood sugar "Hypoglycemia".
- The Glucagon increases the concentration of glucose in the blood, "Hyperglycemia". Then these results produce the state of balance between insulin and Glucagon for normal blood sugar level.
- The catecholamines "Epinephrine" and the "Norepinephrine" cause high blood sugar .
- Integrated regulation

The hormonal function is considered hormone attached to the nervous system in animals, where the hormones have special mechanism that facilitate the integration of these, such as the result of panic and fear that activates the nervous system, the hypothalamus, by neural responses of adrenal then secreted the "Catecholamine" that increase in blood sugar by similar to the mechanism which include glucagon.

- Morphogenetic function
The hormones are required to pass from one phase to another during different growth stages .

- Permissive function
The hormones need other hormones to work, for example, without having Thyroxine the growth hormone can not be influential. The Epinephrine unable to increase oxygen consumption and blood glucose unless the thyroid gland is functioning after taking Thyroxine .

Measuring hormones

The measuring of the concentrations of peptide hormones and non peptide hormones depends on the nature of their chemical structure , such as using iodine, for example, to measure the hormones of the thyroid gland and using the annular structure to measure steroid hormone.

Then many researchers use other ways to measure the concentrations of these hormones, including the use of antibodies , then both "Berson, Yalow" used the immunological method with radiation at the end of fifties " the radio immuno assay". At the same time "Etkins" pointed out in 1960 the possibility of measuring

Thyroxine by studying its association with protein " TBG" due usually to its presence in low concentrations in biological fluids. The lack of precision and sensitivity of these ways has led to a lack of agreement of the amount of hormones concentrations (different methods of measuring hormones, life tests, chemical tests and immunological tests).

In light of the above, there came the need to find more accurate and sensitive ways then to was reached the design the basics of the Radio immuno assay.

Bioassays

The levels of PRL were estimated by using inconvenient bioassays, the classical of bioassays were based on the growth-promoting action of PRL using the pigeon crop sac or the rat mammary gland. There is fast proliferation of the epithelial lining of the crop sac, which contact with basophilic changes in the cytoplasm and they are sensitive to PRL, these changes were used as an indicated to the effect of the hormone.

Radioimmunoassay (RIA)

Radioimmunoassay (RIA) has been a popular method for clinical endocrine determination. RIA is a competitive protein binding (CPB) technique that uses radio-labeled hormone as the tagged hormone and antisera prepared against the specific hormone as a binding site. Competition between unlabeled hormone in the patient sample and the added-labeled hormone for a limited number of antibody–binding sites forms the basis of the assay.

Although RIA is a good technique when high sensitivity is required

Immunoradiometric assay (IRMA)

These methods are similar to (RIA) in that a radio labeled substance is used in an antibody-antigen reaction. However, the radioactive label is attached to the antibody instead of the hormone (Antigen). In addition, an excess of antibody is present in the assay, rather than a limited quantity, is present in the assay. Because the entire unknown antigen becomes bound in IRMA rather than just a portion as in RIA, so that IRMA assays are more sensitive.

Both one-site and two-site IRMAs exist. In the one site assay, the excess antibody that is not bound to the patient sample is removed by addition of a precipitating binder, this binder is antigen bound to some solid support. In the two-site assay (Sandwich technique), a hormone with at least two antibody-binding sites is adsorbed into a solid phase to which one of the antibodies is firmly attached to the walls of the assay tube. After binding to this antibody is completed, a second antibody labeled with 125I is added to the assay. This antibody reacts with the second antibody-binding site to form the Sandwich comprising antibody-hormone-labeled antibody.

Hormones Receptors

The hormones receptors is divided into hormones membrane receptors "Membrane Receptors" and "Nuclear Receptors". Hormones are working through binding membrane receptors through influential proteins accompany the membrane as active paths referring to "Second messenger signalting pathway" of "Cyclic Adenosine Monophosphate (cAMP)" and calcium "Calcium diglycerol", or other transmitters, that stimulates a series of enzymes the "Kinases". hormones bind to the cytoplasm before reaching the nucleus, such as the of the thyroid hormone receptors, which binds to the hormone in the nucleus without the movement hormone step.

Regardless of the class of receptors, there is a certain general principle of how receptors interact with hormones, associated with high affinity and high specificity, in order to allow access of a physiological response (functional) suitable. The Bio response to the hormone is due to the displayed action on the target cell through its unions with relative specificity of protein nature called receptors, and these receptors received a great deal of attention because of its ability to bind hormone and the different their effects in bio chemical reactions that followed by binding of hormone to its receptors, such as glycoproteins, lipoproteins, neurotransmitters, drugs and several antigens.

The isolation and characterization of a small number of the receptors requires many diverse methods of technology.
Criteria of hormone receptor

The receptors are characterized by the different practical basic criteria such as :-
-Saturability
The receptors binding to antigen are characterized by saturability due to the prescence of limited or fixed number of binding sites .
- Affinity
The receptors are characterized by high affinity to bind antigens through the specific centers and the affinity of these recptors association of antigen to another may also vary in the same antigen.

-Specificity
The specificity is the most important criteria for the characterization of the receptors and is susceptible to different antigens for stimulation or inhibition of life response. Hormonal receptors have been extensively studied in ovarian cancer, estrogen and progesterone receptors are reported in approximately 50% of tumors , and 98% of ovarian cancer contains androgen receptors .

Methods of assay of hormone receptors
There are three methods of assay for hormone receptors .
- The conventional biological methods
The conventional biological methods are used and usually set to determine the hormone sites in organs and tissues and include histochemical technique and Immunocytochemical" technique and "Autoradiography". These methods can not be used to measure the receptors concentration or get extra information on the receptors affinity and specificity.
- Radioreceptorassay (RRA)
The radioreceptor assay " RRA" technique is used to study the binding of labeled hormones with specific receptors in vitro.

These studies are usually carried out "In vitro" and the basic needs of these methods the presence of radio labeled hormone and a source of receptors such as tissue homogenate , intact cells and membranes, or any of the parts inside the cell.

The hormone labeled with isotope is bound to its receptor until the equilibrium and a state of balance, then the amount of the labeled hormone receptors complex is calculated , and then the complex (labeled hormone - the recptors) is separating from free hormones by a number of ways including the centrifugation , gel filtration, dialysis at the equilibrium state, differential absorption and is then leaching "Decantation" .

- Antireceptor- antibody technique

A modern techniques which is used to study the role of the receptors to reach a physiological response and to find out the nature of the units for molecule of the receptors .

Chemical Regulator

It believed that any chemical material that impact on the activities of cells or whole organisms can be divided into:

- Exogenous chemical substances

These materials undergone many different responses through its impact on a small number of receptors in the cells and the organism takes these substances or received through injection intravenously, where these materials become part of the cell as co- enzymes, or materials that maintain the ionic composition inside the cells.

The materials that belong to this group can be divided into :-

- Nurient substances

such as carbohydrates, fats, vitamins and minerals and non-metallic elements, calcium, phosphorus, chlorine ... etc.

- Pheromones

This group is composed of attractive sexual material produced by female insects (mites and cockroaches).

- The drugs
- Antihistamines

decrease with histamine that binds to "Receptor sites" in the cells.

- Antagonists of the enzyme "Cholinesterase"

Such as "Anticholinesterase" which combine with the enzyme "Cholinesterase" and inhibits its work.

- Endogenous chemical substances

These materials can be divided into three groups:

- Hormones

A material which are synthesized in special cells that secrete into the blood and then to special cells of specific organs . For example, the hormone testosterone in most somatic cells will increase susceptibility of making the protein in the sperm-generating tubes, this hormone contribute to the making sperms.

- Tissue factors

The materials which are liberated and worked on specific areas such as:
- Histamine
- 5- hydroxyl tryptamine
- kinins
- A set of multiple peptides such as "Brady kinin", " "Kallidin".
- Intracellular – acting substances These materials can be synthesized as a result of influential works on special cells and then increase its activity and the best example is the cyclic adenosine mono phosphate annular "c-AMP."

Endocrine Glands

The glands have different functions such as secretion and unloading, the secretion is beneficial to the body's need, and the unloading process is carried out unless the body is in need to it.

The glands are generally divided into:unloading gland, or external. These glands secrete their materials by specific channels, which are numerous, and the breast is one of them.
Mixed glands:

For example The pancreas is capable of synthesizing enzymes that emptied to the duodenum through the pancreatic duct.
The endocrine glands also synthesize and secrete hormones then effect the body, have a variety of molecular structures that is released from endocrine glands, coordinates many different cell activities and for this reason, the disease of glandular endocrine has a wide range of events that affect many other organs of the human body.

The endocrine glands are characterized by its small size and weight which are situated in different parts of the body, and secrete hormones that have effects on the life of all members of the body and its cells.

Endocrine glands are responsive to the needs of the body from hormones slowly influenced by many factors such as the hormonal factors, emotional and psychological factors and pathogenesis of acute and chronic diseases and others.
Endocrine system works in an integrated manner and proportionate works because of the phenomenon of reverse effect and influence that operate in the two forms long loop and short loop.

Hormones are chemically classified into three categories: peptides, steroids, and amines.
It is functionally classified into two categories:
- Biased hormones stimulate other endocrine gland to synthesize hormones and its secretion.

- The unbiased hormones that affect the non gland tissues.

Classification of Hormones

Hormones can be classified chemically into three groups:
- Protein and peptides hormones Examples insulin, hormones of pituitary gland, parathyroid gland ر.

Glycoprotein Hormones

The glycoprotein hormone of the pituitary (LH, FSH, TSH) and of the placenta (hCG) are composed of two peptide chains, usually referred to as α and β subunits, each with carbohydrate substituent group attached. The carbohydrate moiety, which accounts for 15-31% of the molecular weight include fucose, mannose, galactose, N-acetyl glucose amine, N-acetyl galactose amine and sialic acide.

Each of these hormones consists of two subunits α and β joined by non-covalent bonding.

The α-subunit is almost identical for all of these hormones but the β subunit differs considerably from one hormone to another, this suggests that the β subunit carries the hormonal specificity. Isolated α subunit lack biological activity while isolated β subunit may have little intrinsic biological activity, but full activity is attached when α and β subunits are recombined. This suggests that the presence of both α and β subunit is important for specific receptor recognition and that the β-subunit is responsible for eliciting the specific biological response.

The carboxyl terminal pentapeptide of β subunit is essential for receptor binding but not for α/β association. The feature that distinguishes hormones in the glycoprotein group from hormones of other groups is their glycosylation in each glycoprotein hormone, the α subunit contains two complex asparagine-linked oligosaccharides and the β subunit has either one or two. The glycosylation may be necessary for α/β interaction. The α subunit has five s-s bridges, and the β moiety has six.

The glycoprotein is water soluble molecules circulate in a free form not bound to plasma proteins. The half life of peptide hormones is short, varying between 8-60 minutes. A percentage of intact hormone and fragments of hormone is filtered by the kidney and excreted in urine, they are also degraded by proteases and peptidases in plasma and at the target gland after internalization other organs such as the liver and lunges, which also metabolize glycoprotein hormones.

Second : Hormons and Endocrine gland
. Hormones of the Pituitary gland
The pituitary gland is pea-sized structure located at the base of the brain., it consists of two lobes:
the Anterior Lobe
the Posterior Lobe

The Anterior Lobe
The anterior lobe contains six types of secretory cells, which . All of them secrete their hormone in response to hormones reaching them from the hypothalamus of the brain.
Thyroid Stimulating Hormone (TSH)
TSH (also known as thyrotropin) is a glycoprotein consisting of:
a beta chain of 118 amino acids and
an alpha chain of 92 amino acids.
The alpha chain is identical to that found in two other pituitary hormones, FSH and LH as well as in the hormone chorionic gonadotropin.
The secretion of TSH is stimulated by thyrotrophic releasing hormone (TRH) from the hypothalamus.
 the TSH stimulates the thyroid gland to secrete its hormone thyroxine (T4)

Hormone deficiencies

Follicle-Stimulating Hormone (FSH)

the same alpha chain found in TSH (and LH)
a beta chain of 118 amino acids, which gives it its unique properties.
Synthesis and release of FSH is triggered by the arrival from the hypothalamus of gonadotropin-releasing hormone (GnRH).

FSH in females
In sexually-mature females, FSH (assisted by LH) acts on the follicle to stimulate it to release estrogens.
FSH produced by is available to promote ovulation in women

FSH in males
In sexually-mature males, FSH acts on spermatogonia stimulating (with the aid of testosterone) the production of sperm.

Luteinizing Hormone (LH)
LH is a glycoprotein hormone consists of two polypeptide called and subunit. The -subunit of LH is identical to FSH, TSH and hCG.
The sequence of subunit of human LH and hCG are very similar differing at about 20% of all residues.
The amino acid sequence of the subunit contains 92 amino acid, 5 disulfide bridge and 2-carbohydrate moieties. The subunit is longer contain 6 disulfide bridges and 3 carbohydrate moieties (116).
The molecular weight of the LH is 28KD. Expected values for LH in serum is as shown in table (1).

Table (1): Expected values in serum for LH in normal adults.

	LH (IU/L)
Males, 23-70 years	1.2-7.8
Females	
Follicular phase	1.7-15.0
Midcycle peak	21.9-56.6
Luteal phase	0.6-16.3
Postmenopausal	14.2-52.3

LH is synthesized within the same pituitary cells as FSH and under the same stimulus (GnRH). It is also a heterodimeric glycoprotein consisting of

the same 92-amino acid alpha subunit found in FSH and TSH (as well as in chorionic gonadotropin);
a beta chain of 121 amino acids that is responsible for its properties.
The effects of LH also depend on sex.

LH in females
In sexually-mature females,
LH its release (ovulation) in the middle of the menstrual cycle; stimulates the now-empty follicle to develop into the corpus luteum, which secretes progesterone during the latter half of the menstrual cycle.

Action of LH
In the female LH induces ovulation and stimulates formation of corpus luteum after ovulation has occurred. In response to LH corpus luteum secrete progesterone and estrogens.
In males LH stimulate the interstitial cells in the testis to secrete testosterone.

Hypothalamic Control of Luteinizing Hormone
The release of (LH and FSH) is controlled by one releasing hormone GnRH. This in turn is a primarily regulated by circulating levels of gonadal hormone that reach the hypothalamus. The hypothalamus secretes GnRH in a pulsatile manner lasting for several minutes that occur every 1-3 hrs. The pulsatile release of GnRH also causes pulsatile output of LH.

Estrogen in small amounts has a strong effect in inhibiting the production of both LH and FSH, also when progesterone is available, the inhibitory effect of estrogen is multiplied even through progesterone by itself has little effect. These feedback effects seem to operate mainly directly on the anterior pituitary gland but to a lesser extent on the hypothalamus to decrease secretion of GnRH especially by altering the frequency of the GnRh pulses.

Another negative feedback by a hormone called inhibin, which secreted along with the steroid sex hormones by the granulosa cells of the corpus luteum. This hormone has the same effect in the female as in the male of inhibiting the secretion of FSH by the anterior pituitary gland and LH to a lesser extent as well.

A decline in ovarian hormones secretion during menopause or following castration cause increased secretion of LH and FSH. Ovarian steriod and peptide hormones are also able to exert positive feedback, which is important in the regulation of the LH surge required to induce ovulation and is regulated by sharply rising levels of estrogen in the late first half of the menstrual cycle .

Mechanism of Action of Luteinizing Hormone (LH)

Peptide hormones are too polar to diffuse passively through lipoprotein membranes. Peptide and protein hormones are also too large to pass through membrane pores. Instead these hormones initiate their response by binding to receptors located on or in the cell membrane. This binding interaction results in the generation of an intra cellular signal or "second messenger" that in turn mediates the hormones effects on intra cellular enzymes gene expression and membrane transport. Whereas the "first messenger" of intracellular communication is the hormone the second messenger may be a small organic molecule such as cyclic adenosine monophosphate (cAMP.

After the binding of peptide hormones to their specific plasma membrane receptors on the cell surface, the majority of hormone induced responses are believed to be mediated by the second messenger molecule, cyclic adenosine monophosphate (cAMP). The binding of peptide hormones to their receptors activates the membrane-bound enzyme adenylate cyclase and results in the synthesis of cAMP.

The cAMP synthesis from ATP as the result of adenylate cyclase activation interacts with at least two types of enzyme in the cell, it can bind protein kinase molecules to proceed with its normal action or it can be rapidly degraded to -AMP by another enzyme phosphodiesterase and thus is devoid of any activity as second messenger.

The mature ovaries are paired nodular structures 2.5 – 5 * 2* 1cm, weighing from 4gm to 8gm , the weight varyies during the menstrual cycle . Usually the ovary lies with its long axis vertical but it shares in any movement of the broad ligament and uterus . The human ovaries are attached to the posterior surface of the broad ligament by a peritoneal fold called the mesovarium . Nerves, blood vessels and lymphatics traverse the mesovarium and penetrate the ovary at its hilum. The blood supplied to the ovaries are derived from the ovarian and uterine blood vessels. The medullary and cortical branches supply the entire ovary, with arborizations virtually supplying every follicle. The venous drainage is by the ovarian veins, which enters the inferior vena cava just below the entry of the renal veins.

The ovary is comprised of three distinct regions
- An outer cortex containing the ovarian follicles.
- A central medulla consisting of ovarian stroma.
- An inner hilum around the area of attachment of the ovary to the mesovarian.

The ovarian cortex contains the ovarian follicles in various stages of maturation (Primary, Secondary, Tertiary, Graafian and atretic) together with corpora lutea and corpora albicantia for those that have reached full maturation. Figure (1).

LH acts on the interstitial cells (also known as Leydig cells) of the testes stimulating them to synthesize and secrete the male sex hormone, testosterone.

LH in males is also known as interstitial cell stimulating hormone (ICSH). LH is synthesized within the same pituitary cells as FSH and under the same stimulus (GnRH).

LH in males is also known as interstitial cell stimulating hormone (ICSH).

Prolactin (PRL)

Prolactin is a protein of 198 amino acids. During pregnancy it helps in the preparation of the breasts for future milk production.
After birth, prolactin promotes the synthesis of milk.
Prolactin secretion is stimulated by TRH and repressed by estrogens and dopamine.
Prolactin (PRL) is a poly peptide hormone with a molecular weight of 22000 Dalton and an amino acid sequence of 199 amino acids, it is synthesized and secreted by the lactotrophs of the pituitary. The lactotrophs can be identified by immunoperoxidase staining by using specific antisera and believed to arise directly from the acidophil stem cell line. The relative number and PRL content of

lactotropic cells is increased in fetal pituitary glands and during pregnancy. The increase in the number of lactotropic cells is the result of elevated concentration of estrogens during pregnancy.

Regulation of Prolactin Secretion :

The normal plasma prolactin concentration is approximately 8ng /ml in women and 5ng/ml in men. Secretion of PRL, as for other hormones released by the anterior lobe of the pituitary gland falls under hypothalamic control Figure (1-6). However, PRL is unique among adenohypophyseal hormones in that the primary control of its secretion is Inhibitory (prolactin inhibitory factor, PIF) rather than stimulatory (prolactin-releasing factor, PRF).

In humans, exercise, surgical and psychological stresses, and stimulation of the nipple increase PRL secretion. The PRL level rises during sleep, the rise starting after the onset of sleep and persisting throughout the sleep period. Also PRL secretion is increased during pregnancy, reaching a peak at the time of parturition. Finally, an important influence on PRL regulation appears to be exerted by PRL itself by short feedback loop between the pituitary and hypothalamus. Short feedback-loop inhibition of gonadotrophin releasing hormone (GnRH) secretion by PRL has also been suggested as the cause for inhibition of gonadotrophin (Gn) secretion that occurs in women who are nursing and in patients with PRL- secreting adenomas of the pituitary gland.

The Structure of Prolactin

The structure of human pituitary prolactin contains 199 amino acids residues and has three intramolecular disulfide bridges with a molecular weight of 22000-25000 (Figure 2). Most of studies concerning PRL chemistry were performed on ovine prolactin.

The structure of 0-PRL is similar to that of human growth hormone (hGH) and human chorionic somatomamotropin (hCS). The half- life of PRL, like that of GH, is about 20 minutes. Each of PRL, GH has a single tryptophan residue, and each has two homologous disulfide bonds, GH has disulfide bond between residues 53-165 and 182-189 while PRL has disulfide bonds between residues 4-11, 58-173 and 190-198.

The secondary structure of 0-PRL was determined by using Circular Dichroism Technique (CD), which gave a strong negative band at 233 nm and a second weak band at 209nm. Because of many properties of 0-PRL are similar to those of hGH, therefore spectra of 0-PRL is similar to those of hGH.

Function of Prolactin:
PRL function can be classified into two main types:
Non –Reproductive Function
These functions of PRL have been demonstrated in different organs such as adrenal gland and liver

Growth Hormone (GH)
Human Chorionic Gonadotropin

Human chorionic gonadotropin (hCG) is the signature hormone of the placenta. It's a hetrodimeric glycoprotein hormone that also belongs to the cystine–knot growth factor family. hCG is a member of closely related pituitary glycoprotein hormones (TSH, FSH, and LH), present in mammals, which are important to the correct functioning of the reproductive system. In mammals the glycoprotein hormones consist of a common alpha subunit which in human is encoded by a single gene. Each hormone has different β subunit and this subunit is responsible for the different target specificity of each hormone.

Human growth hormone (HGH; also called somatotropin) is a protein of 191 amino acids. The GH-secreting cells are stimulated to synthesize and release GH by the intermittent arrival of growth hormone releasing hormone (GHRH) from the hypothalamus. GH promotes body growth by binding to receptors on the surface of liver cells.

This stimulates them to release insulin-like growth factor-1 (IGF-1; also known as somatomedin)
IGF-1 acts directly on the ends of the long bones promoting their growth In childhood,
hyposecretion of GH produces a short but normally-proportioned body.
Growth retardation can also result from an inability to respond to GH. GHRH orGH (causing Laron syndrome, a form of dwarfism)
which is part of the "downstream" signaling process after GH binds its receptor.
hypersecretion leads to gigantism
In adults, a hypersecretion of GH or GHRH leads to acromegaly

Structure of hCG

The alpha subunit of hCG, which is identical in sequence to the alpha subunit of the pituitary glycoprotein hormones, is composed of 92 amino acids; while the target-receptor-specific β subunit of hCG has significant sequence homology, with 80% identity to β-LH subunit. The β-hCG subunit contains 145 amino acids with 30 additional amino acids at the carboxyterminus(4). This modification is an essential facet of hCG biology allowing sera concentration in the pregnant mother to reach peak level of 50-100000mIU/ml (1-10mg/l). hCG also varies from LH by virtue of increased glycosylation, internal disulfide bond and sialic acid content.

Carbohydrates constitute approximately 30% by weight of each subunit. Their moieties play a role in the secretion, stability, folding, subunit assembly of hormone and also it seems to be important to maintain the proper conformation of the hormone. Each carbohydrate moiety terminates in sialic acid, which accounts for 10% of the weight, of the molecule and confers a high degree of resistance to degradation and consequently a long plasma half-life. The solid crystal structure of hCG showed that each of the gonadotropin subunits is rich in disulfide bonds (the α subunit have 10 half-cystines residues that form five intrachain disulfide linkages and 12 half-cystines that form six conserved disulfide bridges. It also has structural homology to the disulfide-knot growth factor proteins. The three dimensional model of hCG, proposed a structure which predominantly composed of three helical segments ,two in the α and one in the β subunit forming antiparallel strands in each subunit. They are joined by three hairpin loops giving the hormone only a small hydrophobic core with a large interfacing area,.

Human Chorionic Gonadotropin Forms
The combination of multiple subunits, multiple N-linked and O-linked oligosaccharide side chains causes significant heterogeneity in hCG structure.
hCG free subunits, degraded molecules, molecules with irregular N-and O-linked oligosaccharide side chains and fragments, which well informed, are present in sera, urine and other body fluids.

   a. Blood and Urinary Forms of hCG
Although most measurements of hCG are made in serum, the source of hormone used for clinical treatment of infertility and other medical problems is usually a purified urinary fraction containing the biologically active forms of the hormone. All standard hCG reference preparations are all from the urine of normal pregnant women, that hCG serum measurements were based on.

Although a number of isoforms of hCG circulate in blood, most of these forms vary only by PI differences, due to variable content of the sialic acid residues. Only a small quantity of free subunit and various quantities of nicked hCG circulate in blood; in contrast, a much greater variety of molecular forms of hCG is present in urine due to proteolytic processing of hCG passing through the kidney. Many efforts to separate the several forms of hCG from different variety raw urine sources were evaluated.
Nicked hCG:

The urinary metabolites of hCG and which may also exist to a lesser extent in blood are nicked form of hCG as well as nicked β subunit. Nicked hCG is simply a heterodimer with M.wt.(~36500D), it has peptide bond cleavages in loop 2 of the β subunit (between beta residues 44-49). Loop 2 is known to be exposed to solvent and is easily eliminated by protease. Dissociation of nicked hCG will result in free nicked β subunit, which is also found in a small extent in urine. The existence of nicked hCG was confirmed by isolation and gel electrophoresis. Reliable immunoassays for nicked forms of hCG were not available, due to diminished immunopotency after cleavages in β loop 2,until the development of a fairly specific nicked hCG immunometricassay and immunoassay systems. Anumber of studies suggested that nicked forms of hCG may have clinical significance as markers of certain cancers.

hCG βcf :

Is the main urinary metabolite that produced by proteolysis of hCG or its β-subunit in the kidney since very little hCG βcf is detected in the blood. Core molecules are generally present in urine in much greater molar concentration (2-10 times) than dimeric gonadotrophins or free subunits. Other studies have shown that the molecule may be produced directly by placental cells in tissue culture and pituitary tissue. The structure of this fragment was reported since 1983 and was the first to be studied and discovered upon solution of the crystal structure of hCG. It is derived from hCG β subunit and is composed of residues 6-40 disulfide bridged to residues 55-92 and contaning tirmned carbohydrate groups with no sialic acid and the polypeptide chains are head together by disulfide bridges. Highly specific assays were developed to the hCG βcf and applyed in verious clinical situations. Extensive studies of its compartment distribution and concentration throughout pregnancy were conducted by de Medeirosetal .finding that it is increased in parallel to hCG throughout pregnancy, making its measurement a useful marker of ecto pregnancy.

Furthermore, it has diagnostic applications in cancer tests including ovarian, lung, bladder and various gynecological cancers.

Intact hCG:
It is the major hCG-related molecule present in sera that was separated as purified standard from different raw urine sources. It is composed of a heterodimer with intact polypeptide backbone. Composing of α subunit with 92 amino acids and β subunit with 145 amino acids residue polypeptide, mono and biantennary oligosaccharides and mostly O- linked trisaccharides . Detectable level of intact hCG in sera and urine has been reported, observing that its level in sera of nonpregnant women increase with age, and higher than in men . It rises during the first tirmester (8-10 weeks) to maintain the steroid environment necessary for the pregnancy; following by a rapid decrease until 15 weeks. Intact hCG levels have been also measured in non trophoblastic tumor, observing a slight elevation in its levels as it is compared to a pregnant woman.

freeβ hCG:
It is defined as mono subunit drived from dissociation of hCG. In this form, with M.Wt about (~22000D), only β subunit is present and no α-subunit; having biantennary N-linked oligosaccharides and mostly trisaccharides O-linked oligosaccharide. Using different techniques, the presence of βhCG in sera, urine, tissues, cyst fluid and cell line in different normal, benign and malignant tumor have been demonstrated. Free βhCG was found to be the most secreted form in the non trophoblastic tumors; while its level was found to be low or in normal range in the normal individuals.

free α hCG:

The α-subunit of hCG was found to be synthesized and released by normal placental tissue as a small precursor form that convertes to a larger form prior to secrete; with finding that the free α form do not bind to purified βhCG subunit. It is derived from the dissociation of hCG having identical structure of the combined α subunit excepte its carbohydrate composition which prevents recombination. Many studies were undertaken for studying the presence of the free α subunit in different organs, its immunological and biological property and its anti α monoclonal antibodes. It was found to be in a small extent in urine and sera of healthy individuals. It has also been detected in sera, tissues and cytosol of the pregnant individuals, several endocrine and nonendocrine, and cell line tumors.

The hCG-R belongs to the family of glycoprotein hormone G protein –coupled receptors, which also comprise TSH-R and FSH-R(149).All three receptors are glycoprotiens containing a large extracllular leucin rich repeats (LRR), N-terminal domain with several N-linked glycosylation. A plasma membrane domain composed of seven hydrophobic α-helices ending with an intracellular C-terminus).These three receptors have been shown to transduce hormone binding via coupling to the heterotrimeric Gs protein which activates adenylyl cyclase, or other G proteins that activate phosphlipase C (as assessed by measuring inositol phosphates, diacylglycerol, and / or intracellular Ca^{+2})). The mechanism of the coupling in hCG-R is different from other G protein- coupled receptors whose ligands are much smaller and intercalate among the transmembrance helices(150).

The hCG-R, which is present on gonadal cells, plays a pivotal role in reproductive physiology. Until recently, it was believed that hCG-R were only present in gonadal cells, where they regulate steroidogenesis. However, it is now known that these receptors are also expressed in human and animal uteri, human pituitary ,skin fallopian tubes, human placenta, fetal membranes, decidua, human gestational trophoblastic neoplasms, human breast and breast cancer, endometrial and myometrial blood vessels) and brain. These studies have shown that hCG can directly regulate the functions of nongonodal tissues indicating that the receptor are functional in these tissues.

The evidence for hCG receptors has been obtained by ligand binding measurements or immunocytochemistry using receptor –specific antibodies or evaluation the presence of hCG receptor mRNA. The traditional hCG-R binding studies involved binding a radioactive form of either the hormone itself or a biologically active analog (agonist) to membrane preparations (particulate or solubilized)of target tissues, have been used to identify and characterize the hCG-R. In most hCG-R binding studies,125I-labeled hCG prepared by radio iodination with chloramines or lactoperoxidase has been used. Tritium, has also been used in some studies for labeling.

The binding of LH or hCG to cell surface receptors was thought to be a reversible process, thus saturation binding experiments were performed under steady state conditions. It has now become clear that the hCG-receptor interaction is much more complex and cannot be described as a simple bimolecular reaction. The hormone-receptor complex is not only involved with the stimulation of adenylate cyclase, but also functions in the ligand-induced regulation of the receptors. From different separation techniques of bound from free hormone, the choice was usually between centrifugation and precipitation of the hormone receptor complex.

ACTH — the adrenocorticotropic hormone- Adrenocorticotrophin
ACTH is a peptide of 39 amino acids. It is cut from a larger precursor proopiomelanocortin (POMC).ACTH acts on the cells of the adrenal cortex, stimulating them to produce glucocorticoids, like cortisol;mineralocorticoids, like aldosterone;
androgens (male sex hormones, like testosterone).

In the fetus, ACTH stimulates the adrenal cortex to synthesize a precursor of estrogen called dehydroepiandrosterone sulfate (DHEA-S). which helps prepare the mother for giving birth. Production of ACTH depends on the intermittent arrival of corticotropin-releasing hormone (CRH) from the hypothalamus. Hypersecretion of ACTH is a frequent cause of Cushing's syndrome.

ACTH is a peptide of 39 amino acids. It is cut from a larger precursor proopiomelanocortin (POMC).

ACTH acts on the cells of the adrenal cortex, stimulating them to produce

alpha Melanocyte-Stimulating Hormone (α-MSH)

Alpha MSH is also a cleavage product of proopiomelanocortin (POMC). In fact, α-MSH is identical to the first 13 amino acids at the amino terminal of ACTH.

-Thyroid Stimulating Hormone (TSH)
.This hormone is secreted from anterior pituitary gland located at the bottom of the brain and after the arrival of a signal of their existing in the bottom of the brain . This hormone works to stimulate the entry of the day of the thyroid gland to build hormones (T3 and T4) . The purpose of this analysis is to determine the location and type of the disease, which affects the thyroid gland.

The high level of the hormone TSH after partial eradication of the thyroid gland and in cases of inadequate primary thyroid function and which result in various diseases , as well as in cases of hyperthyroidism and thyroid function as a result of malfunction of hypothalamus and pituitary gland.

Neurohypophysis The Posterior Lobe
The posterior lobe of the pituitary releases two hormones vasopressin and oxytocin
Vasopressin

Vasopressin is a peptide of 9 amino acids (Cys-Tyr-Phe-Gln-Asn-Cys-Pro-Arg-Gly). It is also known as arginine vasopressin (AVP) and the antidiuretic hormone (ADH) acts to reduce the volume of urine formed (giving it its name of antidiuretic hormone).

Oxytocin

Oxytocin is a peptide of 9 amino acids (Cys-Tyr-Ile-Gln-Asn-Cys-Pro-Leu-Gly). It acts on certain smooth muscles, stimulating contractions of the uterus at the time of birth.

Hypothalamus and Factors

Thyrotropin-releasing hormone (TRH)
Gonadotropin-releasing hormone (GnRH)
Growth hormone-releasing hormone (GHRH)
Corticotropin-releasing hormone (CRH)
Somatostatin
Dopamine.

Thyrotropin-releasing hormone (TRH)
TRH is a tripeptide (GluHisPro) it stimulates the release of thyroid-stimulating hormone (TSH) and prolactin (PRL)
GnRH is a peptide of 10 amino acids.

GnRH

is a peptide of 10 amino acids. In both sexes, its secretion occurs in periodic pulses usually occurring every 1–2 hours. after puberty, a hyposecretion of GnRH may result from
intense physical training anorexia nervosa
.Growth hormone-releasing hormone (GHRH)
GHRH is a mixture of two peptides, one containing 40 amino acids, the other 44., GHRH stimulates cells in the anterior lobe of the pituitary to secrete growth hormone (GH).

Corticotropin-releasing hormone (CRH)
CRH is a peptide of 41 amino acids. its acts on cells in the anterior lobe of the pituitary to release adrenocorticotropic hormone (ACTH)

As its name indicates, its acts on cells in the anterior lobe of the pituitary to release adrenocorticotropic hormone (ACTH)

Somatostatin

Somatostatin is a mixture of two peptides, one of 14 amino acids, the other of 28., Somatostatin acts on the anterior lobe of the pituitary to Somatostatin is also secreted by cells in the pancreas and in the intestine where it inhibits the secretion of a variety of other hormones.

The pancreatic gland

Insulin

This hormone is secreted by the pancreas "Pro-insulin" with another hormone called glucagon that contains 81 amino acid depending on the type of source, while effective insulin contains 51 amino acid due to the loss of peptide chain of 30 amino acid.
Chemistry insulin Insulin is a small protein consisting of an alpha chain of 21 amino acids linked by two disulfide (S—S) bridges to abeta chain of 30 amino acids.

However, scattered through the pancreas are several hundred thousand clusters of cells called islets of Langerhans. The islets are endocrine tissue containing four types of cells. In order of abundance, they are the:
beta cells, which secrete insulin and amylin;
alpha cells, which secrete glucagon;
delta cells, which secrete somatostatin, and gamma cells, which secrete pancreatic polypeptide .

However, the pancreas are several hundred thousand clusters of cells called islets of Langerhans. containing four types of cells.

.Beta Cells

Beta cells have channels in their plasma membrane that serve as glucose detectors. Beta cells secrete insulin in response to a rising level of circulating glucose ("blood sugar").

Insulin action

- Fatty acid oxidation component of bilateral carbon vehicles within the crepe cycle and increase its focus when the insulin and Atia making it difficult for the liver converts excess and Aaksdha transforms them into ketone compounds.
- Increase the amount of fat in the blood when the low amount of insulin.
- Low protein-making down the amount of insulin.
- The high amount of carbohydrates which are made down the amount of insulin.

It seems that insulin contributes to multiple effects, including:
- Glucose enter cells.
- Glucose oxidation to CO_2, H_2O.
- Glucose turning redundant to glycogen in the liver and muscles.
- Life operations for structural Kleikojin, protein and fat.

It is believed that the action of insulin President is located in the cell membrane, which contributes to the introduction of sugar and amino acids. And being the AMP annular, and converts glucose to fatty tissue fatty acids that the presence of insulin.

Insulin affects many organs. Itstimulates skeletal muscle fibers to
acts on liver cells
acts on fat (adipose) cells to stimulate the uptake of glucose and the synthesis of fat.

In each case, insulin triggers these effects by binding to the insulin receptor — a transmembrane protein embedded in the plasma membrane of the responding cells.

Insulin formation and storage and excretion

Biosynthesis, storage and releases of insulin

Insulin is made in the network ER "Endoplasmic reticulum" beta cells in "Islets of Langerhans" in the pancreas and in the form of a single insulin-born series "Proinsulin" then turn to insulin losing amino acids chain and called "C-peptide" (C) , where insulin is stored in conjunction with zinc, and in the case of increasing the concentration of glucose in blood.

Glucose main factor in regulating insulin secretion and it, which caused some stimuli of insulin secretion for example, glucose cause an increase in the formation and secretion of insulin and "Growth hormone" due to an increase in the formation of insulin only.

Diabetes is treated by injecting insulin, which regulates how much intake of carbohydrates, where it is not easy every lunch tocalculate carbohydrate content of the weight and then inject the appropriate amount of insulin. The insulin is given in the form of injection under the skin, In spite of the formation of insulin, it is chemically extracted from the pancreas of animals..

Diabetes Mellitus

Diabetes mellitus is an endocrine disorder characterized by :-

a failure of the kidney to efficiently reclaim glucose a resulting increase in the volume of urine

There are three main types of diabetes mellitus:
Type 1
Type 2
Inherited Diabetes Mellitus

Type 1 Diabetes Mellitus

(Insulin-Dependent Diabetes Mellitus or IDDM) is characterized by (hypo) insulin and ;appears in childhood.

Type 2 Diabetes Mellitus

Type 2 is also known as Non Insulin-Dependent Diabetes Mellitus (NIDDM) and adult-onset diabetes. Many people develop Type 2 diabetes mellitus without an accompanying drop in insulin levels (at least at first).

Type 2 diabetes mellitus usually in adults and, particularly often, in overweight people. Several drugs, all of which can be taken by mouth, are useful in restoring better control over blood sugar in patients with type 2 diabetes.

Inherited Diabetes Mellitus

mutations in one or both copies of the gene encoding the insulin receptor. These patients usually have extra-high levels of circulating insulin but defective receptors.

Insulin (receptors)

Binding of insulin to its receptor is the basis for measuring the activity or property adopted of the effectiveness of insulin, where it seems that the degree of this correlation is very high s. In each case, insulin triggers these effects by binding to the insulin receptor — a trans membrane protein embedded in the plasma membrane of the responding cells.

The chemical composition of these compounds is believed that it is a protein concentrated in the plasma membrane of the cell. So its plays a role in regulating life processes of moderation and variety of life the effectiveness of the following enzymes:

- Glucokinase
- Hexokinase
- Phosphofeactokinase
- Pyruvate kinase

As we mentioned by binding to the cell as his future is followed by an increase in the effectiveness of the enzyme "Adenylate cyclase"

A series peptide multiple consists of 29 sour acids are made by alpha cells from of the pancreas and the lifting of glucose in the blood by binding to special receptors in the tissue-affected, followed by increasing the linking process in the activity of the enzyme "Adenylate cyclase" followed by a rise in the level of cyclic AMP, and subsequently break down glycogen in the liver

Alpha Cells

The alpha cells of the islets secrete glucagon, a polypeptide of 29 amino acids.
Glucagon acts principally on the liver where it stimulates the conversion of glycogen into glucose ("glycogenolysis") and Glucagon secretion is stimulated by low levels of glucose in the blood.

Delta Cells

The delta cells secrete somatostatin. This consists of two polypeptides, one of 14 amino acids and one of 28. Somatostatin has a variety of functions.

Gamma Cells

The gamma cells of the islets secrete a 36-amino-acid pancreatic polypeptide, which reduces appetite.

Second:
The amino acid derivatives
Hormones derived from the amino acids derivatives"for example, those hormones derived from tyrosine an adrenaline and Thyroxine

Melatonin and the Pineal Gland

The pineal gland is a tiny structure located at the base of the brain.
Its principal hormone is melatonin, a derivative of the amino acid tryptophan.
Synthesis and release of melatonin is stimulated by darkness and
inhibited by light.

- Adrenal Medulla

Norepinephrine
Epinephrine
Pineal gland Melatonin forming cells
   Nerves serotonin
   5- OH tryptamine
Thyroid hormones
-L-T3,L-T4
  -Para Thyroid

- Calcitonin

The thyroid gland is a double-lobed structure located in the neck. Embedded in its rear surface are the four parathyroid

The thyroid gland synthesizes and secretes:
thyroxine (T4) and triiodothyronine (T3);
calcitonin

The third hormone which is clled called calcitonin will affect the General skeleton of bone the body, where it has major phenomenon of the dynamic interaction of calcium and phosphate ions and decreasing the level of calcium during the deposition of calcium in the bones.

Both hormones are derivatives of the amino acid tyrosine with four atoms of iodine in T4 , three inT3. The thyroid secretes mainly (80%) T4 , but when T4 enters target cells, one atom of iodine is removed from it converting it into T3. T3 is the more potent of the two hormones. It has many effects.

The thyroid cells responsible for the synthesis of T4 and T3 take up circulating iodine from the blood and attach them to tyrosine residues in the protein thyroglobulin. This action, as well as the synthesis of the hormones, is stimulated by the binding of thyroid stimulating hormone (TSH; also known as thyrotropin) to transmembrane receptors at the cell surface.

The thyroid gland is the most important endocrine gland in the body because they play an important role in maintaining the metabolism in the body and the impact on the central nervous system "Central Neural System" general circulation ,pituitary adenomas "Anterior Pituitary gland" and plasma proteins in addition to the different mechanisms of metabolism and also the thyroid is one of the most organs of the body that change the weight, shape, and construction on a microscopic change according to the age, sex and marital status. The thyroid gland consists histologically from component of hypothalamus "Hypothalamus" which produces a hormone

TRH and pituitary adenomas that secrete a hormone TSH and thus the thyroid that secrete hormones L-T3, L-T4. It is stimulated by hormone TSH after its association with its membrane, which works through the activated enzyme Adenylate Cyclase" and then produces the cAMP secondary messenger.

Thyroid gland consists of two lobes each lobe is located on one side of the trachea ,bound to isthmus "Isthmus" and pierce the trachea from the front and from the bottom of the cartilage ring "Circoid Cartilade". The weight of normal thyroid gland up to 20-30 grams, and consists histologically of follicles surrounded by connective tissues, follicular cells is located on the edge of the follicles that are filled with the colloid material ".When the gland is active the follicular cells become vertical and when the gland is ineffective the cells become flat and the meterials of colloid then grouped, as well the gland also contains on outside the follicle cells known as C cells that secrete "calcitonin". The size of the thyroid gland in women is greater than men beside that the size of gland also grow in over the role of adolescence in the case of pregnancy and lactation, and during the last period of the menstrual cycle.

Thyroid in humans
The thyroid is the only organ of the body which has the ability to collect large amounts of iodine to make its hormones , the thyroid is the only gland of the endocrine that store large amounts of inactive iodine with the contents abroad the cells whereas the other glands store small amounts of hormones on the site within the cell and the thyroid be distinct in three differents phases depending on the functional status. The "Resting phase", a stage at which the cells become flat and Acini "becomes

extended and the second phase is the phase of the endocrine" secretory phase ", while the third stage is the "Resorption phase" that represents absorbing iodine from the blood to make hormones and thyroglobulin containing vertical cells and almost light empty cavity.

The essential component of the follicles is the thyroglobulin which represents a"Iodinated " gylco protein that has a molecular weight of 680,000 Dalton produced from the epithelial cells of the tissue of the thyroid as well from cells enhancing the "Neoplasia" in the case of cancers of the thyroid, especially "differentiated thyroid carcinoma. The tumor diagnosis depends, on the high concentration of antibodies in blood such as Grieve Hachmidto .

The advantage of thyroid is its proximity to the two types of cells differ from each other in terms of fetal origin, "Thyroid Follicular cell", which represents the structural unit of the thyroid. As for the other type of thyroid cells are side follicular cells "(C Cell) Para Follicular Cell" or C cells that secrete the hormone "Calcitonin". The clusters or scattered of alveoli arise from neural stem cells "Neuro Crest Cells" and secreted the hormone calcitonin in response to the high level of calcium in blood.The hormone inhibit the absorption of Ca + 2 in stem cells, which is resistant of a hormone secreted by the gland side parathyroid PTH as operating, both to preserve the natural concentration of Ca in blood and fluid overseas cellular .The thyroid gland secrete group of hormones including ,"Tri iodothyronin" which is secreted in very small quantities, also secrete Thyroxine "Tetra iodothyronine"but in larger amount so is a hormone L-T4 is more important than a hormone L-T3 .In addition the thyroid gland secretes Diiodothyronine " DIT" and "Tri iodothyronin" in very small amounts ,these hormones are all an acid derivatives of amino acid "Tyrosine", the nucleus of T4 consist from two benzene rings linked by link ether and there is a direct correlation between the amount of iodine, the amount of the follicles of the thyroid .

The food is the main source of inorganic iodine which is oxidized by the enzyme "Peroxidase", turning them into ions of iodine and associated with proteins follicles making a series of MIT, DIT, L-T4, L-T3 that are associated with Tg and stored with glue until it is needed and released by by the enzyme "protease" which, as shown in Figure (4) and a "half life" for hormone l-T3 one day L-T4 7 days.

It explains physiological functions within the thyroid follicle
The thyroid hormones after their secretion into the blood bind to plasma proteins and small amounts of become free and resultant depends on the effectiveness of hormonal L-T3, L-T4 and the strength of the link between the hormone and plasma protein. The transfer of thyroid hormones are carried out by three proteins are:

- (TBG) Thyroxin Binding Globulin
- (TBPA) Thyroxin Binding Prealbumin
- Albumin

The TBG is the most important functionally and is made in the liver and the transfer of these hormones depend on the chemical affinity with protein with hormone as well as its focus the blood (or plasma) and 99.95% of the L-T4 is linked to these proteins and the TBG transmits 70-75% of L -T4 and TBPA albumin and transmits 15-20% of L-T4.

The hormone L-T3, the 99.7% of which is linked to albumin, and the affinity TBG to L-T3 less than a the affinity of L-T4 . The hormones of the thyroid gland are fundamentals in many processes in the body, after transfer by plasma proteins and reach down to the cells or organs it penetrates the cellular membrane because of its ability to melt fat and then later linked to the surface of the nucleus, leading to motivate many physical processes to increase metabolism and O2 consumption rate there is also a strong

relationship between the catechol Amin that stimulate some enzymes, such as:
- Hexokinase
- Decarbboxylase
- Isocitric dehydrogenase
- Glycerol phosphate drhydrogenase

The hormones of thyroid gland have significant impacts on the vascular system of heart by increasing blood volume and quantity and increase blood pressure, heart rate and increase the tensile strength and muscle tension of heart resulted in occurrence of angina associated with cases of toxic thyroid "Thyrotocosis", and affecting the breathing, gastrointestinal, nervous system, functions of the muscles, sleep, sexual functions and other endocrine glands.

The thyroid gland secrete hormones metabolically active regulate biochemical reactions and metabolic functions, such as L-T3, L-T4, which are responsible for regulating the metabolic rate of the body and the growth of the fetus and the maturity of the bones and normal growth of the body as well as the nervous tissue growth specialization.

The L-T3 and the L-T4 affect the central and peripheral nervous ,the absence these hormones lead to slow the mental maturity and the high level of cerebrospinal fluid (CSF) proteins. The thyroid hormones increase the rate of absorption of carbohydrates in the gut and decrease in in the level of cholesterol in the bloodstream through its impact on the mechanics of liver function.

The pituitary gland controls the degree of thyroid activity mediated by the secretion of the thyroid stimulating hormone "TSH", which is controlled by "Thyrotropin releasing hormone (TRH)", and the hormone, TSH works to stimulate the secretion of hormones the thyroid by "Negative feed - back mechanism "(Figure) and that this mechanism controls the concentration of TSH level in the blood stream as well, as the increase in the level of thyroid hormones in the blood inhibit the secretion of TSH and is affected by the mechanism of inhibition amount of L-T3 and the L-T4 free at serum through a direct impact in the pituitary gland and the hypothalamus directly and so to affect the secretion of each of the TSH and the TRH rate.

The thyroid glandis located in the front of the neck and contain certain cells called follicular cells (follicular cells) that manufacture and secretion of hormones of two basic types:
-T4 Thyroxine
-T3 (triiodothyronine)
These hormones contain iodine, which relies on food as the main source for iodine, and settle most of the iodine intake of food in thyroid gland so that the body includes all mechanism on iodine uptake and reduced it and stored in the thyroid gland.
These hormones (T4 -T3) effects on some of the following operations:
-The effect On carbohydrate metabolism: excess thyroid hormones Please choose a level of glucose in the blood.
-altotar on lipid metabolism: increase the break down of fats which leads to increased formation of ketone bodies, and help the oxidation of cholesterol in the liver, leading to decrease the level of cholesterol in the blood
-The effects on the metabolism of proteins:

The thyroid hormone help the protein synthesis and breaks down of proteins depending on the concentration and muscle weakness and increased excretion of non proteinic nitrogen material in the urine and increase the proportion of creatinine in urine.
-The Hormones play role in physical growth and sexual growth.
-Other effects :The thyroid hormones increase in oxygen consumption in all body tissues except the thyroid gland itself, raise body temperature and to convert carotenoids to vitamin "A" and also help in absorbing vitamin "B".

The hormone TSH include control of several steps inside the thyroid gland, including the synthesis and secretion of the thyroid hormones because of its impact on the metabolic pathways in the body through thyroid hormones, L-T3, L-T4 in blood, there are many functional factors indirectly affect the thyroid gland and its hormones especially for changes in the genital tract "reproductive system". The thyroid tumors afflict women more than men with the ratio of 1: 4 and arise from the follicular epithelium, which produces the hormones L-T3, L-T4 and except medullary tumor arises from shrubs follicular cells.

Because of the significant role played by the thyroid gland and its hormones it is necessary to note "Benign tumors", "Malignant tumors" that afflict important effects in the body on the thyroid gland that lead to a rise in the concentration of the hormone TSH with a significant decrease in the level of hormones of thyroid gland, and an increase in the production of antibodies against antigens of the thyroid, and the case of excessive toxic and "Graves" disease lead to the production of antibodies that are binds to TSH receptors leading to inhibition of its work, as it became clear that measurement of the concentration of the hormone TSH level is sensitive to the hormone level T4, and the "Immuno Radio Metric assay (IRMA)" RIA "Radio Immunoassay" helped in the measurement of the concentration of of thyroid hormones and TSH and note the changes in the incidence of tumors of the thyroid gland and diagnosis of changes in the level of this hormone imbalances depending on the pathological condition that affects the thyroid gland, which facilitated the possibility of identification and detection.

The thyroxine which was considered of the most important hormones of the thyroid gland, was discovered and after twenty-five years old the (L-T3) (3, 5, 3', - triiodothyronine) was discovered and then pointed Rukh and his group to the existence of another form that is reverse (3, 3', 5'- triiodothyronine) which was not effective as L-T3 in the thyroid gland and blood serum.

The movement of hormones of the thyroid gland into the blood stream by two forms free, which constitute a very small percentage of less than 1% of the quantitative amount of hormones the thyroid gland in the blood serum, and the rest is linked to serum proteins. The most important carrier proteins to L-T4 in serum, "Thyroxin - binding globlin"(TBG) a "Glycoprotein" created in the liver andits molecular weight 54000 Dalton " according to Johnson et al., 1980", as well as the (TBPA) "Thyroxin - binding -prealbumine" which has a molecular weight of 13,500 Daltons.

As for the L-T3 it is linked also to TBG but with less affinity , and is also associated with TBPA with a very fast rate of dissociation , difficult to measure it, and low affinity of association . The associatiation of the L-T3 with albumin and in general with the all the carrier proteins is linked with lower affinity than L-T4 . But he nuclear receptors that are associated with L-T3 by high affinity than the L-T4.

The proteins association with thyroid hormones facilitate the supply of these hormones in various tissues, and it was found that the unbound L-T4 to associated protein accumulates in the first cell encountered while in the case of its association with protein- then arriving in all cells so far from the bloodstream in the same amount . Other studies show that about 3% of L-T4 in the human serum blood is associated with (VLDL) "Very low - density lipoproteins", and proteins lipid low-rise density (LDL) "Low - density lipoproteins", as well as the fatty proteins high-density (HDL) "high - density lipoproteins".

The association of the L-T4 with these lipoproteins reflect specialist influence by a weak or medium affinity between the L-T4 and half of the core protein adipose "apolipoprotein (apo) moiety", while the association of the L-T3 has very weak affinity .

The thyroglobulin is brocken into multiple compounds, containing iodine, all derivatives of histidine tyrosine, namely:

"3- monolodo -tyrosine", "3,5 -diiodo -tyrosine", "3, 5, 3' -triiodo -thyronine",
"3,3'-5 -triiodo -thyronine"
"3,3' -diiodo - thyronine"and
"3,3 ', 5,5' -tetraiodothyronine"
"Thyroxin".

Metabolism of L-T3 and L-T4

Liver can through its distinctive role of regulation the level of thyroid hormones then iodine is released from the liver to outside then iodide, reduces the bioactivity of the hormone, in addition to that the thyronine iodide linked to glucuronic acid releasing the product to to gall bladder and also pulling of the amino group from tiiodo thyronine tri-iodine and glucuronic and also releasing certain amounts of free hormones from gallbladder.

Biosynthesis of the hormones of the thyroid gland

There are six stages of biosynthesis of thyroid hormones according to the following :-
- Storage of inorganic iodide in the thyroid gland.
- Bio-synthesis of the protein complex (Thyroglobulin).
- Thyroglobulin Storage.
- Liberation the hormone(L-T4-Thyroxine) and tri-iodo thyronine(L-T3) from Thyroglobulin.
- Transfer the of these hormones mentioned in paragraph 5 to various tissues by the blood.

After entering ion iodide to thyroid gland then it becomes effective in the form of the ion Iodinium "through the oxidation process by enzyme" Iodide peroxidase ".
And then the effective iodine enter to carbon No. 3 of tyrosine constructing unilateral iodine MIT and to the carbon atom 5 to form a binary iodine of tyrosine in the follicles which contain the thyroglobulin by the enzyme "Tyrosine iodinase". After that two molecules of the tyrosine di iodine unite in the thyroglobulin through the oxidation process with the liberation one molecule of alanine and thyroxine and when one molecule of, in the

case of one molecule of di iodine tyrosine DIT with tyrosine unilateral iodine MIT the reaction produce L-T3, and when the two molecules of the tyrosine unilateral iodine are reacted the reaction produce thyronine di iodine . In the final step thyroglobulin break down by enzymes degrading the protein component materials free MIT, as well as the DIT and the L-T3 and the L-T4 the breakdown is followed by distribution of the hormones L-T3,L-T4 the blood, which is associated with:
- Thyroxin binding globulin (TBG)
- Thyroxine that binds pre albumin (TBPA).
- Albumin linked to thyroxine TBA.

Inorganic Iodine transmission

It is not possible to consider the thyroid as only gland that has the ability of iodine concentration, the following glands participated in this process,:
A. salivary gland
B. Skin
C. Mammary gland
D. Choroid

- Halogenation of Tyrosine

Hormones L-T3, and L-T4 and some common medical conditions

There are two common medical conditions as cases resulted from to decrease or increase production of L-T3, and L-T4 by the thyroid gland . These cases are:
- Thyroid deficiency "Hypothyroidism" and Non -toxic goiter".
- "Hyper thyroidism"

Levels of Thyroid hormones
L-T3, and L-T4 in the blood

The levels of the hormone L-T3 in normal human serum are estimated of 1-2 ng / cm 3 in comparison to an average level of the hormone L-T4 of 80 ng / cm 3 (8 micrograms / 100 cm 3) and these values depend on the measurement the specific laboratories.

Unusual levels of the hormone are considered evidence of dysfunction in the thyroid Changes in the levels of L-T3 be in parallel with the same changes in the levels of L-T4 (sometimes be independent). The measurements of both levels are important part in the investigation of the dysfunction in the pituitary / hypothalamus "Hypothalmic / pituitary thyroid axis" and we may find that there is a rise in hormone levels, especially when embryos "Neonates" and healthy babies (generally from the age of one to ten years), an increase from their normal level as due to the need.

Receptors of the thyroid hormones.

The cell membrane carry binding locations to thyroid hormones in each of the tissues of kidney liver, and transplanted cell "Cell line" for the cells of the pituitary tumor, as well as in fat cells "Adipocytes", where it is possible distinguish two different sites and affinity to first, "Primary binding site" and also found that these have saturable binding in the plasma membrane of red blood cells RBCs and in fibroblast cells for implanted cell sites. Other studies have shown an association sites of the L-T4 in the transplanted cells of "Spinal cord" as well as in the "Cerebellum" for newborns also through electron microscope studies, as well as autoradiography also the presence of a "facilitative system" in the red blood cell of man.

Thyroid hormones binding sites in the mitochondria
Several studies in vitro and in vivo to the presence of binding sites to the thyroid hormones in the mitochondria, especially for the L-T3 and showed a number of other studies in vitro susceptibility binding sites in the mitochondria to saturation, where several studies on affinity association, and the capacity of association, and molecular weight in the number of of organisms,. All of these binding sites in the mitochondria membrane is located in the matrix. As it proved scientifically that all sites is limited to the "Inner membrane" of the mitochondria in which the phosphorylation process of oxidizing "Oxidative phosphorylation" is occurring except

in the case of kidney Rat, where it lies outside the membrane "Outer membrane" of the mitochondria.
Binding proteins to hormones the thyroid gland in the cytosol

Although the prescence of available information on cytosol proteins (CTBP) "Cytosolic thyroid hormone binding proteins", however these sites, have physiological significance but is not known precisely. The affinity of L-T3 binding to its "Nuclear T3 receptors" is characterized by greater affinity than withCTBP while MBC capacity is larger in the case of hormone CTBP them L-nuclear T3, derived from the value of high-MBC for these hormones on the role of the CTBP represents a repository "Rcervoir" hormones for thyroid instead of being a specialized regulatory proteins, while trance between him and his group on the functional core of CTBP regulate the level of L-T3, which will be linked to the nuclear receptors.

Receptors of nuclear thyroid hormones

The thyroid hormones carry out vital role and effectiveness through its nuclear receptors and many studies indicated that receptors of L-nuclear T3 be few in non-tissue target, such as the spleen, testes, and these two are non-responsive organs of the hormones of the thyroid gland, but the amount of these receptors are increasing in the tissues with high response to the hormone such as the liver, kidney, brain , pituitary gland, and heart, .
The chemical nature of the-nuclear receptors of L- T3
The chemical characteristics of the nuclear receptors of the L-T3 being non-histonic proteins, with molecular weight up to 50,000 Daltons, and other studies have shown the existence of two types of the receptors , the first recipients of 47,000 Daltons and the second with a molecular weight of 57,000 Daltons.

Inflammation of the thyroid gland is one of the diseases that are caused by a variety of situations, and is considered the most common types of inflammation, which is an auto-immune disease that called inflammation of the thyroid, and in this case, the immune system cells attack the thyroid gland and with antibodies to the cells, and in the few months of infection may be for a temporary period and due to excess of fatty activity, followed by a feeling many of Addison's disease which is one of the diseases that occur as a result of damage to the adrenal cortex and this leads to a lack of the hormone cortisol and this is known as primary adrenal deficiency. Symptoms of Addison's disease characterized by weight loss ,disturbance in the digestive tract, feeling of weakness and fatigue,feeling pain in muscles and joints and decrease in blood pressure ...

Diseases of the thyroid

Hypothyroid diseases; includes the following diseases caused by inadequate production of T3

cretinism: hypothyroidism in infancy and childhood leads to stunted growth and intelligence

. .myxedema: hypothyroidism in adults leads to lowered metabolic rate .

goiter: enlargement of the thyroid gland caused by inadequate iodine in the diet with resulting low levels of T4 and T3;

Hyperthyroid diseases; caused by excessive secretion of thyroid hormones

Graves´ disease. Autoantibodies against the TSH receptor bind to the receptor mimicking the effect of TSH binding. Result: excessive production of thyroid hormones. Graves´ disease is an example of an autoimmune disease. Graves disease is one of the diseases related to immunes system , which is characterized by the presence of antibodies to the future stimulant of the hormone which is secreted by the anterior lobe of the pituitary gland. It is found that increase in the secretion of hormones of the thyroid gland as a result of the antibodies, such as immune globulin that stimulate the thyroid gland with inflation. The most important diseases in hyperthyroidism:

It is also called "Diffuse Toxic Goiter" and it is one of the most important thyroid disease, this illness occurs at any age, and someappears more common in the third and fourth decades, in females than males, often occurs in the same members family. Graves disease is within a group disorders that cause displacement of immune components, and is rarely occurred before reaching the age of ten years.

The disease is characterized by increased secretion of thyroid hormones and some of its symptoms the thyroid rampant inflation and hyperthyroidism and exophthalmos, and the situation could occur in the absence of exophthalmos and without thyroid hyperplasia (this condition is rare). Moreover, some patients suffer from eye-inflammation.

- Toxic nodular goitre

This disease is characterized by the production of too much of the thyroid hormones the by one or a few nodules of the thyroid gland, and those effected by the disease are usually older people and carry out complications such as cardiovascular effects "Angina pectoris" and the "Congestive heart failure" and muscular weakness, with other muscle diseases .

In spite of the hyperthyroidism nodular more common in women, but many men are subjected to this kind more than Grieve disease. The thyroid hormones are excreted in the rare cases of cancers where it leads to hyperthyroidism, and the thyroid gland be solid and irregular and that taking radioactive iodide by the thyroid is high .

Osteoporosis. High levels of thyroid hormones suppress the production of TSH through the negative-feedback mechanism mentioned above. The resulting low level of TSH causes an increase in the numbers of bone-reabsorbing osteoclasts resulting in osteoporosis.

Thyroid gland Tumors

There are two types of tumors of the thyroid gland, benign and malignant tumors.

Benign thyroid gland tumors

Benign thyroid diseases are divided into two types:

- Hyperthyroidism
- Hypothyroidism

Hyperthyroidism

Also called "Thyrotoxicosis" which is characterized by the increased production of thyroid hormones, the symptoms of hyperthyroidism depend on the nature of on the patient's age, and the presence or absence of other diseases in specific organs in the body.

The symptoms of hyperthyroidism are associated with influences of hormones on the metabolism of the tissues, which represents the initial symptoms of hyperthyroidism a decrease in the level of cholesterol in the blood and lower the concentration of lipoproteins in the blood serum, high glucose blood, weight loss, weakness of nerves and heart palpitations strongly and quickly and heat intolerance, sweating, muscle weakness ,tremors, irregular session monthly entrapment or menopause in females, impotence and lack of libido in males.

Treatment of hyperthyroidism

The treatment of hyperthyroidism includes surgery, anti-thyroid drugs and radioactive iodide

- Surgery

The main types of treatment of hyperthyroidism complete removal of thyroid tissue and make the gland unable to produce increasing amounts of its hormones , and in order that the gland easy to be eradicated effectively the surgeon must removes at least 90% of thyroid tissue, and obviously thyroidectomy can be carried out to the cases of thyroid toxicity without control over the case of hyperthyroidism.

- Antithyroid Drugs

These drugs are characterized by their ability to control the symptoms of hyperthyroidism such as "Propyl thiouracil" and "Methimazole" these drugs inhibit the synthesis of thyroid hormones through interference with the iodide incorporation , duplication of tyrosine di-iodine, and the transfer of iodine in the thyroid gland and the impact of these drugs on the toxicity of the disease depends on its ability to reduce or minimize the flow of iodine to the thyroid gland. Anti-thyroid drugs is also characterized by possessing the characteristics of inhibitory factors of immune cells lymphoma.

- Radio active iodide

The iodide radioactive is appropriate treatment for the toxicity of the thyroid, which is not only effective, but also inexpensive and easy to understand and free from centrist and influential effects for a long period of time. The radioactive iodide 131 is the type of most widely used and given orally by using dose of 5-10 mci for Grave disease and 10-15 mci of nodular hyperthyroidism, symptoms begin to disappear after giving radioactive iodide (131) two weeks to four weeks.

- Other therapeutic modulities

Other remedies for hyperthyroidism "Glucocorticoids" and immune inhibitory factors such as the "6-mercaptopurine" and "Cyclosporin". The "Glucocorticoids" have many effects on the function of the thyroid, which inhibit the secretion of TSH and thus thyroid hormones, interfere with the TSH stimulation mediated by TRH, and reduce free protein binding with thyroxine and also interfere with peripheral conversion of L-T4 to L-T3.

Furthermore the "6- mercaptopurine" Interferes link with iodide and lead to the removal of iodine and an increase in its secretion in urine and this medication is toxic too. The "Cyclosporin" is inhibiting the immune function and then effecting lymphocytes of type T more than inhibition of bone marrow and is also an effective and successful drug treatment for people suffering Grave disease.

Hypothyroidism

It is characterized by a decrease in the concentration of one of the thyroid hormones, and the symptoms of this case include fatigue, slow metabolism, a decrease in metabolism base rate and body temperature and high cholesterol levels and a drop in the level of free fatty acid in the blood with a lack of glucose consumption and muscle weakness accompanied by a slowly movement and weight gain and shows the disease in women more than men, especially women at the period of menopause. The hypothyroidism is of two types:

- Primary hypothyroidism

The causes of a primary hypothyroidism include the following :-

- Inflammation of the thyroid due to "Autoimmune thyoiditis" and characterized mostly by chronic infections and include "Hashimoto disease".
- Genetic abnormalities include the lack of a single enzyme controlling the synthesis of thyroid hormones (the absence of any enzyme in the biosynthesis of T3 and T4).
- After treatment hyperthyroidism specifically after removal of the thyroid surgically or after radioactive treatment.
- In rare cases, the cause due to using anti-thyroid drugs resulted in decreasing the thyroid hormones such as (lithium carbonate).
- Iodine deficiency in food causing a decrease in the Production of natural amount of thyroid hormones.

- Secondary hypothyroidism

It is less common than type I and is caused by a lack of TSH secretion of the pituitary gland or due to lack of TRH or may be caused by the inability of the hypothalamus in to secrete adequate amount of TRH.

Some of the hypothyroidism diseases the "diffuse non - toxic Goiter and "nodular Non -toxic Goiter".

Treatment of Hypothyroidism

The treatment of hypothyroidism by taking one of the hormones to the patient and the L-T4 is the hormone of choice and the usual daily dose of 0.15 - 0.2 mg for males and 0.1 - 0.15 mg for females.

The doctors are working to save Thyroxine concentration of serum up to 8-10 mcg / dl. It is recommended to start treatment by mostly up to 0.025 mg / day for a week to two weeks and then increase the daily dose using 0.025 mg every one to two weeks then reaching to the required dose.

Tests that determine the function of the thyroid gland:

- Hormone T-3 and T-4 test

It does not have to reflect the level of thyroxine (t4) kidney his job because physiological levels of thyroxine change depending on the concentration of proteins concentration and these proteins are affected by physiological situations such as pregnancy and taking the pill or any of compounds containing estrogen.

And the level of total t4 in normal blood ranges from 5-12 Microgram / 100 ml (65-16-56 nmol / L).

And the level of natural t3 in the blood ranges between 0.07-0.17 micrograms / 100 ml blood (0.91-2.2 nmol / L) There are cases where t3 and t4 level are rising and other cases at least both of them decrease according to the following :-

Free thyroxine test (t4 free):

The metabolic activity of the hormone t4 on its free concentration (non-mobile protein) and the normal level of the free hormone is between 0.8-2.4 Nano gram / 100 ml (0.1-0.03 nmol / L).

This hormone is rising in its level in the case of hyperthyroidism and thyroid function in the case of its infection and also its level decrease in the case of inadequate thyroid function. This measurement is useful to make sure of the diagnosis of hyperthyroidism when the rise in total thyroxine on the upper limits of normal.

The ratio of t3 absorbed on the (resin)

This is a measurement of the loaded parts is from globulin bearing Thyroxine, if added t3 irradiated to serum patient, the part of it became associated with the protein and remains the other part free, then absorbs this free portion on the (resin) (are similar substance is chemically for industrial purposes).

And it is separated from the serum because the rate of uptake on t3 (resin) is inversely proportional-free part of the carrier protein. This ratio ranging between 25-35% natural.

In the case of excessive thyroid function we get a high value of this hormone and in cases associated with the low level of globulin (carrier of thyroxine) resulted in no change in the function of the thyroid gland.

Furthermore the low value in the case of excessive function of the thyroid gland in cases associated with the high level of globulin without any change in thyroid function.

Measuring the free t4 (FT4i-free thyroxine)

This measurement is a measure of the amount of serum free t4 and we get that by multiplying the value by t3 t4 absorbed on the rt3u)) resin

In cases of hyperthyroidism and thyroid, and we get high values in cases of failure function, regardless of any change in the level .

- Genetic Malfunction of Thyroid gland .
- Discrimination between primary hypothyroidism and secondary hypothyroidisim .
- Proving the primary of hypothyroidism.
- During Test of pituitary gland failure for any reason

The Parathyroid Glands The Parathyroid Glands

The parathyroid glands are 4 tiny structures embedded in the rear surface of the thyroid gland. They secrete parathyroid hormone (PTH) a polypeptide of 84 amino acids. PTH increases the concentration of Ca^{2+} in the blood in three ways. PTH promotes release of Ca^{2+} from the huge reservoir in the bones and reabsorption of Ca^{2+} from the fluid in the tubules in the kidneys.

PTH also regulates the level of phosphate in the blood. Secretion of PTH reduces the efficiency with which phosphate is reclaimed in the proximal tubules of the kidney causing a drop in the phosphate concentration of the blood.

Control of the Parathyroids: the calcium receptor

The cells of the parathyroid glands have surface G-protein-coupled receptors that bind Ca^{2+} . Binding of Ca^{2+} to this receptor depresses the secretion of PTH and thus leads to a lowering of the concentration of Ca^{2+} in the blood. Two classes of inherited disorders involving mutant genes encoding the Ca^{2+} receptor occur:

Hypoparathyroidism

One of Causes of Hypoparathyroidisim is accidental removal of or damage to the parathyroids during neck surgery and the treatment concentrated on give calcium supplements.

Hyperparathyroidism

Tumors in the parathyroids elevate the level of PTH causing a rise in the level of blood Ca^{2+} at the expense of calcium stores in the bones..

Two glands of parathyroid hormone on both sides of the thyroid gland, secrete pth by cells secreting known (chief- cells) ,the (pth) is a protein in nature which consists of the multi-chain peptides. The secretion of the hormone (pth)is regulated by the concentration of calcium ions (ca^{++}) in the blood and the presence of an inverse relationship between the two.

Function of the (pth):

This hormone (pth) is effecting on the concentration of calcium in the body where the concentration of the hormone increases because of its direct impact on the kidneys and bones and indirect impact on the intestinal absorption of calcium and phosphorus concentration decrease because of the direct effect of the hormone on the kidney filtration .

The most important functions of this hormone are:

1-effect on the kidneys:

The (pth) effects the kidney by increasing calcium absorption and increasing the secretion of potassium, phosphorus , carbonic acid and the lack of secretion hydrogen ion and ammonia.

The sites of transferring sodium, calcium located in (distal renal tubule), are effected by increasing the absorption of calcium, while the effect of the hormone on the phosphorus transfer can be inhibited in two locations (distal renal tubule) and (tubule proximal renal) and therefore less concentration phosphorus in the blood in exchange for increasing the concentration of calcium.

2-Effect on bone:
The hormone has four effects on bone, includes all kinds of bone cells.
a-Inhibiting (collagen) biosynthesis in the process of bone formation (osteogenesis), which is done by the constituent (osteobiast cells)
b- Increasing bone's ability to absorb.
c-Increasing decomposition of bone (osteogenesis) by (osteoblast)
d-Increasing the speed of the maturation of the ancestral cells in the decomposition process of bone cells (osteoblast) and the process of biosynthesis of bone cells (osteoblast) ..
As a result of these effects are less than the bone's ability to link and retain calcium and bones begin to fade in the pathological condition.

3-The effect on the intestines (gastrointestinal tract)
The effect on intestine resulted in calcium and phosphorus absorption and then secreted to the blood, through vitamin d effect .
There is a connection between the Para thyroid hormone and the level of calcium in the blood Hiq \ w is considered excessive and inadequate function of the thyroid gland, a neighbor of the main reasons for the low level of calcium in the blood .

The hormone analysis ((PTH)could be used in the following cases:
1. To confirm the diagnosis of (Hyperparathyroidism)
2-To distinguish between Primary hyperthyroidism and the conditions that lead to high calcium in the blood.
Consequently, the diagnosis of hyper thyroid gland first depends on:
1-rising proportion of calcium in the blood.
2-low phosphorus in the blood.
3-high Alosvataz enzyme alkaline (Alkaline phosphatase)

The high calcium in the blood at the same time it is located high hormone (PTH) is hardly to be clear evidence for the diagnosis Verwt thyroid gland first.

The gland neighbor thyroid norepinephrine hormonal Albarrathormon "Parathoromne" and Alcalsetcn "Calcitonin" and the latter is also secreted by glands lemusah "Thymus gland" and the thyroid 'Thyroid gland ". These are located amounting glands in the back of the top section and bottom. The forming cells are divided

Into two types, the Supreme containing glycogen, which secretes parathyroid gland. And the other that does not contain glycogen.
Pathological cases of para thyroid gland

- Hypoparathyroidism
That this situation is rare and clinical manifestations are vague and usually result from the eradication of this gland, and examples of this (cachexia Tetany disease) which is characterized by frequent neuromuscular excitement. And spastic cases, is accompanied by this disorder in the life processes of calcium, where land and weaken the transmission of nerve messages across nerve and muscle can address this disease to raise the level of calcium by mouth, are added and vitamin D2 or by calcium intravenously.
- Hyperparathyroidism
For this case types known as the first and second, where the first is characterized by infecting the bones and kidneys, accompanied by chronic renal lesions with a long-term failure.
Calcitonin
Calcitonin is a polypeptide of 32 amino acids. The thyroid cells in which it is synthesized have receptors that bind calcium ions (Ca^{2+}) circulating in the blood. A rise in its level, such as would occur with the absorption of calcium from a meal, stimulates the cells to release calcitonin. Calcitonin prevents a sharp rise in blood calcium by

inhibiting the uptake of Ca2+ from the small intestine and inhibiting the Ca2+-releasing activity of osteoclasts. Because it slows the loss of Ca2+ from bones, calcitonin has been examined as a possible treatment for osteoporosis, a weakening of the bones that is a leading cause of hip and other bone fractures in the elderly. Being a polypeptide, calcitonin cannot be given by mouth (it would be digested), and giving by injection is not appealing. However, inhaling calcitonin appears to be an effective way to get therapeutic levels of the hormone into the blood. A synthetic version of calcitonin (trade name = Miacalcin) is now available as a nasal spray.

Multi-peptide with a partial weight 5000 manufactured in the thyroid and parathyroid gland effect opposite to the hormone Albarratormon, and it downwards to the level of calcium by increasing the deposition of calcium.
Because it slows the loss of Ca2+ from bones, calcitonin has been examined as a possible treatment for osteoporosis, a weakening of the bones that is a leading cause of hip and other bone fractures in the elderly. Being a polypeptide, calcitonin cannot be given by mouth (it would be digested), and giving by injection is not appealing. However, inhaling calcitonin appears to be an effective way to get therapeutic levels of the hormone into the blood. A synthetic version of calcitonin (trade name = Miacalcin) is now available as a nasal spray.

Steroids hormones
This is secreted by different glands such as the gonads and adrenal glands (cortex). Examples include androgens and estrogens and these hormones are all made up from cholesterol. It can also be classified into two groups of hormones based on the career of operation:
Hormones of the Reproductive System

Females

The ovaries of sexually-mature females secrete: The mature ovaries are paired nodular structures 2.5 – 5 * 2 * 1cm, weighing from 4gm to 8gm, the weight varyies during the menstrual cycle. Usually the ovary lies with its long axis vertical but it shares in any movement of the broad ligament and uterus. The human ovaries are attached to the posterior surface of the broad ligament by a peritoneal fold called the mesovarium. Nerves, blood vessels and lymphatics traverse the mesovarium and penetrate the ovary at its hilum. The blood supplied to the ovaries are derived from the ovarian and uterine blood vessels. The medullary and cortical branches supply the entire ovary, with arborizations virtually supplying every follicle. The venous drainage is by the ovarian veins, which enters the inferior vena cava just below the entry of the renal veins.
The ovary is comprised of three distinct regions
An outer cortex containing the ovarian follicles.
A central medulla consisting of ovarian stroma.
An inner hilum around the area of attachment of the ovary to the mesovarian.
The ovarian cortex contains the ovarian follicles in various stages of maturation (Primary, Secondary, Tertiary, Graafian and atretic) together with corpora lutea and corpora albicantia for those that have reached full maturation. Figure ().
The ovarian stroma consists of three specific cell types:-
Contractile cells, connective tissue cells and interstitial cells.
The hilum consist of specific type of interstitial cells known as the hilus cell. Normal hilus cells have been shown to synthesize and secrete testosterone in response to Luteinizing hormone (LH).

a mixture of estrogens of which 17β-estradiol is the most abundant).
progesterone.
Estrogens

The estrogen is secreted primarily by the ovaries via ovarian follicle and corpus luteum and during pregnancy by the placenta. Adrenal and tests are also believed to secrete small quantities of estrogens.

Only three estrogens are present in significant quantities in the plasma of the human female: B-estradiol, estrone and estriol. The principle estrogen secreted by the ovaries is B-estradiol.

Estrogens stimulate the development of tissue involved in reproduction. Under estrogen stimulation vaginal epithelium proliferates and differentiate uterin endometrium proliferates, the myometrium develop intrinsic rhythmic motility and breast ducts proliferate. Estradiol also has anabolic effects on cartilage so its growth promoting. By the effect on peripheral blood vessels, estrogen typically causes vasodilation and heat dissipation.

Androgen

The adult female produces androgen as well as estrogens and progesterone. About 25% of these androgens are secreted from the ovaries. 25% from the adrenal cortex and 50% from either ovarian or adrenal precurosors. In the female androgens contribute to normal hair growth at puberty).

Progesterone

Progesterone is a C21 steroid secreted by the corpus luteum the placenta and in small amounts by ovarian follicle and small amounts also enter the circulation from the tests and adrenal cortex.

On the uterus. Promote secretory changes in the uterine endometrium preparing the uterus for implantation fertilized ovum, also progesterone decrease the frequency and intensity of uterine contraction helping to prevent expulsion of implanted ovum (28).

On the fallopian tubes. Progesterone promotes secretory changes in the mucosal lining of the fallopian tubes thus maintaining the nutrition of the fertilized ovum (28).

On the breast tissue. Progesterone stimulate the development of lobules and alveoli. It induces differentiation of estrogen prepared ductal tissue and supports the secretory function of the breast during lactation (30, 31).

It has many effects in the body, some having nothing to do with sex and reproduction.

In "target" cells, i.e., cells that change their gene expression in response to the hormone, they bind to receptor proteins located in the cytoplasm and/or nucleus. The hormone-receptor complex enters the nucleus and binds to specific sequences of DNA, called the estrogen (or progesterone) .

The hormone-receptor complex acts as a transcription factor

.Males

The principal androgen (male sex hormone) is testosterone. This steroid is manufactured by the interstitial (Leydig) cells of the testes..

Testosterone is also essential for the production of sperm.

production of testosterone is controlled by the release of luteinizing hormone (LH) from the anterior lobe of the pituitary gland, which is in turn controlled by the release of GnRH from the hypothalamus. LH is also called interstitial cell stimulating hormone (ICSH).

The level of testosterone is under negative-feedback control: a rising level of testosterone suppresses the release of GnRH from the hypothalamus.

The ovaries of sexually-mature females secrete:
a mixture of estrogens of which 17β-estradiol is the most abundant (and most potent).
progesterone.
menopause

The menstrual cycle continues for many years. But eventually, usually between 42 and 52 years of age, the follicles become less responsive to FSH and LH. They begin to secrete less estrogen. Ovulation and menstruation become irregular and finally cease. This cessation is called menopause.

The cessation of menstruation in western women occurs at a median age of 50.8 year . Premature menopause is defined as ovarian failure and menstrual cessation before age of 40 years, this often has a genetic or autoimmune basis . Surgical menopause due to bilateral oophorectomy is common and can cause more severe symptoms owing to the sudden rapid drop in sex hormone levels .

During the menopause, women lose the ability to reproduce and develop symptoms include spontaneously occurring hot flushes, sweat eruptions, irritability, palpitations, lethargy sleeplessness and depression .

During menopause, low estradiol and progesterone concentrations are found in the menopause as an expression of ovarian regression. Thus, circulating levels of LH and FSH are high reflecting primary gonadal failure and the fact that few if any functional follicles remain in the ovary .

The lowered Estradiol concentration results in osteoporosis in roughly one third of women .

Regulation of Estrogen and Progesterone

The synthesis and secretion of estrogens is stimulated by follicle-stimulating hormone (FSH), which is, in turn, controlled by the hypothalamic gonadotropin releasing hormone (GnRH).

High levels of estrogens suppress the release of GnRH (bar) providing a negative-feedback control of hormone levels.

Elevated levels of progesterone control themselves by the same negative feedback loop used by estrogen (and testosterone).

Menstrual Cycle

The term menstrual cycle refers to the series of changes that occur in sexually mature, nonpregnant females and that culminate in menses .

Regular cyclic changes that may be regarded as periodic preparation for fertilization and pregnancy. Its most conspicuous feature is the periodic vaginal bleeding that occurs with the shedding of the uterine mucosa (menstruation).

The mean cycle length is 28 days, but regular cycle lengths of 23 to 35 days may be considered normal.

At the start of each cycle, the primordial follicles enlarge and cavity forms around the ovum (antrum formation). One of the follicles in one ovary starts to grow rapidly on the 6th day, while the others regress At the 14th-16th day of the cycle, the follicle ruptures and the ovum is extruded in to the abdominal cavity. This process is called ovulation. The precise cause of ovulation is not known, but ovulation occurs 1 to 24h after the LH peak. The ovum is picked up by the fimbriated ends of the uterine tubes and transported to the uterus. The follicle that ruptures at the time of ovulation filled with blood. The granulosa and theca cells of the follicles lining promptly proliferate and the clotted blood is replaced with yellowish lipid-rich luteal cells, forming the corpus luteum. If successful fertilization and implantation occurs the corpus luteum function is sustained by hCG produced by the trophoblastic cells of the developing embryo.

The function of the corpus luteum is to support the ovum in the early days after fertilization and implantation into the endometrium, before the placenta is established. If there is no pregnancy, the corpus luteum begins to degenerate about 4 days before the next menses and is replaced by fibrous tissue forming corpus albicans within a few days of the corpus luteum beginning to degenerate, another follicular phase commences in the ovary and the whole process is repeated.

about every 28 days, some blood and other products of the disintegration of the inner lining of the uterus (the endometrium) are discharged from the uterus, a process called menstruation. During this time a new follicle begins to develop in one of the ovaries. After menstruation ceases, the follicle continues to develop, secreting an increasing amount of estrogen as it does so.

The rising level of estrogen causes the endometrium to become thicker and more richly supplied with blood vessels and glands.

A rising level of LH causes the developing egg within the follicle to complete the first meiotic division (meiosis I), forming a secondary oocyte.

After about two weeks, there is a sudden surge in the production of LH.

This surge in LH triggers ovulation: the release of the secondary oocyte into the fallopian tube.

Under the continued influence of LH, the now-empty follicle develops into a corpus luteum (hence the name luteinizing hormone for LH).

Stimulated by LH, the corpus luteum secretes progesterone which

continues the preparation of the endometrium for a possible pregnancy

inhibits the contraction of the uterus

inhibits the development of a new follicle

If fertilization does not occur (which is usually the case), the rising level of progesterone inhibits the release of GnRH which, in turn,

inhibits further production of progesterone.

As the progesterone level drops,

the corpus luteum begins to degenerate;

the endometrium begins to break down, its cells committing programmed cell death (apoptosis);

the inhibition of uterine contraction is lifted, and

the bleeding and cramps of menstruation begin.

Fertilization

Fertilization of the egg is also influenced by progesterone. Sperm swim towards the egg by chemotaxis following a gradient of progesterone secreted by cells surrounding the egg. Progesterone opens CatSper ("cation sperm") channels in the plasma membrane surrounding the anterior portion of the sperm tail. This allows an influx of Ca^{2+} ions which causes the flagellum to beat more rapidly and vigorously.

Pregnancy

As the fertilized egg passes down the fallopian tube, it undergoes its first mitotic divisions. By the end of the week, the developing embryo has become a hollow ball of cells called a blastocyst. At this time, the blastocyst reaches the uterus and embeds itself in the endometrium, a process called implantation. With implantation, pregnancy is established.

The blastocyst has two parts:
the inner cell mass, which will become the baby, and
the trophoblast, which will
develop into the placenta and umbilical cord and
begin to secrete human chorionic gonadotropin (HCG).
HCG is a glycoprotein. It is a heterodimer of
the same alpha subunit (of 92 amino acids) used by TSH, FSH, and LH and
a unique beta subunit (of 145 amino acids).

HCG behaves much like FSH and LH with one crucial exception: it is NOT inhibited by a rising level of progesterone. Thus HCG prevents the deterioration of the corpus luteum at the end of the fourth week and enables pregnancy to continue beyond the end of the normal menstrual cycle.

Because only the implanted trophoblast makes HCG, its early appearance in the urine of pregnant women provides the basis for the most widely used test for pregnancy (which can provide a positive signal even before menstruation would have otherwise begun).

As pregnancy continues, the placenta becomes a major source of progesterone, and its presence is essential to maintain pregnancy. Mothers at risk of giving birth too soon can be given a synthetic progestinto help them retain the fetus until it is full-term.

Birth

Toward the end of pregnancy,
The placenta releases large amounts of CRH which stimulates the pituitary glands of both mother and her fetus to secrete

ACTH, which acts on their adrenal glands causing them to release the estrogen precursor dehydroepiandrosterone sulfate (DHEAS).
This is converted into estrogen by the placenta.
The rising level of estrogen causes the smooth muscle cells of the uterus to
synthesize connexins and form gap junctions. Gap junctions connect the cells electrically so that they contract together as labor begins.
oxytocin.
Oxytocin is secreted by the posterior lobe of the pituitary as well as by the uterus.
Prostaglandins are synthesized in the placenta and uterus.
The normal inhibition of uterine contraction by progesterone is turned off by several mechanisms while both oxytocin and prostaglandins cause the uterus to contract and labor begins.
Three or four days after the baby is born, the breasts begin to secrete milk.
Milk synthesis is stimulated by the pituitary hormone prolactin (PRL), and
its release from the breasts is stimulated by oxytocin.
Milk contains an inhibitory peptide. If the breasts are not fully emptied, the peptide accumulates and inhibits milk production. This autocrine action thus matches supply

Adrenal Glands
The adrenal glands are located on either side of the spine toward the first lumbar vertebra. Structurally adrenergic gland is divided into "Medulla" cortex and (adrenal) "Cortex"..

The adrenal glands are two small structures situated one atop each kidney. Both in anatomy and in function, they consist of two distinct regions: an outer layer, the adrenal cortex, which surrounds the adrenal medulla.
Hormones of the adrenal gland
 For example, the functions of the nervous system raises systolic tension, which has an impact extensively on the

coronary arteries and accelerates the heartbeat and inhibits muscle intestines and relaxes the nervous muscle and stimulates the decomposition of glycogen and the movement of glucose and raises the temperature and increases the life processes essentially without mediation of nervous.

The Adrenal Cortex

Using cholesterol as the starting material, the cells of the adrenal cortex secrete a variety of steroid hormones.
These fall into three classes:
glucocorticoids (e.g., cortisol)
mineralocorticoids (e.g., aldosterone)
androgens (e.g., testosterone)
Production of all three classes is triggered by the secretion ACTH from the anterior lobe of the pituitary.

Glucocorticoids

.The most abundant glucocorticoid is cortisol (also called hydrocortisone).
Cortisol and the other glucocorticoids also have a potent anti-inflammatory effect on the body. They depress the immune response, especially cell-mediated immune responses.

Mineralocorticoids

The mineralocorticoids get their name from their effect on mineral metabolism. The most important of them is the steroid aldosterone.
Aldosterone acts on the kidney promoting the reabsorption of sodium ions (Na^+) into the blood. Water follows the salt and this helps maintain normal blood pressure.
The secretion of aldosterone is stimulated byACTH (as is that of cortisol)and a drop in the level of sodium ions in the blood;

Androgens

The adrenal cortex secretes precursors to androgens such as testosterone.However, excessive production of adrenal androgens can cause premature puberty in young boys.

Addison's Disease: Hyposecretion of the adrenal cortices

Addison's disease has many causes, such as destruction of the adrenal glands by infection;
This disease is characterized by skin tow, as well as muscular weakness, indigestion and a drop in blood pressure. This disease is due, of course, to the lack of the adrenal cortex.

Cushing's Syndrome: Excessive levels of glucocorticoid
In Cushing's syndrome, the level of glucocorticoids, especially cortisol, is too high excessive production of ACTH by the anterior lobe of the pituitary;

The Adrenal Medulla
The adrenal medulla consists of masses of neurons that are part of the sympathetic branch of the autonomic nervous system the adrenal medulla functions as an endocrine gland.
The adrenal medulla releases:
adrenaline (also called epinephrine) and
noradrenaline (also called norepinephrine)
Both are derived from the amino acid tyrosine.
Some of the effects are increase in the rate and strength of the heartbeat resulting in increased blood pressure;

.

Other Hormones
Relaxin
Relaxin is found in pregnant humans but at higher levels early in pregnancy than close to the time of birth. Relaxin plays a more important role in the development of the interface between the uterus and the placenta .
Activins, Inhibins, Follistatin.
Activins increase the action of FSH; inhibins, inhibit it. The important role that activin and follistatin play in the embryonic development of vertebrates.
Oral contraceptives "pill"
The feedback inhibition of GnRH secretion by estrogens and progesterone provides the basis for the most widely-used form of contraception. Their inhibition of GnRH prevents the mid-cycle surge of LH and ovulation.
RU-486

RU-486 (also known as mifepristone) is a synthetic steroid related to progesterone. RU-486 is a progesterone antagonist; that is, it blocks the action of progesterone. It does this by binding more tightly to the progesterone receptor than progesterone itself but without the normal biological effects. The embryo can no longer make chorionic gonadotropin (HCG). Consequently the corpus luteum ceases its production of progesterone.

These properties of RU-486 have caused it to be used to induce abortion of an unwanted fetus. In practice, the physician assists the process by giving a synthetic prostaglandin .

Hormones of the Kidney, Skin, and Heart
Kidney

The human kidney secretes two hormones:
Erythropoietin (EPO)
Calcitriol (1,25[OH]2 Vitamin D3)
Erythropoietin is a glycoprotein. It acts on the bone marrow to increase the production of red blood cells. People with failing kidneys can be kept alive by dialysis. But dialysis only cleanses the blood of wastes. Without a source of EPO, these patients suffer from anemia.
EPO is also synthesized by osteoblasts in mice that have been made anemic; and in in the brain when oxygen becomes scarce there and helps protect neurons from damage.

Calcitriol
Calcitriol is 1,25[OH]2 Vitamin D3, the active form of vitamin D. It is derived from calciferol (vitamin D3) which is synthesized in skin exposed to the ultraviolet rays of the sun

Renin
One of the functions of the kidney is to monitor blood pressure and take corrective action if it should drop. Renin acts on angiotensinogen, a plasma peptide, splitting off a fragment containing 10 amino acids called

angiotensin I.
angiotensin I is cleaved by a peptidase secreted by blood vessels called angiotensin converting enzyme (ACE) — producing
angiotensin II, which contains 8 amino acids.
angiotensin II
constricts the walls of arterioles closing down capillary beds;
stimulates the proximal tubules in the kidney to reabsorb sodium ions;
stimulates the adrenal cortex to release aldosterone.

. Skin

When ultraviolet radiation strikes the skin, it triggers the conversion of dehydrocholesterol (a cholesterol derivative) into calciferol (vitamin D3).Calciferol travels in the blood to the liver where it is converted into 25[OH] vitamin D3.This compound travels to the kidneys where it is converted into calcitriol (1,25 [OH]2vitamin D3). This final step is promoted by the parathyroid hormone (PTH)

Heart

Natriuretic Peptides

In response to a rise in blood pressure, the heart releases two peptides:

A-type Natriuretic Peptide (ANP)

This hormone of 28 amino acids is released from stretched atria (hence the "A").

B-type Natriuretic Peptide (BNP)

This hormone (29 amino acids) is released from the ventricles. (It was first discovered in brain tissue; hence the "B".).The net effect of these actions is to reduce blood pressure by reducing the volume of blood in the circulatory system.

.

Chapter Two

Gynecologic Cancers

The major malignancies of the female reproductive tract are endometrial, cervical, and ovarian cancer (Fig.1).Other gynecologic cancers, such as choriocarcinoma, fallopian tube cancer, vagina cancer, and vulva cancer are rare, but cancers metastasis to these female genitals are more common[1].

Ovarian cancer accounts for 4% of the total cancer in women worldwide. In Iraq, this cancer is the eighth most common malignant neoplasm in women (3.8-4.2% of the total cancer) behind cancer of the breast ,non-Hodgkin's lymphoma ,brain and other CNS, leukemia ,urinary bladder ,bronchus and lung ,and skin. The number of cases for the period (1992-1997) was 828[2].

The incidence rates of corpus uteri cancer (95% endometrial) were found to be higher in the richer countries and urban populations; it is the fifth leading cancer in women at these countries[3].However, there is evidence of some change in the socioeconomic determinants of the disease in developed countries. In Iraq, cancer of corpus uteri accounts for (0.9-2.2%) of the total cancer in women. The number of cases for the period (1992-1997) was 477[2]. The risk of uterine cancer is low before age 40 years and increases sharply thereafter [4].

Most cancers of the vulva, vagina and cervix are detected relatively early because these organs are easily visualized and assessable for examination by cytology and biopsy [1]. Endometrial cancers are also detected early because they normally heralded by vaginal bleeding. In contrast, cancers of the ovary mostly present at a late stage with widespread intra-abdominal metastases. Ovarian cancer remains the most lethal gynecologic malignancy in women [5].

1.2 Endometrial Cancer

Endometrial cancer is the most common type of the corpus uteri (Fig.1c), constituting about 95% of all malignant lesions of the uterine cavity, while uterine sarcomas accounts for fewer than 5 %[3].

The normal endometrium contains two major components: endometrial stroma and glands. The glandular cell produces cytoplasmic receptors for estradiol and, when stimulated, results in proliferation of glandular epithelium, producing larger and more numerous glands. In the premenopausal cycling endometrium, the maturational effects of progesterone from the corpus leteum counter balance the proliferative influence of estrogen. If conception does not occur, the lining is shed and a new cycle is initiated. When ovulation ceases at menopause, the endometrial becomes quiescent and the glands atrophy. Tumors arising from the glandular epithelium, either de novo or as a result of abnormal estrogenic stimulation; are adenocarinomas[6]. The transition from benign proliferative endometrium to well differentiated adenocarcinoma probably proceeds through several intermediary steps, collectively termed hyperplasia. Hyperplasias are characterized by an overgrowth of the glandular component of the endometrial lining, whereas cancers are defined by the size of the cells, irregular nuclear membranes, coarse chromatin clumping, loss of glandular pattern, increasing nuclear atypia, and mitotic activity[6].

1.2.1 Histology

About 90% of endometrial cancers are endometriosis type (adenocarcinomas), which contain glands similar to those seen in the normal endometrium. These tumors are subclassified by their degree of differentiation into three grades:well (grade I, 50%), moderately (grade II, 35%), and poor differentiated (grade III, 15%). The International Federation of Gynecology and Oncology (FIGO) grading scheme is based on the growth pattern (relative proportion of glandular and solid areas):less

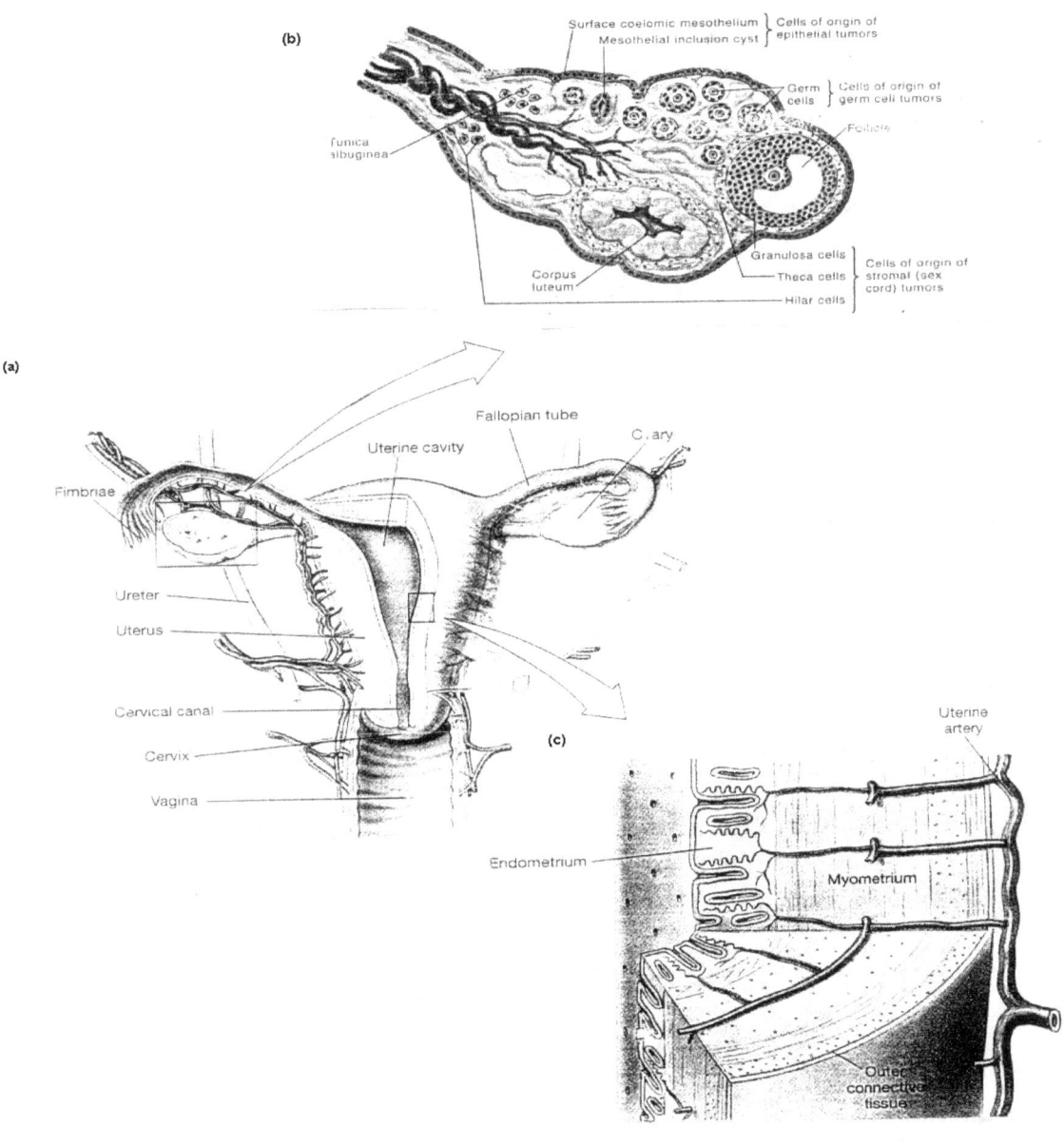

Fig.1 [(9)] (a):Female reproduction tract

(b): Embryologic origins of ovarian cancer.

(c):Tissue layers of uterus

than 5% solid growth is grade I, 6-50% is grade 2, and more than 50% is grade 3[3].   Endometriosis adenocarcinoma accounts for 60-65% of the adenocarcinomas, it is characterized by the disappearance of stroma between abnormal glands, necrosis, and hemorrhage[1].Some typical adenocarcinoma contains areas of squamous metaplasia, which can have a benign or malignant appearance, such tumors are termed adenocarcinoma with squamous differentiation[6].

A variant of endometriosis adenocarcinoma is the villoglandular type, which appears to have more aggressive behavior [7].Uncommon cell types (clear, papillary serous, mucinous, squamous and undifferentiated) account for the remaining 10% of endometrial cancers [6].

1.2.2 Etiology

Endometrial cancer is one of the few malignancies for which the etiology is well understood.Epidemiologic, endocrinologic, and clinical studies have shown that the association between endometrial cancer and most of the identified risk factors can be explained by increased exposure to endogenous or exogenous estrogens. This hypothesis is supported by the increased risk of endometrial cancer (related to obesity, early age at menarche, late age at menopause, low parity or nullparity, certain types of ovarian tumors, history of menstrual disorders, history of infertility, and use of estrogen replacement therapy (ERT) or sequential oral contraceptives). It is also supported by the decreased risk of endometrial cancer (related to the use of combined oral contraceptives, early age at menopause, high parity, and smoking)[4].

Medical conditions such as diabetes, hypertension, and thyroid diseases have been suggested to increase the risk of endometrial cancer, but these finding are not consistent[8]. It remains, unclear whether these finding are the result of the association of these medical conditions with obesity or with other biologic mechanisms.

1.2.3 Clinical Evaluation and Treatment

The initial growth phase of most endometrial cancers consists of polyploidy expansion within the endometrial lining. Small areas of necrosis or surface breakdown in the tumor produce abnormal vaginal bleeding. Consequently, the premenopausal women usually present with metrorrhagia, and the older woman has postmenopausal bleeding. Patient with advanced disease may have pelvic pain, bleeding, or bloating[3]. The diagnosis of endometrial can be reliably established by office endometrial biopsy. Dilatation and fractional curettage or hysteroscopy with biopsy may be helpful diagnostic procedures in cases in which out patient biopsy is non diagnostic[1].

The workup on the woman with endometrial cancer should include a physical examination, endometrial biopsy, laboratory studies, and a chest radiograph. Endometrial cancer is typically a disease of women in their sixties and seventies who frequently have coexisting medical problems, so the evaluation should focus on an assessment of operative risk. Additional more sophisticated studies such as proctoscopy, barium enema, ultrasound, computed tomography (CT), or magnetic resonance imaging should be reserved for clinical situations in which advanced disease is suspected[6].

Schematic approach to treatment endometrial cancer is illustrated in (Fig. 2)[6].

Surgical excision of the primary tumor via total abdominal hysterectomy with bilateral salpingo –oophorectomy (TAH-BSO); sometimes accompanied by removal of the pelvic and periaortic lymph nodes; is the mainstay of therapy for endometrial cancer. The selection of additional therapy is based on the result of the staging biopsies of uterine tumors, which was adopted in 1988 (Table 1) by FIGO surgical staging[3].

1.3 Epithelial Ovarian Cancer

The ovary is composed of an outer cortex and an inner medulla. The cortex containing the coelomic surface mesothelium, germ cell (oocytes), specialized hormone producing stroma (granulose and thecal cell), and unspecialized supporting stromal cells. Primary ovarian cancers are classified; according to the structure of the ovary from which the tumor is derived (Fig.1b); into three major categories, epithelial, germ cell, and stromal[9]. The ovary may also be the site of metastasis spread from

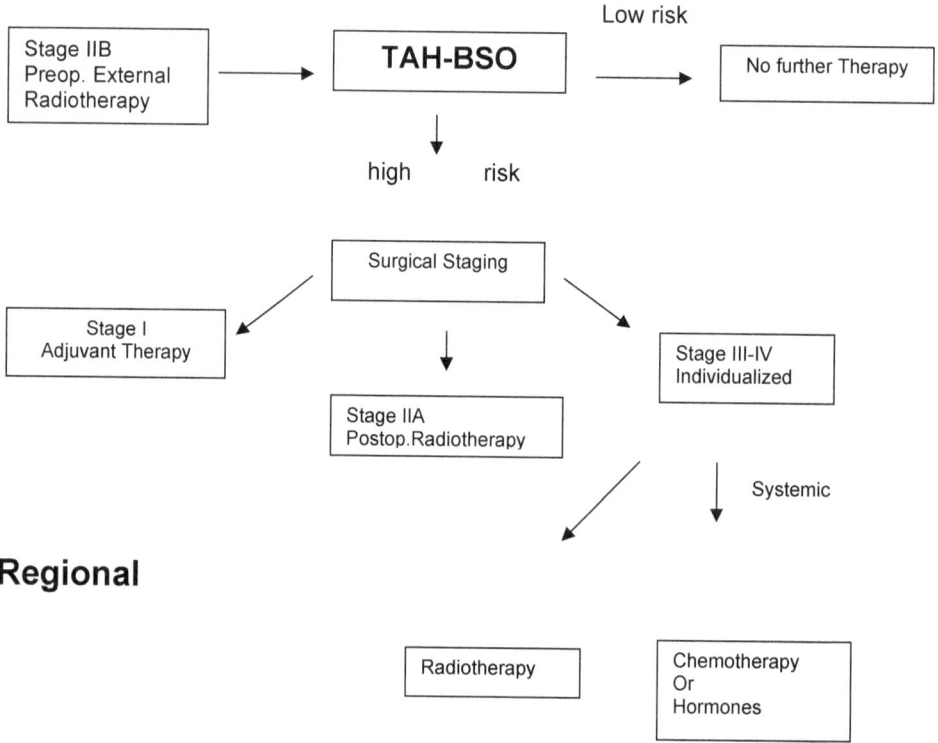

Fig.2 Basic treatment strategy for endometrial cancer[6].

Table (1) Surgical staging system for cancer of the uterine body[3].

Stage	Extent
I	Cancer confined to the corpus
Ia	Tumor limited to endometrial
Ib	Invasion to less than one half the myometrium

Ic	Invasion to more than one half the myometrium G_1: Well differentiated (≤5% of a nonsquamous or nonmorular solid growth pattern) G_2: Moderately differentiated (6%-50%) G_3: Predominantly solid or undifferentiated
II	Cancer involves corpus and cervix but does not extend outside the uterus
IIa	Endocervical glandular involvement only
IIb	Cervical stromal invasion
III	Cancer extends outside the uterus but not outside the true pelvis
IIIa	Tumor invades serosa and/or adnexa or positive peritoneal cytology
IIIb	Vaginal metastases
IIIc	Positive pelvic and/or paraaortic nodes
IV	Cancer extends outside true pelvis or invades bladder or rectal mucosa
IVa	Tumor invasion of bladder and/or bowel mucosa
IVb	Distant metastases including intra-abdominal and/or inguinal lymph nodes

gynecologic and other tumors. Epithelial ovarian tumors are the most common, accounting for 80-90% of cases[5].

Epithelial ovarian cancers are derived from the coelomic surface mesothelium. During ovulation, the dominant ovarian follicle ruptures, releasing the ovum. To heal the ruptured follicle, surface mesothelial cells proliferate and fill the defect. Inclusion cysts may form and are lined with these mesothelial cells. Over years of repeated ovulation and repeated growth of mesothelia cells, the regulation of this proliferation may become disrupted, allowing the development of tumors (Fig.1b)[9].

Numerous factors can stimulate –or fail to control mesothelial proliferation including growth factors, oncogenes, and tumor suppressor genes[10].

1.3.1 Histology

Epithelial ovarian tumors classified according to cell type, pattern of growth (cystic, solid, and surface), atypia and invasiveness (benign, borderlines, malignant). The ovarian surface epithelium, when involved in neoplastic conditions, often undergoes a mullerian differentiation as a result, it may produce any of the adult epithelia, including tube, endometrial and, endocervical mucosa, singly or in combination. So the histological types of the epithelial tumors are serous, mucinous, endometriosis, clear cell, Brenner, mixed epithelial, undifferentiated, and unclassified[3]. Serous tumors make up about 20% of all benign ovarian tumors and 75% of epithelial ovarian cancers. Malignant serous tumors are usually multicystic, bilateral and resembling fallopian tube epithelium. Serous carcinoma may have a complex mixture of cystic and solid areas with extensive papillations, or they may contain a predominantly solid mass with areas of necrosis and hemorrhage[11]. The second common type of epithelial ovarian cancer is endometriosis tumor (10%) which is histological indistinguishable from endometrial uterine cancer. About 6-10% of epithelial ovarian cancers are mucinous, which characterized by epithelial – mucin producing cell resembling those of the endocervix[9]. Clear cell tumors account for<5% of all primary ovarian cancers. The tumor cell is large and their cytoplasm is clear. These tumors have more aggressive behavior than other epithelial cancers. Brenner tumors, these unusual lesions are though to originate from the ovarian mesothelium despite their urothelial –like transitional cell structure. They are solid and almost benign.

1.3.2 Etiology

Several factors have been associated with an increased risk of epithelial ovarian cancer, while two factors appear to reduce the incidence of this malignancy – pregnancy and use of oral contraceptives – presumably because they suppress ovulation. The risk of epithelial ovarian cancer is high in ovulating nulliparous women, women using of fertility drug, women exposuring to hormones (particularly gonadotropins) and carcinogens, such as infectious agents and chemical carcinogens[12].

The large majority of ovarian cancer presumably related to environmental factors, but 5-10% of women with epithelial tumors have a predisposing familial syndrome. Three autosomal dominant syndromes have been identified: site –specific ovarian cancer, hereditary breast – ovarian cancer syndrome, and lynch syndrome II (adenocarcinomas of multiple sites, including breast, ovary, and endometrial, gastrointestinal tract). All three types of familial ovarian cancer develop at a younger age than sporadic ovarian cancer[9].

Other factors, such as obesity, high-fat diet, early menarche, late menopause, single women, may predispose women to develop this malignancy. The hypothesis is that the continual ovulation, uninterrupted by pregnancy may develop the disease[4].

1.3.3 Clinical Evaluation and Treatment

Epithelial ovarian cancer is seldom detected at an early stage; up to 80% of these tumors have metastasized by any of the four routes (peritoneal seeding, lymphatic embolization, direct extension and hematogenous dissemination). The first sign of early ovarian cancer is usually an asymptomatic mass found during a pelvic, rectal, or abdominal examination. A benign tumor or slow-growing cancer may grow quite large and cause abdominal distention and may compress the rectum or bladder, producing constipation, urinary frequency or nocturia. Ovarian tumor occasionally undergoes torsion, resulting in localized pain, nausea, and possibly low fever, and may also cause symptoms of inappropriate estrogen or androgen synthesis[9].

The diagnosis of epithelial ovarian cancer is surgical, usually by a laparotomy, but occasionally at laparoscopy. Ultrasound of adnexal pelvic mass is useful prior to the surgical diagnosis. Most patients with a pelvic mass require ultrasonography, chest radiography, CT, and aspiration of pleural effusion to check for metastases. Mammography should be performed to exclude primary breast cancer[3].

The management of primary epithelial ovarian cancer may be divided into three phases: primary definitive surgery, postoperative adjuvant therapy usually chemotherapy[13], and systematic follow up and reevaluation. The initial surgical procedure (usually TAH-BSO) is used to determine the extent of the disease (staging) and to remove as much gross disease as possible (cytoreduction). In ovarian cancer, metastases adhere superficially to peritoneal surfaces without deep invasion. Therefore, surgical resection of metastasis ovarian cancer is standard care[9]. The tumor must be staged to facilitate planning of adjuvant therapy, the FIGO staging system of 1987 (Table 2)[3] is based on the finding at surgical exploration. Following the cytoreductive surgery or debulking, patients should be treated, generally with chemotherapy and some time with radiation or endocrine therapy[14][15].The majority of women with advanced epithelial ovarian cancer will ultimately relapse and develop drug-resistant disease.Thus,there is a common need for second-line treatment including cytotoxic drugs,or secondary cytoreductive surgery or palliative surgery[16].

Table (2) Surgical staging system for epithelial ovarian cancers[3].

Stage	Extent (proportion of cases)
I	Cancer limited to ovaries (15%)
Ia	Limited to one ovary, no ascites
Ib	Both ovaries involved, no ascites
Ic	Ia or Ib with ascites or positive peritoneal washings
II	Cancer of one or both ovaries with extension limited to pelvic tissue(15%)

IIa	Extension to uterus or tubes
IIb	Extension to other pelvic tissues
IIc	IIa or IIb with ascites pr positive peritoneal washings
III	Cancer involving one or both ovaries with peritoneal implants outside the pelvis and/or positive retroperitoneal or inguinal nodes. Tumor is limited to the true pelvis but with histological proved extension to small bowel or omentum (65%)
IIIa	Tumor grossly limited to the true pelvis with negative nodes but with histological confirmed microscope seeding of abdominal peritoneal surface
IIIb	Same as IIIa, but abdominal peritoneal implants do not exceed 2 cm in diameter
IIIc	Abdominal implants greater than 2cm in diameter and/or positive retroperitoneal or inguinal lymph nodes
IV	Distant metastases present (including cytology-positive pleural effusion metastasis liver parenchyma or peripheral superficial lymph nodes(5%).

1.4 Tumor Markers in Gynecologic Cancers

Tumor markers; which may be considered as any biological aberration that indicates the presence of a tumor; are required for primary diagnosis, histological identification, assessment of the extent of the disease pre- and post operatively, therapy guidance, and early detection of recurrence. Ideally, the tumor marker assessed in serum, or in another body compartment, should accurately reflect the presence and the amount of residual tumor, and should also distinguish benign from malignant disease[17]. In gynecological practice, it is well documented that human chorionic gonadotropin(hCG), alpha-feto protein(AFP), estrogens, and androgens have been contributed greatly to the diagnosis and management of gestational trophoblastic disease, gram cell tumor,and granulose cell tumor, respectively[17][18].

By the introduction of polyclonal and monoclonal antibody technology, variety of tumor carbohydrate associated antigens have been identified as a result of glycosylation changes in the carbohydrate moieties of glycoproteins and glycolipids by tumor cells, some of tumor associated antigens also represent differentiation antigens and receptors that may reexpressed or over expressed[18][19].

1.4.1 Antigenic Markers

1- Squamous Cell Carcinoma (SCC) Antigen: The SCC antigen was described in 1977 as one of 14 subfractions of tumor associated antigen found in human cervical cancer tissue [20]. SCC antigen have been reported to be elevated (>2 ng/ml) in 35% of stage I patients, and increasing to 91% in stage IV for cervical cancer [21][22]. While serum SCC antigen level were increased in 8-29% of endometrial cancer. In vulvar and vaginal squamous cell cancers, the reported frequencies of SCC antigen elevation are lower (42% and 17%, respectively) than in cervical cancer [17].

2- Lipid-Associated Sialic Acid (LSA): Elevated levels of LSA are found in a variety of malignancies, including cancers of the vulva, vaginga, uterus and ovary [18]. Due to its lack of specification, LSA used alone had not been useful in gynecologic cancers.

3- CA-125: Since its first report in 1983 [23], CA-125 has been used widely for monitoring epithelial ovarian cancer. Among healthy individuals, 99% will have been serum level <35U/ml [24]. Approximately 85% of ovarian cancer patients have elevated CA-125 level (> 35 U/ml) for all histological types of epithelial ovarian cancers but more frequently for nonmucinous tumors than for mucinous types[25][26]. It is also expressed in cancer of epithelium female reproductive tract[27] and can be detected in 22% of patients with nongynecological cancers (Breast, lung, Colon, Pancreas), also

detected in benign gynecological tumors (endometriosis, adenomyosis, ovarian cystadenomas)[28][29].

The low specificity hampers the use of CA-125 as a screening test for ovarian cancer. Combined with pelvic examination and/or ultra sound, the specificity can be increased[30]. The sensitivity of the marker might be also increased by examining peritoneal fluid, where the CA-125 levels are higher than in serum (Cutoff at 200u/ml)[31]. CA-125 is very rarely elevated in endometrial cancer confined to the uterus, whereas elevation is found in 78-100% of cases with extra uterine spread[27].

4- Other Antigens Defined by Monoclonal Antibodies:
Following the report of CA-125, a number of monoclonal

antibodies such as CA 19-9 and CA 15-3 were developed in colorectal cancer[32], breast cancer[33], ovarian cancer and endometrial cancer[34] as immunnogens, and this explain their lack of specificity.

Various monoclonal antibodies have been raised against epithelial mucin like caner antigen MCA , CA M26 , and CA M29. These cancer antigens have been reported in cancers of breast, colon, ovary, endometrial and cervix [35][36], but non of them, either alone or in combination , has been shown to be as useful as CA-125 .

Another marker for ovarian cancer is NB/70K which seen to be a marker for all histological types of epithelial ovarian cancer[17]. Serum NB/70K is elevated in all stages of ovarian malignancies, including more than 50% of early stage cases[37][38]. 5- Carcinoembryonic Antigen (CEA): This oncofetal protein can be demonstrated in most gynecological cancers immunohistologically, while plasma levels are often too modestly elevated to be useful for diagnosis or for disease monitoring[17] . Slightly elevated plasma CEA levels are seen in approximately 50% of ovarian cancer patients[39][26], and mucinous tumors tend to have highest levels. Measuring CEA with CA-125 and CA19-9 can be useful to differentiate between ovarian and colorectal cancer. In the uterus, adenocarcinomas that arise in the endocervix express greater amount of CEA than adenocarcinomas of the endometrial. This is reflected in the reported frequency of elevated plasma CEA levels : 68% in cervical cancer, and 34% in endometrial cancer[40]. Squamous cell cancer, which represent 85% of the cervical cancer, also express CEA, but at more modest level. In a study on 205 patients, only 28% had plasma level above 5ng/ml. High levels indicated advance disease or lymph node metastases[41].

1.4.2 Genomic Markers

By cytogenetic examination, using a chromosome spread technique or cell flow cytometry, general genetic alteration have been identifying. Aneuploidy is a common finding in endometrial cancer and ovarian cancer[42][43]. There is a correlation between DNA ploidy abnormalities and grade of differentiation (more common in moderately and poorly differentiated lesions). Trisomy of chromosomes 1,7 and 10 and allelic loss at loci on chromosomes 3p (71%), 9q(38%) 10q(35%) and 17p(35%) have also reported in endometrial cancer[44]. Most ovarian and endometrial cancers over express of macrophage colony stimulating factor (M-CSF) and c-fms which encode the M-CSF receptor[45][46]. The macrophages secrete cytokines (tumor necrosis factor α and interleukins 1 and 6) that stimulate the growth of tumors[47]. Epidermal growth factor (EGF) and polo-like kinase (PLK) protein (contribute to regulation of the cell cycle) have been found to be expressed in most ovarian tumor, while the suppressor gene p53 defective[9][48]. However, the clinical significance of these genomic markers has not yet been established.

1.4.3 Enzymatic Markers

Lactate dehydrogenase (LDH) and heat stable alkaline phosphatase (HSAP) have been also considered as potential tumor markers in gynecology cancers.

The glycolytic enzyme LDH aroused a considerable interest in oncology since Warburg[49] reported increased rate of glucose utilization producing more lactate by tumor cells. LDH is often elevated in epithelial ovarian cancer[50], as in many solid malignancies[51], making it a non-specific tumor markers. In germ cell tumor, the two isoenzymes (LDH-1 and LDH-2) are increased and their activity appear to parallel the response to therapy [52]. Serum LDH levels were documented to be elevated in cervical cancer even at the early stage of the disease and in combination with other biomarker was more beneficial for diagnosis and treatment monitoring of patients[53][54].

Alkaline phosphates (ALP) was one of the first examples of tumor associated enzymes[55]. The heat stable alkaline phosphatase (HSAP, Regan isoenzyme) has been found in a variety of solid tumors and in 6-64% of ovarian cancer patients[56][57]. The marker level did not correlated to tumor burden or prognosis and half of the patients lost the marker during progression.

Chapter Three

Hormonal Markers

1.4.4 Hormonal Markers

Gonadotropin hormones (LH, FSH) can function as tumor markers in gonadal stromal tumor in addition to estrogen and androgen[17]. Inhibins, which are produced in ovarian granulosa cells, cause a specific suppression of pituitary

FSH release. Radioimmunometric determination of inhibin has proven to be reliable marker in the monitoring of granulosa cell tumors[58].

1.4..4.1 Human Chorionic Gonadotropin (hCG)

Since the discovery of human chorionic gonadotropin (choriogonadotropin) hormone by Hirose[59]] and Ascheim and Zondek[60], its measurement has been the basis of pregnancy diagnosis and, a marker for many trophoblastic and nontrophoblastic tumors.

This hormone is synthesized by the trophoblastic cells of the placenta during the early weeks of pregnancy. Human chorionic gonadotropin stimulates the ovarian corpus luteum to produce progesterone until the placenta it self acquires the ability to produce this pregnancy sustaining steroid[61]. An increasing evidence supports the synthesis of hCG in small quantities by the pituitary[62].

Human chorionic gonadotropin belongs to the glycoprotein hormones family (also includes hLH, hTSH, hFSH) whose members share a common α-subunit (contains 92 amino acids) and vary in their β-subunit. Each subunit is produced by a separate gene, the α subunit is encoded by a single gene present on chromosome 6q21.1-q23 whereas cluster of seven nonallelic genes located on chromosome 19q13.3 encoded for β-subunit of hCG(hCGβ). (One or more of these genes may preferentially expressed during pregnancy and tumor genesis)[63][64]. The β-subunit confer biological specificity; in humans, hCGβ and hLHβ are closly related subunits with 82% identity, and this reflects a common biological function for these two hormones.The hCGβ (contains 145 amino acids) further differs from that of hLHβ (contains 114 amino acids) by the 24-amino acid carboxy terminal polypeptide extension (CTP). It has been suggested that this glycosylated extension may impart extra solubility and in vivo circulation life time to hCG[65]].

Human chorionic gonadotropin could be purified from pregnancy urine by a combination of organic precipitation, ion exchange chromotography, and gel filtration[66].

A- Chemical Structure of hCG

The primary structure of α and β subunits for hCG were determined in 1973[67][68]. There are five disulfide bridges in the α-subunit and six in the β-subunit (these are conserved across the family). Human chorionic gonadotropin (38,633 Dalton) is approximately 30% carbohydrate by weight and contains both complex biantennary N-asparagine linked and simple O-serine linked. Each carbohydrate moiety terminates in sialic acid, but considerable carbohydrate heterogeneity result in a wide isoelectric point (pI) distribution of the hormone (pI generally 3-6 and as high as 10 for asialo forms of the hormone). The hCG produced in trophoblaste disease exhibits greater carbohydrate heterogeneity (such as triantennary structures) and some cancers produce only asialo hCG[66]. Removal of the terminal sialic acid residues markedly reduce the half-life of hormone in the circulation by enhancing binding to hepatocyte lectin and thus clearance from the blood stream. The carbohydrate moieties of hCG also played a role in hormone secretion, stability, folding and subunit assembly[99]. In vitro, studies with deglycosylated hormone generally indicate a greater loss of biological response than receptor binding. It is believed that the carbohydrate chains bind to a lectin-like membrane component to give the biological response. It was further found that the carbohydrate in the α-subunit was more important in the hormone function than that present in the β-subunit, which seems to be important to maintain the proper conformation of the hormone[70][71].

Various attempts have been made to construct three dimensional (3D) model of hCG[72-78]. Based on amino acid sequence and information; accumulated from chemical modifications (oxidization, deglycossylation, desialyation, reduction), enzymatic modifications (nicks, fragmentation and deletion of fragments), cross reactions, disulfide pairings, molecular biology studies (site directed mutagenesis, chimeric hormone constructs) with monoclonal antibodies(MAb) mapping of surface epitopes, spectral analyses (especially circular dichroism (CD)) and diffraction analyses (X-ray diffraction of hydrogen floride-hCG crystals), multi wavelength anomalous diffraction

(MAD)); the proposed structure of hCG is predominantly composed of β structure with three helical segments, two in the α and one in the β-subunit. The antiparallel β-strands in each subunits are joined by three hairpin loops. Both subunits structure like some protein growth factors (NGF, TGF-β and PDGF-ββ) contain the so-called cysteine knot motif by three disulphide bonds (cys I-IV, cys II-V, cys III-VI). The heterodimer is stabilized by a segment of the β-subunit which wraps around the α-subunit and is covalently linked like a seat belt by cys 26-cys110. This hetrodimer has a large area of interface with only a small hydrophobic core. The overall topology of hCG subunit is shown in (Fig. 3).

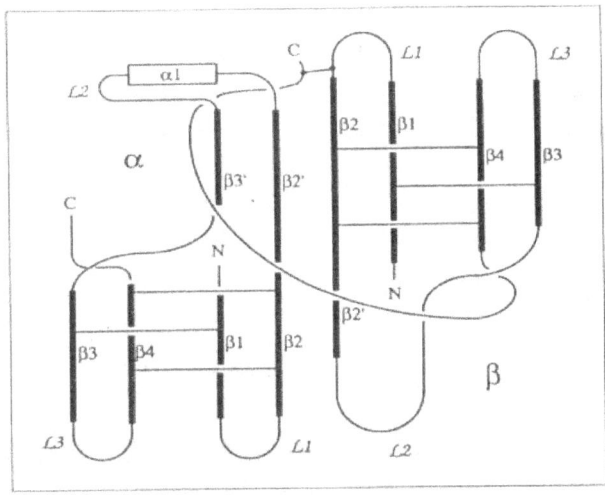

fig.3 A schematic drawing of the hCG dimer topology

B- Molecular Forms of hCG

Various molecular forms of hCG (intact or modified), free subunits (intact or modified), and degradation products are present in biologic fluid, tissues of normal and tumoral organs and reference preparations[79].

Intact hCG is defined as heterodimer comprising the two mature α and β subunits. Its levels in the sera of normal nonpregnant women increase with age, and higher than in men[80][81]. In women <50 year the upper reference limit is

8.6 pmol/l (2.9 Iu/L), and in those >50 years is 15-5 pmol/L (5.3 Iu/L), where the level of hCG in urine is < 2 pmol/mmol creatinine. In contrast, hCG levels are 2-6 pmol/L (0.7-2.1 Iu/L) in sera of healthy men. Plasma intact hCG rises exponentially during the first trimester of pregnancy, reaching a peak of 25-30 µg/ml at 8-10 weeks. Following a rapid decreases until 15 weeks, a slower but continuous fall is observed up to delivery. Interestingly, amniotic hCG levels are <20% of maternal serum level at all times of gestation[79].

Free hCGβ refers to the non-combined β-subunit consisting of a glycosylated 23KD single chain polypeptide. This form appears to be devoid of any biologic role during pregnancy (the peak level reaches 50-70 ng/ml between 8-10 week)[79]. In normal individuals, free hCGβ serum values are extremely low (never excceding 4.5 pmol/L) while this form is most often secreted in nontrophoblastic tumors[82].

Free hCGα refers to non combined α-subunit. Although it appears to have a structure identical to that of the combined α subunit of hCG, its carbohydrate composition is quite different which prevents combination[83]. This form may have a biologic role in pregnancy independent form that of hCG, where its level in serum slowly increase up to 18 weeks to reach a peak of 30-50ng/ml. In healthy individuals, levels up to 3000 pg/ml are present in sera originate in the pituitary gland[62]. Free hCGα may increase in pituitray tumors.

Beta -Core Fragment of hCGβ (hCGβcf), is small forms (10KD) immunoreactivity identify. Its two disulfide-bridged peptides lacking the amino acid sequences 1-5, 41-54 and 93-145 and corbohydrate structure changes of hCGβ[84]. It has been demonstrated that HCGβcf is primarily product of renal degradation of hCGβ, although some fragments may originate from either placenta or cleavage in the circulation. hCGβcf is the most abundant fragment of hCGβ in pregnancy urine and is also found in the urine of patients with a broad spectrum of malignancies and benign tumors. Circulating hCGβcf is almost undetectable, while levels of 0.51-1.25 pmol/mmol creatinine are found in urine of healthy women[69]

Nicked hCG (hCGn or hCGβn), are detectable in a significant proportion of both hCG and its free β subunit of normal and tumoral origins, as a result of intrachain proteolysis nicking within the β44-49 region [85][86]. The hCG appears to be uns Table and rapidly dissociates and weakly recognized by MAb. Furthermore, some results indicate that hCGβn is an intermediary form in the metabolism of hCG, and further cleavage to hCGβcf.

Several other molecules are detectable mostly in tumors, due to the modification of peptide structures both in hCGα and hCGβ such as N-terminal heterogeneity hCGα, mutant hCGα, large hCGα and CTP, hyperglycosylated hCG[87].

C- Production and Metabolism of hCG

Both normal placental cells as well as neoplastic trophoblastic cells,and sperm cells in the male synthesize and secrete hCG and its free subunits[88]. Human chorionic gonadotropin plays a critical role in the maintenance of the corpus luteum during the first 4 to 6 week of gestation to produce progesterone to keep the endometrial intact where the blastocyst is implanted, hCG also stimulates the testosterone production by the developing tests in males Fetuses[79]. The physiological processes regulating hCG production are not well understood. It is likely that its biosynthesis is dependent on the stage of differentiation of the trophoblastic cells and regulated by several autocrine and paracrine factors acting through cellular signals on the regulatory elements of the α- and β-subunit genes[89].

About 80% of hCG is metabolized in liver and renal tissue and excreted predominately as hCGβcf, whereas 20% of circulated hCG is excreted unmodified. In normal individuals, the plasma half-life of hCG is about 24-36 hours. Its clearance includes an initial, fast component ($t_{1/2}$ of 5-6h) and a second, slower component ($t_{1/2}$ of over 24 h)[69].

D- Clinical Use of hCG Determinations

Diagnosis of Pregnancy and Related Disorders: Immunochemical determination of hCG in serum or more commonly in urine, is the main method for diagnosis of pregnancy. In serum a discrimination limit of 10 Iu/L is often used [79].

Spontaneous abortion is the most common complication of pregnancy, and abnormally low serum hCG concentrations are often observed in these patients. Serial determinations of hCG in serum in combination with sonography are much used for diagnosis this condition [66].

Decreasing or slowly increasing hCG concentration during early pregnancy may also indicate ectopic pregnancy (implantation of the blastocyst outside the uterus)[66].

The potential usefulness of either hCG or its free subunits(in combination with AFP) in prenatal screening for chromosomal abnormalities (particularly Down's syndrome), was the subject of intense debate[90][91]. While the mean hCGβ value in cases of trisomy 21 is significantly higher than in unaffected cases, the median value in trisomy 18 was found to be significantly lower.

Placental and Testicular Trophoblastic Tumors: Infact hCG and free hCGβ are ideal tumor marker for gestational trophoblastic disease (GTD ....choriocarcinoma,and hydatidiform mole[92]....)and testicular germ cell tumors[93]. The ratio of hCGβ to intact hCG is higher in patients with malignant than with benign trophoblastic tumors, which used for diagnosis. Measurement of blood intact hCG can be utilized for the therapeutic monitoring of GTD . Molecular heterogeneity has been demonstrated in the carbohydrate structure of hCG in choriocarcinoma, hCG is substantially diminished in sialic acid content and is composed of tri and tetra-antennary carbohydrate moieties[94].

Human chorionic gonadotropin or its free subunits presents in (80%) of sera patients with nonseminomatous testicular tumors, where hCG is elevated to more than 20,000 pg/ml[93]. Indeed, hCG, free hCG, AFP are the Most useful markers for the diagnosis, prognosis, and follow up of these malignancies[95]. Measurement of hCG in CSF is done when cerebral metastasis are suspected (ratios of hCG in CSF to hCG in serum exceed 1:60). In seminoma testicular cancer, hCG is elevated in 16% of patients[96].

Nontrophoblastic Tumors : The presence of hCG and its subunits in the tumor itself or in the serum and malignant effusions of patients with nontrophoblastic and nongonadal tumors were investigated. Numerous investigators, using immunochemical procedures, have demonstrated that cancers of the breast, uterus, lung, ovary, cervix, gastrointestinal tract, bladder, liver produce elevated concentrations of hCG or its subunits.

a-Blood studies: Early results employing radioimmuno assay(RIA) for hCG indicated that a great variety of malignant tumors expressed hCG[97-102]. The overall incidence was in the range of 19%-30% of all tumors studied. The levels of hCG were only moderately increased and shown variable correlation to tumor stage, histological grading or clinical course. Therefore, the clinical value of determine hCG in serum for diagnosing and monitoring patients with non-trophoblastic malignancies remains controversial. Studies with highly specific sandwich procedures and effective MAb showed that free hCGβ is mainly elevated while intact hCG and hCGα are slightly elevated or within the normal range in the sera of nontrophoblastic cancers. The malignancies that exhibited the greatest incidence are gynecologic cancers, tumor of the head, neck, lung, gastrointestinal tract tumors, and melanoma[102-111]. These studies suggested that free hCGβ is highly diagnostic of malignancy in general and defines a subgroup of aggressive nongonadal malignancies.

b-Urine studies: hCGβcf, free hCGβ, and asialo hCG have been reported in urine of different malignancies employing two soild phase capture antibodies[112][104][105]. Using a cutoff of 0.2ng/ml for hCGβcf, 59% of the cervical and endometrial and 70% of the ovarian cancers patients exceeded this level[113]. It has been suggested that the hCGβcf in the urine of patients is due to the degradation of the free hCGβ produced by these cancers.

c-Effusion fluids studies: Free hCGβ immunoreactive material has been reported in ascetic and tumor cyst fluids of patients with non trophoblastic cancers (including gynecologic cancers) more than intact hormone[106][107]. Free hCGβ levels have shown to be higher in malignant ascites as compared to the corresponding plasma values of the same patients. Determining free hCGβ in malignant effusions enhanced the sensitivity of this marker as compared to serum.

d-Tissues studies: Human chorionic gonadotropin in the cell is present in two pools, a secretary pool and membrane pool. The latter becomes an intrinsic part on determination hCG on the cancer cell surface. Various investigations on cultured cancer cells (including cells form nasopharynx, lung, cervix, bladder, breast, endometrial, colon, prostate, oral cancers, leukemia's, lymphomas, and retinoblastomas), using flow cytomatry technique, demonstrated the expression of membrane associated hCG and its subunit[114-124]. These studies showed high expression of free hCGβ than intact hCG and there was no relationship between the hCGβ positivity and the histological type or morphological features of tumor, but a number of authors have suggested that tumor which expresses hCG will pursue amore aggressive course and has a worse prognosis[125]. Analyses also done on extracted of media of cultured caner cells, reflecting the secretary pool or total cell pools. Other techniques such as immunohistochemical techniques were used to identify hCG in tumor and normal tissues[126-137]. Positive staining for hCG were found in, ovarian, cervical , colorectal , gastric , hepatoblastoma , large bowel, urothelial and breast cancers tissue. Authors also reported the stimulation of the growth of several tumor cell lines by hCG[138-141]. These findings with other results which confirmed that hCG has also chemical and physiologic properties of growth factor[76][77], have provided a scientific basis for studies of, prevention from, and control of the cancer by active or passive immunization against hCG and its subunits[142]. Some studies used anti hCG vaccines which originally was developed for fertility control[143][144]. In addition to that, studies on cluster of genes encoding the hCGβ demonstrated that five nonallelic genes are transcribed in vivo with highly variable level of expression in first trimester placenta tissue, whereas normal nontrophoblastic tissue express 2 allelic hCGβ genes. In tumor tissue collected from patient with breast bladder, prostate, thyroid cancer found that up to 61% of these tumors expressed four nonallelic hCGβ genes different from that of normal tissue. These findings provide the basis for a simple test that may be useful for the molecular diagnosis of nontrophsblastic cancer[64][115].

E- Determination Methods of hCG

Immunological methods are nearly exclusively used for clinical determination of hCG[82], while assessment of the potency of hCG preparations to be used for pharmaceutical purposes should be performed by bioassays.

Biological Methods

a-<u>In vivo methods</u>: These bioassays measure the response of gonadal tissue to hCG in various animals (mice, rabbits, frogs). These methods are no longer widely used because its susceptible to yielding spurious results due to variation in the metabolic clearance rates of the forms of hCG being assessed and are based on animal –

derived rather than Human chorionic gonadotropin receptor[69].

b-<u>In vitro methods</u>: Two methods have been widly used for assessing the hLH-like bioactivity of hCG, one is based on the rat interstitial cell testosterone response (RICT assay)[145] and the other on the mouse interstitial cell response (MICT assay)[146]. These methods have the advantage of the low detection limit but the limitation is their lack of specificity (hLH activity, factors modulate cellular responses) and heterologous nature of the systems. Homologous bioassays have been constructed with the modern techniques of molecular biology, where the human hCG receptor gene has been cloned and transfected into cells to produce human receptors[147].

Binding Methods

Binding assay can utilize antibodies or receptor as binding protein.

a. <u>Competitive Immunoprocedures</u>

Radioimmuno assay(RIA), and Receptor methods: Pregnancy test is the first specific immunoprocedueres (qualitative agglutination inhibition methods) in which, latex particles or sheep red blood cell coated with the hCG, mix with the urine sample and then with a solution containing antibodies to the hCG. Presence of hCG in urine inhibits agglutination. This method is still widely utilized, especially as in-house methods[69].

The quantitative specific immunoprocedures for hCG were typical competitive immunoprocedure in which the radio active iodine ^{125}I-hCG competes with sample analyze for binding to anti-hCG. Increased hCG in sample, decreased bound radio activity. These methods are still used widely for measurement of serum and urine hCG levels[148][69].

Another type of assays systems available for serum hCG determination is a radioreceptor assay (RRA). This method is based on the competition between the hCG in the serum and a ^{125}I-hCG for binding to hCG receptor present in tissue [69].

b. <u>**Non – competitive immunoprocedures**</u>:

Non competitive immunometric with labeled antibody (sandwich or two-site immunometric methods)are quantitative methods for hCG and its subunits with low detection limit, large measurement range. The labeled anti hCG reacts with sample hCG bound to solid phase anti hCG. Diffferent labels are used such as enzyme, fluorescent, and luminescent. The specificity of sandwich methods is defined by the reactivity and recognition ability of the antibodies (β •→ α, β •→ β, αβ •→ β, βcf •→ β, βCTP •→ α, α •→ α) where •-- capture antibody and → tracer antibody[79](Fig.4).

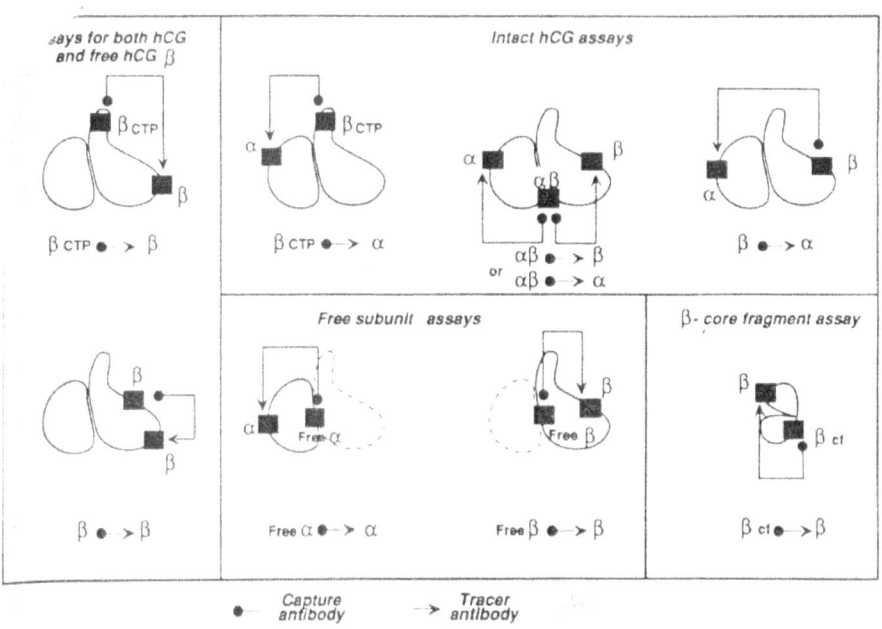

Fig.4 Schematic representation of the sandwich methods used for the measurement of hCG, its free subunits, or hCGcf[79]

1.5 Human Chorionic Gonadotropin Receptor (hCG-R)

Human chorionic gonadotropin receptor is a cell surface protein, present on testicular leydig cells and on ovarian theca, interstitial, luteal, and mature granulosa cells. In both males and females, the hCG-R recognizes the pituitary hormone LH. In the pregnant female, the same receptor also recognizes hCG[149].

The hCG-R belongs to a subfamily of glycoprotein hormone receptor within the G protien-coupled receptor family (whose members are characterized by the common structural feature of seven transmembrance domains)[150].

Although there are differences in the response duration, affinity and dissociation rate of hCG for the hCG-R as compared to LH, it is generally believed that both hormones bind to and activate the receptor similarity. Studies to date indicate that the cAMP second messenger system is the predominant pathway stimulated by either LH or hCG. Yet, it has been reported that stimulation with LH or hCG leads to activation of phospholipase C ,resulting in formation of inositol phosphates and elevation in intracellular Ca^{2+} $[Ca^{2+}]$[151][152]. The responsiveness of a given target cell to hCG or LH can be modulated by alteration in the number of cell surface receptor such as homologous and hetrologous down regulation, as a results of internalization and lysosomal degradation of hormone-receptor complex[149]. Desensitization (reduction in response intensity) has also been observed for hCG-R it has been hypothesized that the desentization occurs as a result of receptor phosphorylation[153].

1.5.1 Structure of the hCG-R and the Nature of Hormone –Receptor Interaction

Structural studies have been hampered by the low abundance of hCG receptor, nonetheless, researchers have made sufficient progress in this area. Data from chemical crosslinking of labed hCG to the receptor, immunoprcipitation of a biosynthetically labeled hCG-R, purification of the receptor, have led to the estimation of the overall size and structure of this receptor[154-158]. Studies on the cDNA and genomic clones for hCG-R have yielded more detailed information on the structure and function[147][159-161]. The antibodies developed to the receptor have confirmed the topology of this receptor and are being used to address structure function relationship[149][154]. All these studies established that hCG-R is a single chain glucoprotien of 85-93 KDa. Its composed of two halves of equal size., the N-terminal extracellular hydrophilic, exodomain of ~ 350 amino acid residues and the membrane associated intracellular C-terminal endodomain of ~334 amino acids residues, which includes seven hydrophobic transmembrane helices and three cytoplasmic loops with three extracellular loops that link the seven transmembrane domains (Fig.5) The hCG-R is encoded by 11 exons located on chromosome 2 p21, the first 10 of which code for the exodomain and the 11th for the endodomain.

hCG Receptor

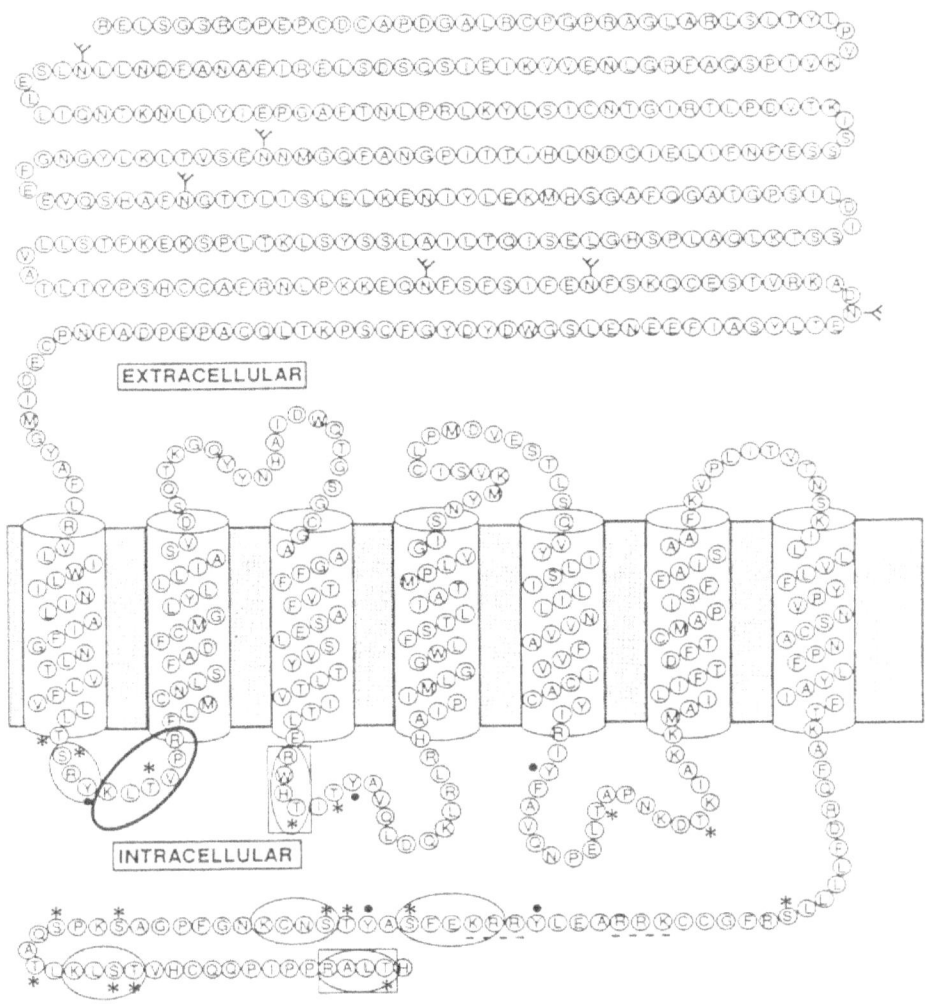

Fig.5 Amino acid sequence, orientation, and proposed topology of the hCG receptor in the plasma membrane. Potential sites for N-linked glycosylation are shown by the branch-like structures. The sequences underlined with dashes in the cytoplasmic tail mark two

clusters of basic amino acids which might represent potential tryptic cleavage sites.Potential intracellular sites for phosphorylation and denoted by asterisks (serine and threonine residues) or dark dots (tyrosine).The rectangles denote weak consensus sequences for cAMP-dependent protein kinase – catalyzed phosphorylation.The ovals and heavy ovals denotes weak and strong consensus sequences for c kinase-catalyzed phosphorylation[154].

Results from mutational analysis[162-164], chimeric construction[165], synthetic peptide studies[166] and chemical modification of the receptor[154] showed that the hormone binding and activation processes are separable.The large extacellular domain of hCG with six sites for N-linked glycosylation and (8-9) leucine –rich repeats (LRR) is responsible for high selectivity and affinity hormone binding without hormone action[167-170]. LRR are thought to form a crescent with concave inner surface consisting of β sheets which may bind hormone (Fig.6)

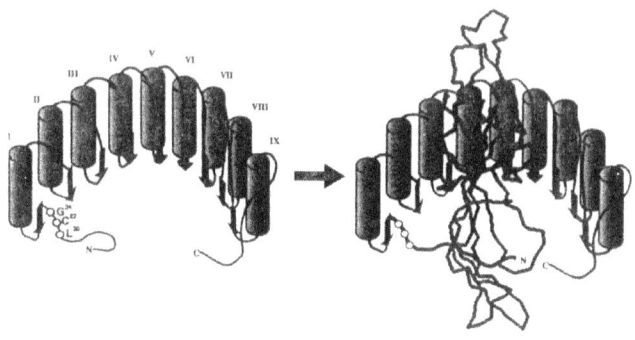

Fig.6 Schematic model for the interaction of the hCG receptor exodomain with hCG. Hormone interacts with the inner face of the crescent structure of the receptor[173].

Activation, in contrast, has been identified with transmembrance domain [171]. The molecular mechanism involves the initial high affinity contact of hCG with receptor exodomain, then the resulting complex (may not be thermodynamically stable) undergoes conformational changes and makes secondary low affinity interactions with exoloops of membrane associated domain(Fig.7), causing an allosteric structural change in the endodomain which leads to signal generation and hormone action[172][150].

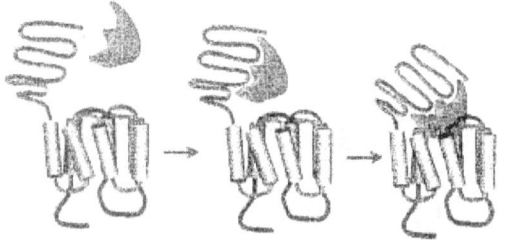

Fig.7 Schematic presentation of the receptor-hCG interaction.
hCG bind to the N-terminal segment of the receptor and the hCG-N-terminal segment interacts with exo loops to generate a signal[150].

The initial high affinity interaction includes multiple contacts between the exdomain and both subunits of the hormone[173][174]. The precise hormone contact sites in the exodomain are unclear. The most putative contact sites are αC-terminal region, N-oligosaccharide at αAsn^{52}, and unique αhelix in the α-subunit, as well as unusual loop (seat belt) in the β-subunit and the peripheral β-hair pin loop of both subunits[175]. Between the initial interaction and the signal generation, hCG and receptor undergo conformational changes. Studies including photochemical crosslinking of hCG, mutational analysis of the receptor and serial truncation has been demonstrated a structural changes of hCG involving the intersubunit interaction[162][164][176-177].

1.5.2 The Role of hCC-R in Nongonadal Tissues and Tumor Tissues

It has long believed that the hCG-Rs were present only in gonads. But the studies, using immunocytochemistry method and MAb for receptor or reverse transcription polymerase chain reaction technique, have demonstrated that various female as well as male nongonadal tissues (uterus, placenta, fetal membrane, deciduas, fallopian tubes, brain, breast, skin) contain low levels of hCG-R protein and hCG-R transcripts (cancer tissues are among these tissues)[178-187]. Moreover the direct action of ectopically synthesis hCG in regulating the growth of various cell types and its role in various cancers, implies the existence of the corresponding receptors in these tissues.

Studies on nonpregnant human uteri, demonstrated not only the presence but also different cellular distribution of hCG-R, where epithelial cells contain more receptors than that stromal cells of endometrial in turn, contain more than myometrial and vascular smooth muscle[178][188]. The hCG-R are increased from the proliferation to the secretary phase of the cycle. These differences suggest that other hormones regulate the receptor, which regulated different functions. Evidences from in vitro studies supported these suggestion, for example, hCG was reported to have a relaxing effect on porcine myometrium[189] and can increase cAMP levels and progesterone synthesis in rat uterus[190]. The expression of receptor mRNA had been found higher in myometrial then in human endometrial blood vessels ,and there is possibility that hCG could an directly increase uterine blood flow [187].Human chorionic gonadotropin may regulate epithelial cell by increasing the local synthesis of steroid hormones[190]. In vitro , hCG can directly regulate proliferation of human myometrial smooth muscle cells causing hyperplasia as well as hypertrophy[188]. Increasing expression of receptor gene in human endometrial cancer has also been obtained [184].

Studies on human placenta , fetal membranes, decidua have been shown the presence of receptors as well as cellular localization differences and changes from mid to term pregnancy suggesting that hCG may regulate their function. In fact, there are data from in vitro studies support this possibility, in placenta, high hCG-R concentration have been shown to increase cAMP formation[191], glycogen breakdown[192], aromatization of androgens to estrogens[193], and interconversion of estron to estradiol[194]. In decidua, high hCG concentration have been shown to directly stimulate prolactin synthesis, which in turn, has been inhibited placental hCG synthsis.This finding indicates the presence of hCG- influenced short loop positive and negative regulatory mechanisms within the human fetoplacental unit[195][196]. Investigation for the presence of functional hCG-R in human breast, revealed that normal epithelial cell, benign lesion, and cancer biopsies content hCG-R[183][197]. The intensity of the immunolabeling of hCG-R varied in individual biopsies. The presence of receptor mRNA was also confirmed.In the breast, it has been proposed that hCG provokes differenentiation, which in turn renders the cells less susceptible to neoplastic transformation[198]. This mechanism would explain the decreased occurrence of breast cancer in women who had on early pregnancy. Studies on human breast cell lines showed the functional presence of hCG-R that can bind hCG to exert a direct antiproliferative effect on human breast epithelial cells through secretion of inhibin and decreasing estrogen receptor[199]. Indeed hCG has been shown to exert either stimulatory or inhibitory effects on the growth of various cancer in vitro. In bladder, lung and endometrial cancer as well as choriocarcinoma, hCG and/or its subunits promote the growth of these cancers while in breast and prostate cancers are growth inhibiting[115].

Part B
Human Chrionic (hCG) in Gynecologic Cancers

Sami A. AL-Mudhaffar
Methal Ahmad Abd Ali
Zainab Mohammad A.AL-Azzawi

Summary

1- The concentration of intact human chorionic gonadotropin (hCG) and free β subunit (hCGβ) in sera and tissues of patients with gynecologic cancers was determined by radioimmunoassay. Intact hCG and free hCGβ levels were found to be elevated above the reference value (5mIU/ml) in sera of 50% of patients with epithelial ovarian cancer and in sera of 44% of patients with endometrial cancer. The sera intact hCG and free hCGβ levels were significantly higher in malignant then in non-malignant gynecologic tumors ($P<0.05$). Elevation intact hCG and free hCGβ above 5mIU/ml in tissues was found in 100% of patients with epithelial ovarian and endometrial cancers.

2- The characteristics of the binding of ^{125}I-hCG to hCG receptor in ovarian and uterine tumors homogenates were investigated using the radio receptor assay (RRA). Different factors affecting this binding were studied and the result showed that the maximum binding was achieved by using 500 and 700 µg protein of ovarian and uterine homogenate respectively, 17 and 4 Pmol of ^{125}I-hCG concentration for ovarian and uterine tumors respectively, pH(7-7.4) for both tumors. Association kinetic studies of the binding reaction revealed the time and temperature dependency. The ^{125}I-hCG binding to the hCG receptor was specifically displaced by low concentration of unlabeled hCG and LH, where IC_{50} was 2.4ng for hCG and 4ng for LH. The RRA appeared that homogenates of malignant ovarian tumor have a specific hCG-binding activity 2 fold higher than that of homogenates of benign. Endometrial cancer homogenates have binding activity lower than that in malignant and benign ovarian tumors.

3- Method was developed for the purification of hCG receptor from detergent soluble membranes extracts of epithelial ovarian cancer by gel filtration chromatography using Sepharose CL-6B column (1.5 x 87cm) and affinity chromatography using hCG-Sepharose affinity column (2 x 8cm). The purification fold was 8,538 fold.

Physicochemical properties of purified HCG receptor were investigated and the

results showed that the Mw 204 KD and the stock's radius 54^0A under non reducing conditions, while the reducing conditions showed three protein component with Mw 80, 350 D, 53,703D, 42,657D respectively, PI 6.4, carbohydrate content 8%. Binding studies demonstrated that the purified hCG receptor retained all the specific binding characterized expected for the hCG receptor. Association kinetic indicated the pseudo first order kinetic of the binding of ^{125}I-hCG to hCG-R, while the kinetic of dissociation was biphasic and characterized by an initial rapid phase followed by a much slower rate of dissociation. The Hill-plot data (n-1) indicated that there was no cooperation between the hCG –R binding sites.

4- The carcinoembryonic antigen (CEA) was investigated in the sera and tissues of patients with ovarian and uterine tumors. The results revealed a slightly elevation of levels of CEA above the reference value (5ng/ml) in the sear of 33% of patients with endometrial cancer and in 43% of patients with epithelial ovarian cancer. CEA immunoreactivities were significantly higher in malignant ovarian homogenates than in corresponding sera samples while not significantly elevated in malignant uterine tissues.

5-Serum LDH, total ALP, and HSAP activity were measured in ovarian and uterine tumors. The LDH activity was higher in patients with epithelial ovarian and endometrial cancers than those of healthy individuals but not statistical significant. The electrophoresis analysis of LDH isoenzymes showed extra band for serum of epithelial ovarian cancer. Total AIP and HSAP activity in the sera of cancer patients were significantly higher than in normal individuals.

6-The elements (copper, zinc, calcium) concentration were measured in the sera of patients with ovarian and uterine tumors. The results revealed an increment in the concentration of zinc in patients with ovarian and endometrial cancer rather than control group, while copper and calcium concentration did not alter in cancer investigated patients

LIST of ABBREVIATIONS

AFP	Alpha –Fetoprotein
ALP	Alkaline phosphatase

B	Bound
BSA	Bovine serum albumin
CA_{125}	Carbohydrate antigen 125
CA_{15-3}	Carbohydrate antigen 15-3
CA_{19-9}	Carbohydrate antigen 19-9
cAMP	3',5' –cyclic adenosine mono phosphate
CD	Circular dichroism
cDNA	Complementary DNA
CNS	Central nervous system
CPM	Counts per minute
CSF	Cerebrospinal fluid
CT	Computed tomography
CTP	Carboxy terminal polypeptide
DNA	Deoxy ribonucleic acid
D	Dalton
EDTA	Ethylene diamine tetra acetic acid
EGF	Epidermal growth factor
ERT	Estrogen replacement therapy
F	Free
FIGO	International federation of gynecology and oncology
FSH	Follicle –stimulating hormone
GTD	Gestational trophoblastic disease
hCG	Human chorionic gonadotropin
hCGβ	β Subunit of human chorionic gonadotropin
hCGα	α Subunit of human chorionic gonadotropin
hCGβcf	Beta –core fragment of hCG
HCG-R	hCG receptor
HSAP	Heat stable alkaline phosphatase

LDH	Lactate dehydrogenase
LH	Luteinzing hormone
LRR	Leucine rich repeats
LSA	Lipid associated sialic acid
MAB	Mono clonal antibodies
MAD	multi wavelength anamalous diffraction
MCA	Mucin like cancer antigen
M-CSF	Macrophage clonny stimulating factor
mRNA	Massenger ribonucleic acid
NGF	Nerve growth factor
NS	Not significant
NSB	Non specific binding

PDGF-ββ	Platelet –derived growth factor – ββ
PAGE	Poly acrylamide gel electrophoresis
PI	Isoelectric point
P^{53}	Tumor suppressor gene
PLK	Polo –like kinase
PMSF	Phenyl methyl sulfonyl flouride
PTH	Parathyroid hormone
RIA	Radioimmuno assay
RRA	Radio receptor assay
SB	Specific binding
SCC	Squamous cell carcinoma
SDS	Sodium dodecyl sulfate
S-phase	Synthesis phase
TAH – BSO	Total abdominal hysterectomy with bilateral salping –oophorectomy
TEMED	N,N,N,N-tetra methyl ethylene diamine
TGF-β	Tumor growth factor - β
TSH	Thyroid –stimulating hormone
UV	Ultra –violet
WGA	Wheat germ agglutinin

Contents

Subject	Page No.
Summary	

	ListofAbbreviation		3
	Contents		5
	ChapterOne:	Introduction	7 13
1.1		Gynecologic cancers	14
1.2		Endometrial cancer	15
1.2.1		Histology	15
1.2.2		Etiology	17
1.2.3		Clinical evaluation and treatment	17
1.3		Epithelial ovarian cancer	18
1.3.1		Histology	20
1.3.2		Etiology	21
1.3.3		Clinical evaluation and treatment	22
1.4		Tumor markers in Gynecologic cancers	24
1.4.1		Antigenic markers	24
		Squamous cell carcinoma (SCC) antigen	24
		Lipid associated sialic acid (LsA)	24
		CA-125	25
		Other antigens defined by monoclonal antibodies	25

		Carcinoembryonic antigen (CEA)	25
	1.4.2	Genomic markers	26
	1.4.3	Enzymatic markers	26
	1.4.4	Hormonal markers	27
	1.4.4.1	Human chorionic gonadotropin (hCG)	27
	A-	Chemical structure of hCG	28
	B-	Molecular forms of hCG	29
		Intact hCG	29
		Free hCGβ	30
		Free hCGα	30
		Beta – Core fragment of hCGβ (hCGβcf)	30
		Nicked hCG (hCGn or hCGβn)	30
	C-	Production and metabolism of hCG	31
	D-	Clinical use of hCG determinations	31
		Diagnosis of pregnancy and related disorders	31
		Placental and testicular trophblastic tumors	32
		Nontrophoblastic tumors	32
		a- blood studies	32
		b- urine studies	33
		c- effusion fluids studies	33
		d- tissues studies	33
	E-	Determination methods of hCG	34
		Biological methods	34
		a- In vivo methods	34

		b- In vitro methods	34
		Binding methods	34
		a- competitive immuno procedures	35
		b- non- competitive immuno procedures	35
	1.5	Human chrionic gonadotropin receptor (hCG-R)	36
	1.5.1	Structure of the hCG-R and the nature of hormone – Receptor Introduction	37
	1.5.2	The role of hCG-R nongonadal tissue and tumor tissue	40
		The aim of the work	42

Chapter Two: Human Chorionic Gonadotropin in Sera and Tissues of Patients with Gynecologic Cancers 43		
Abstract		43
Introduction		44
Materials and methods		44
2.1	Chemicals	44
2.2	Apparatus	45
2.3	Patients	45
2.4	Collection of specimens and preparation of tissue homogenates	45
2.5	Quantitative measurement of intact hCG and free hCGβ	46
2.6	Estimation of protein contents	47
2.7	Statistical analysis	47
	Results and discussion	48
	Preoperative Data	48
	Postoperative Data	53
	Malignant and non malignant tissues	53

ChapterThree: Development of a Radio Receptor Assay for Detection and Analysis of hCG – Receptor in Benign and Malignant Gynecologic Tumors		64
Abstract		64
Introduction		65
Materials and methods		66
3.1	Buffer and reagents	66
3.2	Patients and specimens	66
3.3	Binding studies of ^{125}I-hCG with Its receptor in Benign and malignant ovarian and uterine tumors	66
3.3.1	Preliminary test of hCG-R binding	66
3.3.2	The effect of ^{125}I-hCG concentration on the binding	67
3.3.3	The effect of receptor concentration on the binding	67
3.3.4	The effect of pH on the receptor binding	67
3.3.5	Temperature dependency of the binding	67
3.3.6	Time course of receptor binding	68
3.3.7	Stability of receptor homogenate and hormone receptor complex	68
3.3.8	Determination of the equilibrium binding constants of hCG – R	69
3.3.9	Specification of binding and determination of IC_{50} and K_i	69
Results and discussion		70

Membrane preparation		70
Human chorionic gonadotropin receptor in ovarian and uterine tumors		70
Factors effecting the binding of ^{125}I-hCG with Its receptor		72
1-	The effect of ^{125}I-hCG concentration on the binding	73
2-	The effect of receptor concentration on the binding	73
3-	The effect of pH on the receptor binding	74
4-	Temperature dependency of the binding	74
5-	Time course of receptor binding	75
6-	Stability of receptor homogenate and hormone receptor complex	75
7-	Determination of the equilibrium binding constants and maximum binding capacity	76
8-	Specificity of binding	77
Chapter Four: Purification and Characterization of hCG – R from Ovarian Cancer 90		
Abstract		90
Introduction		91
Materials and methods		92
4.1	Buffers and reagents	92
4.2	Patients and specimens	92
4.3	Protein determination	92
4.4	Binding assay	92
4.5	Purification of hCG-R	93
4.5.1	Gel filtration chromatography onSepharose CL-6B	93

4.5.2	Affinity chromatography on Sepharose-hCG	94
4.6	Analytical and characterization of purified hCG receptor	95
4.6.1	Determination of the association constant	95
4.6.2	Analysis of purified receptor by conventional PAGE	95
4.6.3	Determination of the molecular weight by SDS-PAGE	96
4.6.4	Determination of the molecular weight and stocks radius by gel filtration chromatography	98
4.6.5	Determination of the isoelectric point (pI) of purified receptor	99
4.6.6	Specificity of binding for purified receptor	99
4.6.7	UV spectrum of hCG, and hCG- receptor complex	100
4.6.8	Stability of crude and purified solubilized receptor	100
4.6.9	Binding kinetics of purified receptor	100
4.6.10	Estimation of hill coefficient (n) of purified receptor	101
	Results and discussion	101
	Purification of the receptor	101
	Analytical and characterization of purified receptor	104
	Human chorionic Gonadotropin binding properties of purified receptor	104
	Polyacrylamide gel electrophoresis	105

Molecular weight of the purified receptor		105
Isoelectric point (pI) of purified receptor		107
Carbohydrate composition of the purified receptor		107
Human chorionic Gonadotropin binding specificity of purified receptor		108
The UV – spectrum of hCG and of hCG-Receptor complex		109
Stability of purified receptor		109
Kinetic of purified receptor binding and the determination of Hill coefficient		110
Chapter Five: Evaluation of some Biochemical Constituents in Gynecologic Cancers		123
Abstract		123
Introduction		124
Material and methods		126
5.1	CEA , LDH, and ALP kits	126
5.2	Apparatus	126
5.3	Patients and specimens	126
5.4	CEA quantitative measurement	126
5.5	Determination of LDH activity	127
5.6	Analysis of LDH by electrophoresis	129
5.7	Determination of alkaline phosphatase activity (ALP)	130
5.8	Determination of heat stable alkaline phosphatase activity (HSAP)	131
5.9	Determination of the elements ($Cu^{++}, Zn^{++}, Ca^{++}$) concentrations in the sera of tumor patients	131
Results and discussion		131

Biochemical constituents in ovarian and endometrial cancer patients	131
Copper , zinc , calcium in the sera of ovarian and endometrial cancer patients	134
Conclusions	140
The future work	142
References	143

Chapter One

Introduction

1.1 Gynecologic Cancers

The major malignancies of the female reproductive tract are endometrial, cervical, and ovarian cancer (Fig.1).Other gynecologic cancers, such as choriocarcinoma, fallopian tube cancer, vagina cancer, and vulva cancer are rare, but cancers metastasis to these female genitals are more common[1].

Ovarian cancer accounts for 4% of the total cancer in women worldwide. In Iraq, this cancer is the eighth most common malignant neoplasm in women (3.8-4.2% of the total cancer) behind cancer of the breast ,non-Hodgkin's lymphoma ,brain and other CNS, leukemia ,urinary bladder ,bronchus and lung ,and skin. The number of cases for the period (1992-1997) was 828[2].

The incidence rates of corpus uteri cancer (95% endometrial) were found to be higher in the richer countries and urban populations; it is the fifth leading cancer in women at these countries[3].However, there is evidence of some change in the socioeconomic determinants of the disease in developed countries. In Iraq, cancer of corpus uteri accounts for (0.9-2.2%) of the total cancer in women. The number of cases for the period (1992-1997) was 477[2]. The risk of uterine cancer is low before age 40 years and increases sharply thereafter[4].

Most cancers of the vulva, vagina and cervix are detected relatively early because these organs are easily visualized and assessable for examination by cytology and biopsy [1].Endometrial cancers are also detected early because they normally heralded by vaginal bleeding. In contrast, cancers of the ovary mostly present at a late stage with widespread intra-abdominal metastases. Ovarian cancer remains the most lethal gynecologic malignancy in women [5].

1.2 Endometrial Cancer

Endometrial cancer is the most common type of the corpus uteri (Fig.1c), constituting about 95% of all malignant lesions of the uterine cavity, while uterine

sarcomas accounts for fewer than 5 %[3].

The normal endometrium contains two major components: endometrial stroma and glands. The glandular cell produces cytoplasmic receptors for estradiol and, when stimulated, results in proliferation of glandular epithelium, producing larger and more numerous glands. In the premenopausal cycling endometrium, the maturational effects of progesterone from the corpus leteum counter balance the proliferative influence of estrogen. If conception does not occur, the lining is shed and a new cycle is initiated. When ovulation ceases at menopause, the endometrial becomes quiescent and the glands atrophy. Tumors arising from the glandular epithelium, either de novo or as a result of abnormal estrogenic stimulation; are adenocarinomas[6]. The transition from benign proliferative endometrium to well differentiated adenocarcinoma probably proceeds through several intermediary steps, collectively termed hyperplasia. Hyperplasias are characterized by an overgrowth of the glandular component of the endometrial lining, whereas cancers are defined by the size of the cells, irregular nuclear membranes, coarse chromatin clumping, loss of glandular pattern, increasing nuclear atypia, and mitotic activity[6].

1.2.1 Histology

About 90% of endometrial cancers are endometriosis type (adenocarcinomas), which contain glands similar to those seen in the normal endometrium. These tumors are subclassified by their degree of differentiation into three grades:well (grade I, 50%), moderately (grade II, 35%), and poor differentiated (grade III, 15%). The International Federation of Gynecology and Oncology (FIGO) grading scheme is based on the growth pattern (relative proportion of glandular and solid areas):less

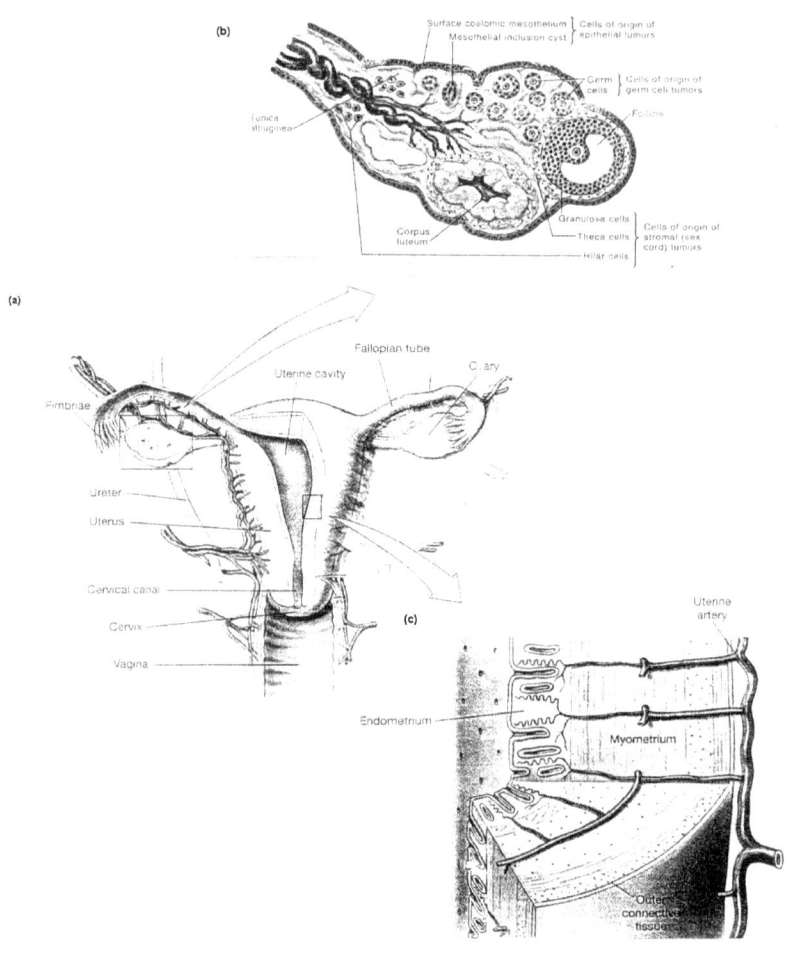

Fig.1 [9] (a):Female reproduction tract
(b): Embryologic origins of ovarian cancer.
(c):Tissue layers of uterus

than 5% solid growth is grade I, 6-50% is grade 2, and more than 50% is grade 3[3].   Endometriosis adenocarcinoma accounts for 60-65% of the adenocarcinomas, it is characterized by the disappearance of stroma between abnormal glands, necrosis, and hemorrhage[1].Some typical adenocarcinoma contains areas of squamous metaplasia, which can have a benign or malignant appearance, such tumors are termed adenocarcinoma with squamous differentiation[6].

A variant of endometriosis adenocarcinoma is the villoglandular type, which appears to have more aggressive behavior [7]. Uncommon cell types (clear, papillary serous, mucinous, squamous and undifferentiated) account for the remaining 10% of endometrial cancers [6].

1.2.3 Etiology

Endometrial cancer is one of the few malignancies for which the etiology is well understood. Epidemiologic, endocrinologic, and clinical studies have shown that the association between endometrial cancer and most of the identified risk factors can be explained by increased exposure to endogenous or exogenous estrogens. This hypothesis is supported by the increased risk of endometrial cancer (related to obesity, early age at menarche, late age at menopause, low parity or nullparity, certain types of ovarian tumors, history of menstrual disorders, history of infertility, and use of estrogen replacement therapy (ERT) or sequential oral contraceptives). It is also supported by the decreased risk of endometrial cancer (related to the use of combined oral contraceptives, early age at menopause, high parity, and smoking)[4].

Medical conditions such as diabetes, hypertension, and thyroid diseases have been suggested to increase the risk of endometrial cancer, but these finding are not consistent[8]. It remains, unclear whether these finding are the result of the association of these medical conditions with obesity or with other biologic mechanisms.

1.2.3 Clinical Evaluation and Treatment

The initial growth phase of most endometrial cancers consists of polyploidy expansion within the endometrial lining .Small areas of necrosis or surface breakdown in the tumor produce abnormal vaginal bleeding. Consequently, the premenopausal women usually present with metrorrhagia, and the older woman has postmenopausal bleeding. Patient with advanced disease may have pelvic pain, bleeding, or bloating[3]. The diagnosis of endometrial can be reliably established by office endometrial biopsy. Dilatation and fractional curettage or hysteroscopy with biopsy may be helpful diagnostic procedures in cases in which out patient biopsy is non diagnostic[1].

The workup on the woman with endometrial cancer should include a physical examination, endometrial biopsy, laboratory studies, and a chest radiograph. Endometrial cancer is typically a disease of women in their sixties and seventies who frequently have coexisting medical problems, so the evaluation should focus on an assessment of operative risk. Additional more sophisticated studies such as proctoscopy, barium enema, ultrasound, computed tomography (CT), or magnetic resonance imaging should be reserved for clinical situations in which advanced disease is suspected[6].

Schematic approach to treatment endometrial cancer is illustrated in (Fig. 2)[6].

Surgical excision of the primary tumor via total abdominal hysterectomy with bilateral salpingo –oophorectomy (TAH-BSO); sometimes accompanied by removal of the pelvic and periaortic lymph nodes; is the mainstay of therapy for endometrial cancer. The selection of additional therapy is based on the result of the staging biopsies of uterine tumors, which was adopted in 1988 (Table 1) by FIGO surgical staging[3].

1.3 Epithelial Ovarian Cancer

The ovary is composed of an outer cortex and an inner medulla. The cortex containing the coelomic surface mesothelium, germ cell (oocytes), specialized hormone producing stroma (granulose and thecal cell), and unspecialized supporting stromal cells. Primary ovarian cancers are classified; according to the structure of the ovary from which the tumor is derived (Fig.1b); into three major categories, epithelial, germ cell, and stromal[9]. The ovary may also be the site of metastasis spread from

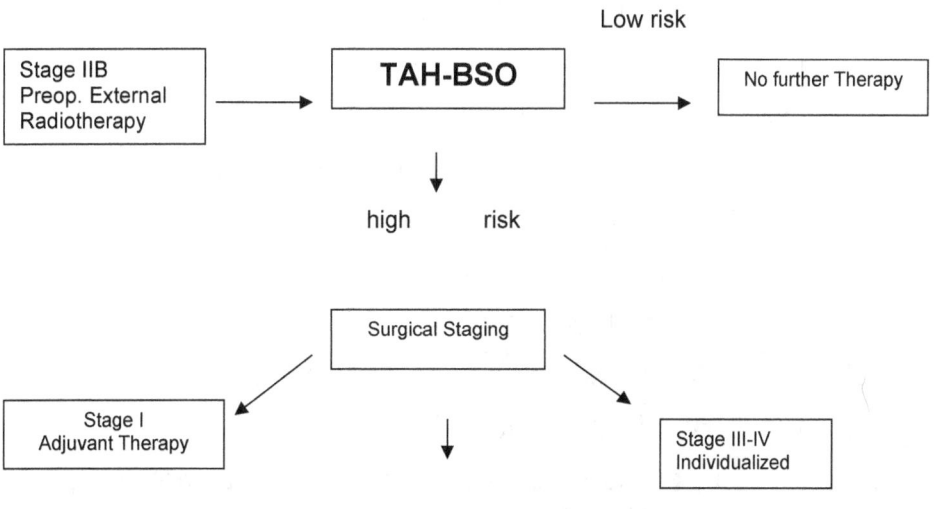

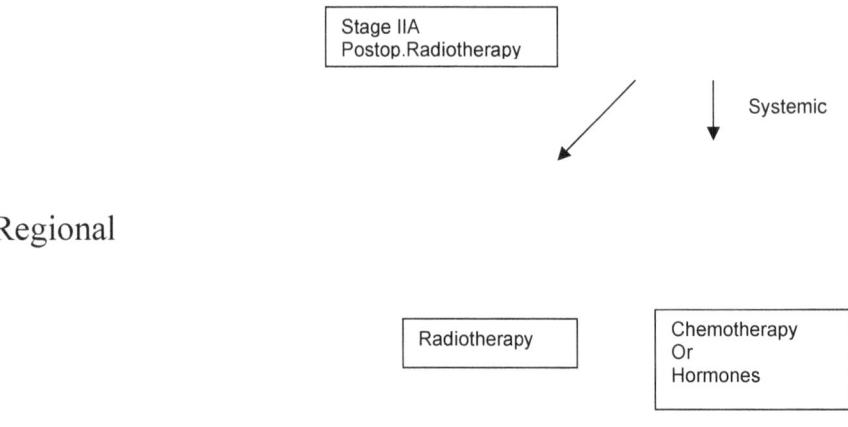

Fig.2 Basic treatment strategy for endometrial cancer[6].

Table (1) Surgical staging system for cancer of the uterine body[3].

Stage	Extent
I	Cancer confined to the corpus
Ia	Tumor limited to endometrial
Ib	Invasion to less than one half the myometrium
Ic	Invasion to more than one half the myometrium G_1:Well differentiated (≤5% of a nonsquamous or nonmorular solid growth pattern) G_2:Moderately differentiated (6%-50%) G_3:Predominantly solid or undifferentiated
II	Cancer involves corpus and cervix but does not extend outside the uterus
IIa	Endocervical glandular involvement only
IIb	Cervical stromal invasion
III	Cancer extends outside the uterus but not outside

	the true pelvis
IIIa	Tumor invades serosa and/or adnexa or positive peritoneal cytology
IIIb	Vaginal metastases
IIIc	Positive pelvic and /or paraaortic nodes
IV	Cancer extends outside true pelvis or invades bladder or rectal mucosa
IVa	Tumor invasion of bladder and/or bowel mucosa
IVb	Distant metastases including intra-abdominal and/or inguinal lymph nodes

gynecologic and other tumors. Epithelial ovarian tumors are the most common, accounting for 80-90% of cases[5].

Epithelial ovarian cancers are derived from the coelomic surface mesothelium. During ovulation, the dominant ovarian follicle ruptures, releasing the ovum. To heal the ruptured follicle, surface mesothelial cells proliferate and fill the defect. Inclusion cysts may form and are lined with these mesothelial cells. Over years of repeated ovulation and repeated growth of mesothelia cells, the regulation of this proliferation may become disrupted, allowing the development of tumors (Fig.1b)[9].

Numerous factors can stimulate –or fail to control mesothelial proliferation including growth factors, oncogenes, and tumor suppressor genes[10].

1.3.1 Histology

Epithelial ovarian tumors classified according to cell type, pattern of growth (cystic, solid, and surface), atypia and invasiveness (benign, borderlines, malignant). The ovarian surface epithelium, when involved in neoplastic conditions, often undergoes a mullerian differentiation as a result, it may produce any of the adult epithelia, including tube, endometrial and, endocervical mucosa, singly or in combination. So the histological types of the epithelial tumors are serous, mucinous, endometriosis, clear cell, Brenner, mixed epithelial, undifferentiated, and unclassified[3]. Serous tumors make up about 20% of all benign ovarian tumors and 75% of epithelial ovarian cancers. Malignant serous tumors are usually multicystic, bilateral and resembling fallopian tube epithelium. Serous carcinoma may have a complex mixture of cystic and solid areas with extensive papillations, or they may contain a predominantly solid mass with areas of necrosis and hemorrhage[11]. The second common type of epithelial ovarian cancer is endometriosis tumor (10%) which is histological indistinguishable from endometrial uterine cancer. About 6-10% of epithelial ovarian cancers are mucinous, which characterized by epithelial – mucin producing cell resembling those of the endocervix[9]. Clear cell tumors account for<5% of all primary ovarian cancers. The tumor cell is large and their cytoplasm is clear. These tumors have more aggressive behavior than other epithelial cancers. Brenner tumors, these unusual lesions are though to originate from the ovarian mesothelium despite their urothelial –like transitional cell structure. They are solid and almost benign.

1.3.2 Etiology

Several factors have been associated with an increased risk of epithelial ovarian cancer, while two factors appear to reduce the incidence of this malignancy – pregnancy and use of oral contraceptives – presumably because they suppress ovulation. The risk of epithelial ovarian cancer is high in ovulating nulliparous women, women using of fertility drug, women exposuring to hormones (particularly gonadotropins) and carcinogens, such as infectious agents and chemical carcinogens[12].

The large majority of ovarian cancer presumably related to environmental factors, but 5-10% of women with epithelial tumors have a predisposing familial syndrome. Three autosomal dominant syndromes have been identified: site –specific ovarian cancer, hereditary breast –ovarian cancer syndrome, and lynch syndrome II (adenocarcinomas of multiple sites, including breast, ovary, and endometrial, gastrointestinal tract). All three types of familial ovarian cancer develop at a younger age than sporadic ovarian cancer[9].

Other factors, such as obesity, high-fat diet, early menarche, late menopause, single women, may predispose women to develop this malignancy. The hypothesis is that the continual ovulation, uninterrupted by pregnancy may develop the disease[4].

1.3.4   Clinical Evaluation and Treatment

Epithelial ovarian cancer is seldom detected at an early stage; up to 80% of these tumors have metastasized by any of the four routes (peritoneal seeding, lymphatic embolization, direct extension and hematogenous dissemination). The first sign of early ovarian cancer is usually an asymptomatic mass found during a pelvic, rectal, or abdominal examination. A benign tumor or slow- growing cancer may grow quite large and cause abdominal distention and may compress the rectum or bladder, producing constipation, urinary frequency or nocturia. Ovarian tumor occasionally undergoes torsion, resulting in localized pain, nausea, and possibly low fever, and may also cause symptoms of inappropriate estrogen or androgen synthesis[9].

The diagnosis of epithelial ovarian cancer is surgical, usually by a laparotomy, but occasionally at laparoscopy. Ultrasound of adnexal pelvic mass is useful prior to the surgical diagnosis. Most patients with a pelvic mass require ultrasonography, chest radiography, CT, and aspiration of pleural effusion to check for metastases.

Mammography should be performed to exclude primary breast cancer[3].

The management of primary epithelial ovarian cancer may be divided into three phases: primary definitive surgery, postoperative adjuvant therapy usually chemotherapy[13], and systematic follow up and reevaluation. The initial surgical procedure (usually TAH-BSO) is used to determine the extent of the disease (staging) and to remove as much gross disease as possible (cytoreduction). In ovarian cancer, metastases adhere superficially to peritoneal surfaces without deep invasion. Therefore, surgical resection of metastasis ovarian cancer is standard care[9]. The tumor must be staged to facilitate planning of adjuvant therapy, the FIGO staging system of 1987 (Table 2)[3] is based on the finding at surgical exploration. Following the cytoreductive surgery or debulking, patients should be treated, generally with chemotherapy and some time with radiation or endocrine therapy[14][15]. The majority of women with advanced epithelial ovarian cancer will ultimately relapse and develop drug-resistant disease. Thus, there is a common need for second-line treatment including cytotoxic drugs, or secondary cytoreductive surgery or palliative surgery[16].

Table (2) Surgical staging system for epithelial ovarian cancers[3].

Stage	Extent (proportion of cases)
I	Cancer limited to ovaries (15%)
Ia	Limited to one ovary, no ascites
Ib	Both ovaries involved, no ascites
Ic	Ia or Ib with ascites or positive peritoneal washings

II		Cancer of one or both ovaries with extension limited to pelvic tissue(15%)
	IIa	Extension to uterus or tubes
IIb		Extension to other pelvic tissues
	IIc	IIa or IIb with ascites pr positive peritoneal washings
III		Cancer involving one or both ovaries with peritoneal implants outside the pelvis and/or positive retroperitoneal or inguinal nodes. Tumor is limited to the true pelvis but with histological proved extension to small bowel or omentum (65%)
IIIa		Tumor grossly limited to the true pelvis with negative nodes but with histological confirmed microscope seeding of abdominal peritoneal surface
IIIb		Same as IIIa, but abdominal peritoneal implants do not exceed 2 cm in diameter
IIIc		Abdominal implants greater than 2cm in diameter and/or positive retroperitoneal or inguinal lymph nodes
	IV	Distant metastases present (including cytology-positive pleural effusion metastasis liver parenchyma or peripheral superficial lymph nodes(5%).

1.4 Tumor Markers in Gynecologic Cancers

Tumor markers; which may be considered as any biological aberration that indicates the presence of a tumor; are required for primary diagnosis, histological identification, assessment of the extent of the disease pre- and post operatively, therapy guidance, and early detection of recurrence. Ideally, the tumor marker assessed in serum, or in another body compartment, should accurately reflect the presence and the amount of residual tumor, and should also distinguish benign from malignant disease[17]. In gynecological practice, it is well documented that human chorionic gonadotropin(hCG), alpha-feto protein(AFP), estrogens, and androgens have been contributed greatly to the diagnosis and management of gestational trophoblastic disease, gram cell tumor, and granulose cell tumor, respectively[17][18].

By the introduction of polyclonal and monoclonal antibody technology, variety of tumor carbohydrate associated antigens have been identified as a result of glycosylation changes in the carbohydrate moieties of glycoproteins and glycolipids by tumor cells, some of tumor associated antigens also represent differentiation antigens and receptors that may reexpressed or over expressed[18][19].

1.4.1 Antigenic Markers

1- Squamous Cell Carcinoma (SCC) Antigen: The SCC antigen was described in 1977 as one of 14 subfractions of tumor associated antigen found in human cervical cancer tissue [20]. SCC antigen have been reported to be elevated (>2 ng/ml) in 35% of stage I patients, and increasing to 91% in stage IV for cervical cancer [21][22]. While serum SCC antigen level were increased in 8-29% of endometrial cancer. In vulvar and vaginal squamous cell cancers, the reported frequencies of SCC antigen elevation are lower (42% and 17%, respectively) than in cervical cancer[17].

2- Lipid-Associated Sialic Acid (LSA): Elevated levels of LSA are found in a variety of malignancies, including cancers of the vulva, vaginga, uterus and ovary[18]. Due to its lack of specification, LSA used alone had not been useful in gynecologic cancers.

3- CA-125: Since its first report in 1983[23], CA-125 has been used widely for monitoring epithelial ovarian cancer. Among healthy individuals, 99% will have been serum level <35U/ml[24]. Approximately 85% of ovarian cancer patients have elevated CA-125 level (> 35 U/ml) for all histological types of epithelial ovarian cancers but more frequently for nonmucinous tumors than for mucinous types[25][26].It is also expressed in cancer of epithelium female reproductive tract[27] and can be detected in 22% of patients with nongynecological cancers (Breast, lung, Colon, Pancreas), also

detected in benign gynecological tumors (endometriosis, adenomyosis, ovarian cystadenomas)[28][29].

The low specificity hampers the use of CA-125 as a screening test for ovarian cancer. Combined with pelvic examination and/or ultra sound, the specificity can be increased[30]. The sensitivity of the marker might be also increased by examining peritoneal fluid, where the CA-125 levels are higher than in serum (Cutoff at 200u/ml)[31].CA-125 is very rarely elevated in endometrial cancer confined to the uterus, whereas elevation is found in 78-100% of cases with extra uterine spread[27].

4- Other Antigens Defined by Monoclonal Antibodies: Following the report of CA-125, a number of monoclonal antibodies such as CA 19-9 and CA 15-3 were developed in colorectal cancer[32], breast cancer[33], ovarian cancer and endometrial cancer[34] as immunnogens, and this explain their lack of specificity.

Various monoclonal antibodies have been raised against epithelial mucin like caner antigen MCA , CA M26 , and CA M29. These cancer antigens have been reported in cancers of breast, colon, ovary, endometrial and cervix [35][36], but non of them, either alone or in combination , has been shown to be as useful as CA-125 .

Another marker for ovarian cancer is NB/70K which seen to be a marker for all histological types of epithelial ovarian cancer[17]. Serum NB/70K is elevated in all stages of ovarian malignancies, including more than 50% of early stagecases[37][38].

5- Carcinoembryonic Antigen (CEA): This oncofetal protein can be demonstrated in most gynecological cancers immunohistologically, while plasma levels are often too modestly elevated to be useful for diagnosis or for disease monitoring[17] . Slightly elevated plasma CEA levels are seen in approximately 50% of ovarian cancer patients[39][26], and mucinous tumors tend to have highest levels. Measuring CEA with CA-125 and CA19-9 can be useful to differentiate between ovarian and colorectal cancer. In the uterus, adenocarcinomas that arise in the endocervix express greater amount of CEA than adenocarcinomas of the endometrial. This is reflected in the reported frequency of elevated plasma CEA levels : 68% in cervical cancer, and 34% in endometrial cancer[40]. Squamous cell cancer, which represent 85% of the cervical cancer, also express CEA, but at more modest level.In a study on 205 patients, only 28% had plasma level above 5ng/ml. High levels indicated advance disease or lymph node metastases[41].

1.4.2 Genomic Markers

By cytogenetic examination, using a chromosome spread technique or cell flow cytometry, general genetic alteration have been identifying. Aneuploidy is a common finding in endometrial cancer and ovarian cancer[42][43]. There is a correlation between DNA ploidy abnormalities and grade of differentiation (more common in moderately and poorly differentiated lesions). Trisomy of chromosomes 1,7 and 10 and allelic loss at loci on chromosomes 3p (71%), 9q(38%) 10q(35%) and 17p(35%) have also reported in endometrial cancer[44]. Most ovarian and endometrial cancers over express of macrophage colony stimulating factor (M-CSF) and c-fms which encode the M-CSF receptor[45][46]. The macrophages secrete cytokines (tumor necrosis factor α and interleukins 1 and 6) that stimulate the growth of tumors[47]. Epidermal growth factor (EGF) and polo-like kinase (PLK) protein (contribute to regulation of the cell cycle) have been found to be expressed in most ovarian tumor, while the suppressor gene p53 defective[9][48]. However, the clinical significance of these genomic markers has not yet been established.

1.4.3 Enzymatic Markers

Lactate dehydrogenase (LDH) and heat stable alkaline phosphatase (HSAP) have been also considered as potential tumor markers in gynecology cancers.

The glycolytic enzyme LDH aroused a considerable interest in oncology since Warburg[49] reported increased rate of glucose utilization producing more lactate by tumor cells. LDH is often elevated in epithelial ovarian cancer[50], as in many solid malignancies[51], making it a non-specific tumor markers. In germ cell tumor, the two isoenzymes (LDH-1 and LDH-2) are increased and their activity appear to parallel the response to therapy [52]. Serum LDH levels were documented to be elevated in cervical cancer even at the early stage of the disease and in combination with other biomarker was more beneficial for diagnosis and treatment monitoring of patients[53][54].

Alkaline phosphates (ALP) was one of the first examples of tumor associated enzymes[55]. The heat stable alkaline phosphatase

(HSAP, Regan isoenzyme) has been found in a variety of solid tumors and in 6-64% of ovarian cancer patients[56][57]. The marker level did not correlated to tumor burden or prognosis and half of the patients lost the marker during progression.

1.4.4 Hormonal Markers

Gonadotropin hormones (LH, FSH) can function as tumor markers in gonadal stromal tumor in addition to estrogen and androgen[17]. Inhibins, which are produced in ovarian granulosa cells, cause a specific suppression of pituitary FSH release. Radioimmunometric determination of inhibin has proven to be reliable marker in the monitoring of granulosa cell tumors[58].

1.4..4.1 Human Chorionic Gonadotropin (hCG)

Since the discovery of human chorionic gonadotropin (choriogonadotropin) hormone by Hirose[59]] and Ascheim and Zondek[60], its measurement has been the basis of pregnancy diagnosis and, a marker for many trophoblastic and nontrophoblastic tumors.

This hormone is synthesized by the trophoblastic cells of the placenta during the early weeks of pregnancy. Human chorionic gonadotropin stimulates the ovarian corpus luteum to produce progesterone until the placenta it self acquires the ability to produce this pregnancy sustaining steroid[61]. An increasing evidence supports the synthesis of hCG in small quantities by the pituitary[62].

Human chorionic gonadotropin belongs to the glycoprotein hormones family (also includes hLH, hTSH, hFSH) whose members share a common α-subunit (contains 92 amino acids) and vary in their β-subunit. Each subunit is produced by a separate gene, the α subunit is encoded by a single gene present on chromosome 6q21.1-q23 whereas cluster of seven nonallelic genes located on chromosome 19q13.3 encoded for β-subunit of hCG(hCGβ). (One or more of these genes may preferentially expressed during pregnancy and tumor genesis)[63][64]. The β-subunit confer biological specificity; in humans, hCGβ and hLHβ are closely related subunits with 82% identity, and this reflects a common biological function for these two hormones. The hCGβ (contains 145 amino acids) further differs from that of hLHβ (contains 114 amino acids) by the 24-amino acid carboxy terminal polypeptide extension (CTP). It has been suggested that this glycosylated extension may impart extra solubility and in vivo circulation life time to hCG[65)].

Human chorionic gonadotropin could be purified from pregnancy urine by a combination of organic precipitation, ion exchange chromotography, and gel filtration[66].

A- Chemical Structure of hCG

The primary structure of α and β subunits for hCG were determined in 1973[67][68]. There are five disulfide bridges in the α-subunit and six in the β-subunit (these are conserved across the family). Human chorionic gonadotropin (38,633 Dalton) is approximately 30% carbohydrate by weight and contains both complex biantennary N-asparagine linked and simple O-serine linked. Each carbohydrate moiety terminates in sialic acid, but considerable carbohydrate heterogeneity result in a wide isoelectric point (pI) distribution of the hormone (pI generally 3-6 and as high as 10 for asialo forms of the hormone). The hCG produced in trophoblaste disease exhibits greater carbohydrate heterogeneity (such as triantennary structures) and some cancers produce only asialo hCG[66]. Removal of the terminal sialic acid residues markedly reduce the half-life of hormone in the circulation by enhancing binding to hepatocyte lectin and thus clearance from the blood stream. The carbohydrate moieties of hCG also played a role in hormone secretion, stability, folding and subunit assembly[99]. In vitro, studies with deglycosylated hormone generally indicate a greater loss of biological response than receptor binding. It is believed that the carbohydrate chains bind to a lectin-like membrane component to give the biological response. It was further found that the carbohydrate in the α-subunit was more important in the hormone function than that present in the β-subunit, which seems to be important to maintain the proper conformation of the hormone[70][71].

Various attempts have been made to construct three dimensional (3D) model of hCG[72-78]. Based on amino acid sequence and information; accumulated from chemical modifications (oxidization, deglycossylation, desialyation, reduction), enzymatic modifications (nicks, fragmentation and deletion of fragments), cross reactions, disulfide pairings, molecular biology studies (site directed mutagenesis, chimeric hormone constructs) with monoclonal antibodies(MAb) mapping of surface epitopes, spectral analyses (especially circular dichroism (CD)) and diffraction analyses (X-ray diffraction of hydrogen floride-hCG crystals), multi wavelength anomalous diffraction (MAD)); the proposed structure of hCG is predominantly composed of β structure with three helical segments, two in the α and one in the β-subunit. The antiparallel β- strands in each subunits are joined by three hairpin loops. Both subunits structure like some protein growth factors (NGF, TGF-β and PDGF-ββ) contain the so-called cysteine knot motif by three disulphide bonds (cys I-IV, cys II-V, cys III-VI). The heterodimer is stabilized by a segment of the β-subunit which wraps around the α-subunit and is covalently linked like a seat belt by cys 26-cys110. This hetrodimer has a large area of interface with only a small hydrophobic core. The overall topology of hCG subunit is shown in (Fig. 3).

B- Molecular Forms of hCG

Various molecular forms of hCG (intact or modified), free subunits (intact or modified), and degradation products are present in biologic fluid, tissues of normal and tumoral organs and reference preparations[79].

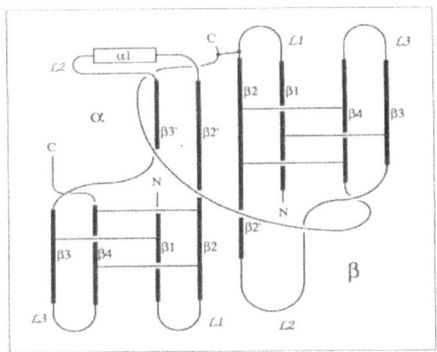

fig.3 A schematic drawing of the hCG dimer topology

Intact hCG is defined as heterodimer comprising the two mature α and β subunits. Its levels in the sera of normal nonpregnant women increase with age, and higher than in men[80][81]. In women <50 year the upper reference limit is 8.6 pmol/l (2.9 Iu/L), and in those >50 years is 15-5 pmol/L (5.3 Iu/L), where the level of hCG in urine is < 2 pmol/mmol creatinine. In contrast, hCG levels are 2-6 pmol/L (0.7-2.1 Iu/L) in sera of healthy men. Plasma intact hCG rises exponentially during the first trimester of pregnancy, reaching a peak of 25-30 µg/ml at 8-10 weeks. Following a rapid decreases until 15 weeks, a slower but continuous fall is observed up to delivery. Interestingly, amniotic hCG levels are <20% of maternal serum level at all times of gestation[79].

Free hCGβ refers to the non-combined β-subunit consisting of a glycosylated 23KD single chain polypeptide. This form appears to be devoid of any biologic role during pregnancy (the peak level reaches 50-70 ng/ml between 8-10 week)[79]. In normal individuals, free hCGβ serum values are extremely low (never exceeding 4.5 pmol/L) while this form is most often secreted in nontrophoblastic tumors[82].

Free hCGα refers to non combined α-subunit. Although it appears to have a structure identical to that of the combined α subunit of hCG, its carbohydrate composition is quite different which prevents combination[83]. This form may have a biologic role in pregnancy independent form that of hCG, where its level in serum slowly increase up to 18 weeks to reach a peak of 30-50ng/ml. In healthy individuals, levels up to 3000 pg/ml are present in sera originate in the pituitary gland[62]. Free hCGα may increase in pituitray tumors.

Beta -Core Fragment of hCGβ (hCGβcf), is small forms (10KD) immunoreactivity identify. Its two disulfide-bridged peptides lacking the amino acid sequences 1-5, 41-54 and 93-145 and corbohydrate structure changes of hCGβ[84]. It has been demonstrated that HCGβcf is primarily product of renal degradation of hCGβ, although some fragments may originate from either placenta or cleavage in the circulation. hCGβcf is the most abundant fragment of hCGβ in pregnancy urine and is also found in the urine of patients with a broad spectrum of malignancies and benign tumors. Circulating hCGβcf is almost undetectable, while levels of 0.51-1.25 pmol/mmol creatinine are found in urine of healthy women[69]

Nicked hCG (hCGn or hCGβn), are detectable in a significant proportion of both hCG and its free β subunit of normal and tumoral origins, as a result of intrachain proteolysis nicking within the β44-49 region [85][86]. The hCG appears to be uns Table and rapidly dissociates and weakly recognized by MAb. Furthermore, some results indicate that hCGβn is an intermediary form in the metabolism of hCG, and further cleavage to hCGβcf.

Several other molecules are detectable mostly in tumors, due to the modification of peptide structures both in hCGα and hCGβ such as N-terminal heterogeneity hCGα, mutant hCGα, large hCGα and CTP, hyperglycosylated hCG[87].

C- Production and Metabolism of hCG

Both normal placental cells as well as neoplastic trophoblastic cells, and sperm cells in the male synthesize and secrete hCG and its free subunits[88]. Human chorionic gonadotropin plays a critical role in the maintenance of the corpus luteum during the first 4 to 6 week of gestation to produce progesterone to keep the endometrial intact where the blastocyst is implanted, hCG also stimulates the testosterone production by the developing tests in males Fetuses[79]. The physiological processes regulating hCG production are not well understood. It is likely that its biosynthesis is dependent on the stage of differentiation of the trophoblastic cells and regulated by several autocrine and paracrine factors acting through cellular signals on the regulatory elements of the α- and β-subunit genes[89].

About 80% of hCG is metabolized in liver and renal tissue and excreted predominately as hCGβcf, whereas 20% of circulated hCG is excreted unmodified. In normal individuals, the plasma half-life of hCG is about 24-36 hours. Its clearance includes an initial, fast component ($t_{1/2}$ of 5-6h) and a second, slower component ($t_{1/2}$ of over 24 h)[69].

D- Clinical Use of hCG Determinations

Diagnosis of Pregnancy and Related Disorders: Immunochemical determination of hCG in serum or more commonly in urine, is the main method for diagnosis of pregnancy. In serum a discrimination limit of 10 Iu/L is often used [79].

Spontaneous abortion is the most common complication of pregnancy, and abnormally low serum hCG concentrations are often observed in these patients. Serial determinations of hCG in serum in combination with sonography are much used for diagnosis this condition [66].

Decreasing or slowly increasing hCG concentration during early pregnancy may also indicate ectopic pregnancy (implantation of the blastocyst outside the uterus)[66].

The potential usefulness of either hCG or its free subunits(in combination with AFP) in prenatal screening for chromosomal abnormalities (particularly Down's syndrome), was the subject of intense debate[90][91]. While the mean hCGβ value in cases of trisomy 21 is significantly higher than in unaffected cases, the median value in trisomy 18 was found to be significantly lower.

Placental and Testicular Trophoblastic Tumors: Infact hCG and free hCGβ are ideal tumor marker for gestational trophoblastic disease (GTD ""choriocarcinoma,and hydatidiform mole[92]"")and testicular germ cell tumors[93].The ratio of hCGβ to intact hCG is higher in patients with malignant than with benign trophoblastic tumors, which used for diagnosis. Measurement of blood intact hCG can be utilized for the therapeutic monitoring of GTD . Molecular heterogeneity has been demonstrated in the carbohydrate structure of hCG in choriocarcinoma, hCG is substantially diminished in sialic acid content and is composed of tri and tetra-antennary carbohydrate moieties[94].

Human chorionic gonadotropin or its free subunits presents in (80%) of sera patients with nonseminomatous testicular tumors, where hCG is elevated to more than 20,000 pg/ml[93]. Indeed, hCG, free hCG, AFP are the Most useful markers for the diagnosis, prognosis, and follow up of these malignancies[95]. Measurement of hCG in CSF is done when cerebral metastasis are suspected (ratios of hCG in CSF to hCG in serum exceed 1:60). In seminoma testicular cancer, hCG is elevated in 16% of patients[96].

*Nontrophoblastic Tumors :*The presence of hCG and its subunits in the tumor itself or in the serum and malignant effusions of patients with nontrophoblastic and nongonadal tumors were investigated.

Numerous investigators, using immunochemical procedures, have demonstrated that cancers of the breast, uterus, lung, ovary, cervix, gastrointestinal tract, bladder, liver produce elevated concentrations of hCG or its subunits.

a-Blood studies: Early results employing radioimmuno assay(RIA) for hCG indicated that a great variety of malignant tumors expressed hCG[97-102]. The overall incidence was in the range of 19%-30% of all tumors studied. The levels of hCG were only moderately increased and shown variable correlation to tumor stage, histological grading or clinical course. Therefore, the clinical value of determine hCG in serum for diagnosing and monitoring patients with non-trophoblastic malignancies remains controversial. Studies with highly specific sandwich procedures and effective MAb showed that free hCGβ is mainly elevated while intact hCG and hCGα are slightly elevated or within the normal range in the sera of nontrophoblastic cancers.The malignancies that exhibited the greatest incidence are gynecologic cancers, tumor of

the head,neck,lung,gastrointestinal tract tumors,and melanoma[102-111]. These studies

suggested that free hCGβ is highly diagnostic of malignancy in general and defines a subgroup of aggressive nongonadal malignancies.

b-Urine studies: hCGβcf, free hCGβ, and asialo hCG have been reported in urine of different malignancies employing two soild phase capture antibodies[112][104][105]. Using a cutoff of 0.2ng/ml for hCGβcf,59% of the cervical and endometrial and 70% of the ovarian cancers patients exceeded this level[113]. It has been suggested that the hCGβcf in the urine of patients is due to the degradation of the free hCGβ produced by these cancers.

*c-Effusion fluids studies:*Free hCGβ immunoreactive material has been reported in ascetic and tumor cyst fluids of patients with non trophoblastic cancers (including gynecologic cancers) more than intact hormone[106][107]. Free hCGβ levels have shown to be higher in malignant ascites as compared to the corresponding plasma values of the same patients. Determining free hCGβ in malignant effusions enhanced the sensitivity of this marker as compared to serum.

d-Tissues studies: Human chorionic gonadotropin in the cell is present in two pools, a secretary pool and membrane pool. The latter becomes an intrinsic part on determination hCG on the cancer cell surface. Various investigations on cultured cancer cells (including cells form nasopharynx, lung, cervix, bladder, breast, endometrial, colon, prostate, oral cancers, leukemia's, lymphomas, and retinoblastomas), using flow cytomatry technique, demonstrated the expression of membrane associated hCG and its subunit[114-124]. These studies showed high expression of free hCGβ than intact hCG and there was no relationship between the hCGβ positivity and the histological type or morphological features of tumor, but a number of authors have suggested that tumor which expresses hCG will pursue amore aggressive course and has a worse prognosis[125]. Analyses also done on extracted of media of cultured caner cells, reflecting the secretary pool or total cell pools. Other techniques such as immunohistochemical techniques were used to identify hCG in tumor and normal tissues[126-137]. Positive staining for hCG were found in, ovarian, cervical, colorectal, gastric, hepatoblastoma, large bowel, urothelial and breast cancers tissue. Authors also reported the stimulation of the growth of several tumor cell lines by hCG[138-141]. These findings with other results which confirmed that hCG has also chemical and physiologic properties of growth factor[76][77], have provided a scientific basis for studies of, prevention from, and control of the cancer by active or passive immunization against hCG and its subunits[142]. Some studies used anti hCG vaccines which originally was developed for fertility control[143][144]. In addition to that, studies on cluster of genes encoding the hCGβ demonstrated that five nonallelic genes are transcribed in vivo with highly variable level of expression in first trimester placenta tissue, whereas normal nontrophoblastic tissue express 2 allelic hCGβ genes. In tumor tissue collected from patient with breast bladder, prostate, thyroid cancer found that up to 61% of these tumors expressed four nonallelic hCGβ genes different from that of normal tissue. These findings provide the basis for a simple test that may be useful for the molecular diagnosis of nontrophsblastic cancer[64][115].

E- Determination Methods of hCG

Immunological methods are nearly exclusively used for clinical determination of hCG[82], while assessment of the potency of hCG preparations to be used for pharmaceutical purposes should be performed by bioassays.

Biological Methods

a-<u>In vivo methods</u>: These bioassays measure the response of gonadal tissue to hCG in various animals (mice, rabbits, frogs). These methods are no longer widely used because its susceptible to yielding spurious results due to variation in the metabolic clearance rates of the forms of hCG being assessed and are based on animal –derived rather than Human chorionic gonadotropin receptor[69].

b-<u>In vitro methods</u>: Two methods have been widly used for assessing the hLH-like bioactivity of hCG, one is based on the rat interstitial cell testosterone response (RICT assay)[145] and the other on the mouse interstitial cell response (MICT assay)[146]. These methods have the advantage of the low detection limit but the limitation is their lack of specificity (hLH activity, factors modulate cellular responses) and heterologous nature of the systems. Homologous bioassays have been constructed with the modern techniques of molecular biology, where the human hCG receptor gene has been cloned and transfected into cells to produce human receptors[147].

Binding Methods

Binding assay can utilize antibodies or receptor as binding protein.

a. <u>Competitive Immunoprocedures</u>

Radioimmuno assay(RIA), and Receptor methods: Pregnancy test is the first specific immunoprocedueres (qualitative agglutination inhibition methods) in which, latex particles or sheep red blood cell coated with the hCG, mix with the urine sample and then with a solution containing antibodies to the hCG. Presence of hCG in urine inhibits agglutination. This method is still widely utilized, especially as in-house methods[69].

The quantitative specific immunoprocedures for hCG were typical competitive immunoprocedure in which the radio active iodine ^{125}I-hCG competes with sample analyze for binding to anti-hCG. Increased hCG in sample, decreased bound radio activity. These methods are still used widely for measurement of serum and urine hCG levels[148][69].

Another type of assays systems available for serum hCG determination is a radioreceptor assay (RRA). This method is based on the competition between the hCG in the serum and a ^{125}I-hCG for binding to hCG receptor present in tissue [69].

b. <u>Non – competitive immunoprocedures:</u>

Non competitive immunometric with labeled antibody (sandwich or two-site immunometric methods) are quantitative methods for hCG and its subunits with low detection limit, large measurement range. The labeled anti hCG reacts with sample hCG bound to solid phase anti hCG. Diffferent labels are used such as enzyme, fluorescent, and luminescent. The specificity of sandwich methods is defined by the reactivity and recognition ability of the antibodies (β •→ α,β •→ β, αβ

•→ β, βcf •→ β, βCTP •→ α, α •→ α) where •-- capture antibody and → tracer antibody[79](Fig.4).

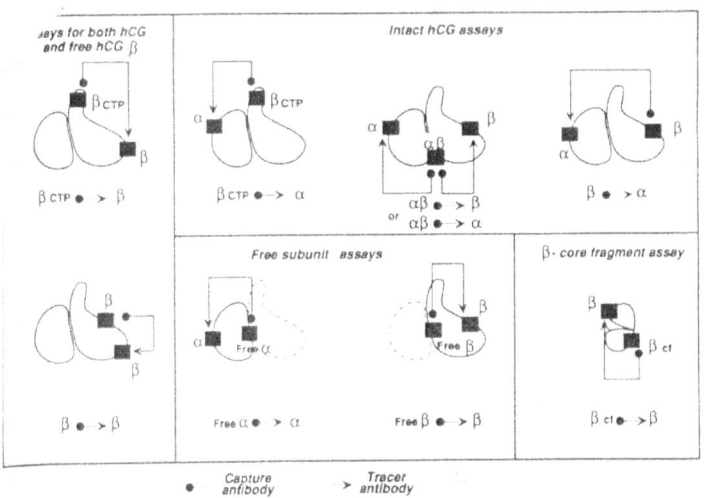

Fig.4 Schematic representation of the sandwich methods used for the measurement of hCG, its free subunits, or hCGcf[79]

1.5 Human Chorionic Gonadotropin Receptor (hCG-R)

Human chorionic gonadotropin receptor is a cell surface protein, present on testicular leydig cells and on ovarian theca, interstitial, luteal, and mature granulosa cells. In both males and females, the hCG-R recognizes the pituitary hormone LH. In the pregnant female, the same receptor also recognizes hCG[149].

The hCG-R belongs to a subfamily of glycoprotein hormone receptor within the G protien-coupled receptor family (whose members are characterized by the common structural feature of seven transmembrance domains)[150].

Although there are differences in the response duration, affinity and dissociation rate of hCG for the hCG-R as compared to LH, it is generally believed that both hormones bind to and activate the receptor similarity. Studies to date indicate that the cAMP second messenger system is the predominant pathway stimulated by either LH or hCG. Yet, it has been reported that stimulation with LH or hCG leads to activation of phospholipase C, resulting in formation of inositol phosphates and elevation in intracellular Ca^{2+} $[Ca^{2+}]$[151][152].

The responsiveness of a given target cell to hCG or LH can be modulated by alteration in the number of cell surface receptor such as homologous and hetrologous down regulation, as a results of internalization and lysosomal degradation of hormone- receptor complex[149]. Desensitization (reduction in response intensity) has also been observed for hCG-R it has been hypothesized that the desentization occurs as a result of receptor phosphorylation[153].

1.5.1 Structure of the hCG-R and the Nature of Hormone – Receptor
Interaction

Structural studies have been hampered by the low abundance of hCG receptor, nonetheless, researchers have made sufficient progress in this area. Data from chemical crosslinking of labed hCG to the receptor, immunoprcipitation of a biosynthetically labeled hCG-R, purification of the receptor, have led to the estimation of the overall size and structure of this receptor[154-158]. Studies on the cDNA and genomic clones for hCG-R have yielded more detailed information on the structure and function[147][159-161]. The antibodies developed to the receptor have confirmed the topology of this receptor and are being used to address structure function relationship[149][154]. All these studies established that hCG-R is a single chain glucoprotien of 85-93 KDa. Its composed of two halves of equal size., the N-terminal extracellular hydrophilic, exodomain of ~ 350 amino acid residues and the membrane associated intracellular C-terminal endodomain of ~334 amino acids residues, which includes seven hydrophobic transmembrane helices and three cytoplasmic loops with three extracellular loops that link the seven transmembrane domains (Fig.5) The hCG-R is encoded by 11 exons located on chromosome 2 p21, the first 10 of which code for the exodomain and the 11th for the endodomain.

hCG Receptor

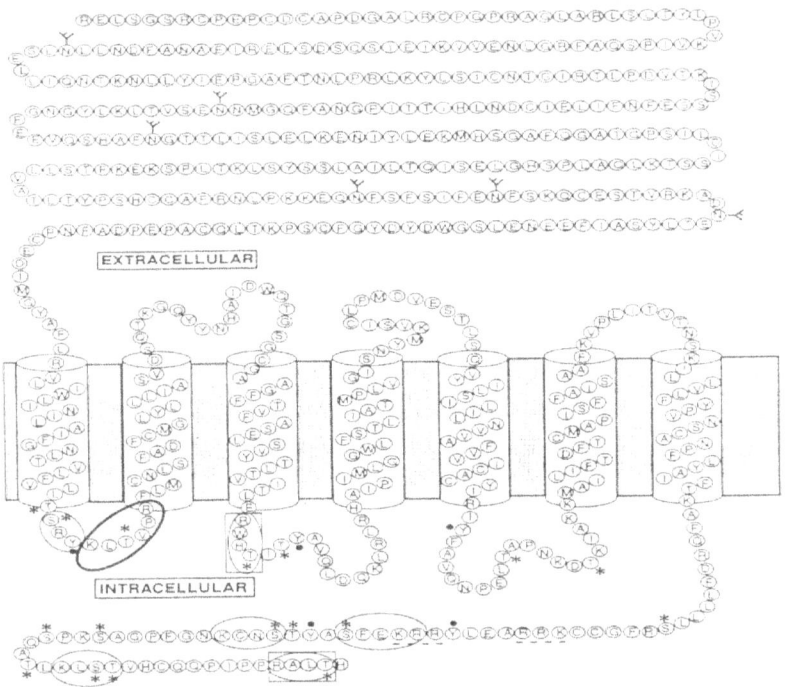

Fig.5 Amino acid sequence, orientation, and proposed topology of the hCG receptor in the plasma membrane.Potential sites for N-linked glycosylation are shown by the branch-like structures .The sequences underlined with dashes in the cytoplasmic tail mark two clusters of basic amino acids which might represent potential tryptic cleavage sites.Potential intracellular sites for phosphorylation and denoted by asterisks (serine and threonine residues) or dark dots (tyrosine).The rectangles denote weak consensus sequences for cAMP-dependent protein kinase –catalyzed phosphorylation.The ovals and heavy ovals denotes weak and strong consensus sequences for c kinase-catalyzed phosphorylation[154].

Results from mutational analysis[162-164], chimeric construction[165], synthetic peptide studies[166] and chemical modification of the receptor[154] showed that the hormone binding and activation processes are separable. The large extacellular domain of hCG with six sites for N-linked glycosylation and (8-9) leucine –rich repeats (LRR) is responsible for high selectivity and affinity hormone binding without hormone action[167-170]. LRR are thought to form a crescent with concave inner surface consisting of β sheets which may bind hormone (Fig.6)

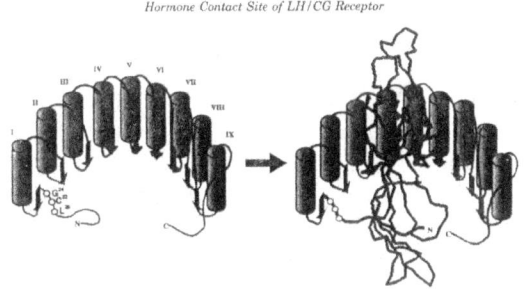

Fig.6 Schematic model for the interaction of the hCG receptor exodomain with hCG. Hormone interacts with the inner face of the crescent structure of the receptor[173].

Activation, in contrast, has been identified with transmembrance domain [171]. The molecular mechanism involves the initial high affinity contact of hCG with receptor exodomain, then the resulting complex (may not be thermodynamically stable) undergoes conformational changes and makes secondary low affinity interactions with exoloops of membrane associated domain(Fig.7), causing an allosteric structural change in the endodomain which leads to signal generation and hormone action[172][150].

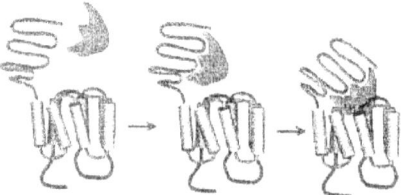

Fig.7 Schematic presentation of the receptor-hCG interaction. hCG bind to the N-terminal segment of the receptor and the hCG-N-terminal segment interacts with exo loops to generate a signal[150].

The initial high affinity interaction includes multiple contacts between the exdomain and both subunits of the hormone[173][174]. The precise hormone contact sites in the exodomain are unclear. The most putative contact sites are αC-terminal region, N-oligosaccharide at αAsn^{52}, and unique αhelix in the α-subunit, as well as unusual loop (seat belt) in the β-subunit and the peripheral β-hair pin loop of both subunits[175]. Between the initial interaction and the signal generation, hCG and receptor undergo conformational changes. Studies including photochemical crosslinking of hCG, mutational analysis of the receptor and serial truncation has been demonstrated a structural changes of hCG involving the intersubunit interaction[162][164][176-177].

1.5.2 The Role of hCC-R in Nongonadal Tissues and Tumor Tissues

It has long believed that the hCG-Rs were present only in gonads. But the studies, using immunocytochemistry method and MAb for receptor or reverse transcription polymerase chain reaction technique, have demonstrated that various female as well as male nongonadal tissues (uterus, placenta, fetal membrane, deciduas, fallopian tubes, brain, breast, skin) contain low levels of hCG-R protein and hCG-R transcripts (cancer tissues are among these tissues)[178-187]. Moreover the direct action of ectopically synthesis hCG in regulating the growth of various cell types and its role in various cancers, implies the existence of the corresponding receptors in these tissues.

Studies on nonpregnant human uteri, demonstrated not only the presence but also different cellular distribution of hCG-R, where epithelial cells contain more receptors than that stromal cells of endometrial in turn, contain more than myometrial and vascular smooth muscle[178][188]. The hCG-R are increased from the proliferation to the secretary phase of the cycle. These differences suggest that other hormones regulate the receptor, which regulated different functions. Evidences from in vitro studies supported these suggestion, for example , hCG was reported to have a relaxing effect on porcine myometrium[189] and can increase cAMP levels and progesterone synthesis in rat uterus[190]. The expression of receptor mRNA had been found higher in myometrial then in human endometrial blood vessels ,and there is possibility that hCG could an directly increase uterine blood flow [187].Human chorionic gonadotropin may regulate epithelial cell by increasing the local synthesis of steroid hormones[190]. In vitro , hCG can directly regulate proliferation of human myometrial smooth muscle cells causing hyperplasia as well as hypertrophy[188]. Increasing expression of receptor gene in human endometrial cancer has also been obtained [184].

Studies on human placenta , fetal membranes, decidua have been shown the presence of receptors as well as cellular localization differences and changes from mid to term pregnancy suggesting that hCG may regulate their function. In fact, there are data from in vitro studies support this possibility, in placenta, high hCG-R concentration have been shown to increase cAMP formation[191], glycogen breakdown[192], aromatization of androgens to estrogens[193], and interconversion of estron to estradiol[194]. In decidua, high hCG concentration have been shown to directly stimulate prolactin synthesis, which in turn, has been inhibited placental hCG synthsis.This finding indicates the presence of hCG- influenced short loop positive and negative regulatory mechanisms within the human fetoplacental unit[195][196]. Investigation for the presence of functional hCG-R in human breast, revealed that normal epithelial cell, benign lesion, and cancer biopsies content hCG-R[183][197]. The intensity of the immunolabeling of hCG-R varied in individual biopsies. The presence of receptor mRNA was also confirmed.In the breast, it has been proposed that hCG provokes differenentiation, which in turn renders the cells less susceptible to neoplastic transformation[198]. This mechanism would explain the decreased occurrence of breast cancer in women who had on early pregnancy. Studies on human breast cell lines showed the functional presence of hCG-R that can bind hCG to exert a direct antiproliferative effect on human breast epithelial cells through secretion of inhibin and decreasing estrogen receptor[199]. Indeed hCG has been shown to exert either stimulatory or inhibitory effects on the growth of various cancer in vitro. In bladder, lung and endometrial cancer as well as choriocarcinoma, hCG and/or its subunits promote the growth of these cancers while in breast and prostate cancers are growth inhibiting[115].

The Aim of the Work

The aim of this work includes the following :

1-Investigation the presence of intact hCG and free hCGβ in sera and tissues of patients with endometrial and epithelial ovarian cancers.

2-Development a quantitative radiorecptor assay for detection and analysis of membrane-associated hCG-Rs of human cancer tissues of endometrial and epithelial ovarian cancers.

3-Purification and characterization the hCG-Rs of epithelial ovarian cancer from detergent soluble membrane extracts .

4-Determination of the kinetic parameters of the binding reaction of hCG with it's receptor in epithelial ovarian cancer.

5-Investigation the CEA in the sera and tissues of endometrial and epithelial ovarian cancers patients.

6-Evaluation the levels of LDH, and ALP enzymes activities in sera of endometrial and epithelial ovarian cancers patients and analysis the different isoenzymes and forms of them.

7-Measuring the concentration of Copper, Zinc, and Calcium elements in sera of patients with endometrial and epithelial ovarian cancers.

Chapter Two

Human Chorionic Gonadotropin in Sera and Tissues of Patients with Gynecologic Cancers

Abstract

In the present study, the presence of intact hCG and free hCGβ in sera and tissues of patients with gynecologic cancers was investigated by radioimmunoassy for these two forms of hCG. Intact hCG and free hCGβ level was found to be elevated above the reference value(>5mIU/ml) in the sera of 7 out of 14 patients (50%) with epithelial ovarian cancer and 4 out of 9 (44%) patients with endometrial cancer. There was no significant difference between the mean age of hCG positive and hCG negative cancer patients, but hCG-positive patients were significantly older in the benign group. The intact hCG and free hCGβ in malignant conditions (87.6 ± 47.9mIU/ml) were higher than in benign conditions (11.2 ± 2.2mIU/ml) ($P< 0.05$).Circulating intact hCG and free hCGβ could be demonstrated in early as well as in advanced stage of gynecologic cancers. After radical surgery,4 out of 6 hCG-positive patients turned to be normal, and 2 of them remained positive.

In malignant tissue extracts, the intact hCG and free hCGβ were significantly higher than in the corresponding sera samples of benign tumor samples. Elevation intact hCG and free hCGβ in tissues(>5 mIU/mI) was found in 100% of patients with gynecologic cancers. The level of intact hCG and free hCGβ exceeded 10mIU/ml in all of 23 malignant tissue samples whereas not exceeded 10mIU/ml in all of 25 benign tissue samples. In this respects, measurement of hCG in tumor tissue may be of higher diagnostic value as serum measurement.

Introduction

Human chorionic gonadotropin is a clinically relevant marker of trophoblastic and nontrophoblastic cancers. Most studies have focused on serum determination of hCG, which have been reported to be elevated in a significant portion of patients with various nontrophoblastic cancers including gynecologic cancers [97-111]. The elevation of hCG reported vary considerably for similar cancer. This is possibly due to differences in patient selection and assay characteristic. Recently, the utilization of monoclonal antibodies, highly specific for intact hCG or its subunits have been re-evaluated the presence of hCG in the sera of nontrophoblastic tumors [102-111]. Several studies have shown that either hCG or its subunits may be localized to a variety of nontrophoblastic tissues and was extrac Table from malignant and normal human tissues [126-137]. Expression of membrane-associated hCG was found to be a phenotypic marker characteristic of all evaluated cultured human cancer cell lines, irrespective to their type or origin [114-125]. A stimulation of the growth of several tumor cells by hCG had also been reported[138-141].

In present work, a re-evaluation of presence of intact hCG and free hCGß in sera samples of patients with endometrial and epithelial ovarian cancer was carried out using a highly sensitive and specific radioimmunoassay (RIA) for these forms of hCG. Determination of intact hCG and free hCGß in these malignant tissues had been carried out to further definition of the role of hCG as tumor marker,

Materials and Methods:

2.1 Chemicals

All laboratory chemicals and reagents were of analar grade. PMSF (phenyl methyl salfonyl flouride), Tris (hydroxymethylaminaomethane), Glycerol, HCL, EDTA, and BSA were obtained from Fluka. NaOH, Na_2CO_3, $CuSO_4$, $MgCl_2$ and Folin reagent were obtained from BDH.

Intact hCG and free hCGß ^{125}I RIA kit was provided by Diasorin which contain one vial of ^{125}I-hCG tracer (Iµg/ml); one vial of rat-monoclonal hCGß antibody recognizes hCGß within holo-hCG as well as in its free form; one vail of precipitating goat anti-mouse serum (second antibody) in polymer solution, one vial of 90 mIU/ml hCG as control; one vial of hCG blank (0 mIU/ml); five vials of hCG standards with concentrations ranging of (5-200 mIU/ml); one vial of hCG calibrator at 50 mIU/ml.

2.2 Apparatus

The apparatus used during this study were, LKB gamma counter type 1270 Rack, cooling centrifuge type Hettich, LKB UV-Visible record spectrophtometers, Unicam pH meter, Memmert water bath, Memmert incubator, homogenizer type HO and SM shaker.

2.3 Patients

The study groups consisted of 82 women with gynecologic disease (age range between 22-80 years).Histological diagnosis was available in all cases. Malignant tumor was present in 23 patients (age range between 38-80), the tumor site was the endometrial adenocarcinoma in 9 and the epithelial ovary in 14 {(serous11), mucinous(3)}. Staging was performed according to the FIGO. The benign condition was present in 59 patients (age range between 22-67),this group consisted of adenomyosis (4 cases), fibromyoma (20 cases), endometriosis (3 cases), follicular cyst(10 cases), serous cystadenom tumors (9 cases), mucinous tumors(3 cases) polycystic (6 cases) and dermiod cyst (4 cases). Thirty healthy female between the age of 20-65 served as the control group, it was confirmed that the individuals were not suffering from any disease at time of the study and did not receive any medication.

All patients were admitted for treatment to aL-Yarmook Hospital, Saddam Medical City, Saddam Collage of Medicine, Al-Arabi Private Hospital.

2.4 Collection of Specimens and Preparation of Tissue Homogenates.

Serum: All blood samples were collected before treatment in all cases using venipuncture.Additional sample were obtained from 6 patients postoperatively while they were still in hospital. The samples were allowed to clot before separation by centrifugation at 1500xg for 15 min. The sera were subjected to RIA measurement.

Tissue: Tumor tissue specimens were obtained from patients during surgical operating. For comparison two term pregnancy placentas from spontaneous normal vaginal deliveries were used . The tissues were brought to the laboratory on crushed ice and washed extensively with ice–cold physiological saline to remove blood then stored frozen until analysis. The frozen tissues were weighted, minced finely then, homogenized in 50 mM Tris/HCL, pH 7.4 containing (0.25 M sucrose, 1mM $MgCL_2$, 5 mM EDTA, 1mM PMSF) with a ratio of 1:4 (w:v) by homogenizer at setting (2.5) using three 20s periods of homogenization. The homogenate was then filtered through a nylon mesh sieve in order to eliminate fibers and cell debris then centrifuged at 600xg for 15min at 4^0c to remove the nuclei. The pellet was again homogenized and centrifuged. The above supernatants were pooled and subjected to RIA measurement and protein determination.

2.5 Quantitative Measurement of Intact hCG and Free hCGß

Intact hCG anf free hCGß of sera and tissues samples were measured using highly sensitive and specific RIA Kit. The assay protocol was described in Table (2.1)

Table (2.1) Assay protocol of intact hCG and free hCGß

	Total count (tracer)	hCG Blank (0mIU/ml)	hCG Standards (mIU/ml)						hCG Control (mIU/ml)	Patient sample
Tube number	T_1,T_2	B_0 1,2	5 3,4	10 5,6	25 7,8	50 9,10	100 11,12	200 13,14	90 15,16	X 17,18
Standards, control, blank	←---------------------- 200 µL ----------------------→									

patients samples (µL)	
125I-hCGß red tracer (µL)	←-------------------------- ------ 100 µL -------------------------- -------------------------→
125I-hCGß blue antibody	←-------------------- 100 µL -------------------- ------------------------→
	1. vortex mixing, incubation for 1h at room temp., total count set aside until counting in gamma counter
Precipitating antiserum	←----------------------1ml----------------------- ----------------------→
	1. Incubation for 5 min at room temp. 2. Centrifugation at 40^0C for 15 min at 1500xg 3. Decantation 4. Counting for radioactivity in a gamma counter for 1min

Calculation:

1. The standard curve was drawn by plotting the CPM bound for hCG standards versus their concentration(Fig 2.1)

2. The samples values of intact hCG and free hCGβ were read directly from the curve. Patients samples with assay values greater than 200 mIU/ml must be diluted.

2.6 Estimation of Protein Contents:

Protein was measured by the method of Lowry et.al.,[200] using bovine serum albumin as standard.

2.7 Statistical Analysis: Results were expressedas mean ± S.E.M.Student's t - test
and one way analysis of variance were used for statistical comparisons and P values of less than 0.05 were considered statistically significant.

Results and Discussion:

Preoperative Data:

Results of hCG determination depend largely on the selection of antibody and the type of technique used. In present study, re-evaluation of the presence of intact hCG and free hCGβ subunit immunoreactivity in serum of patients with ovarian and endometrial cancer, using RIA assay with the lower limit of sensitivity of (1.5 mIU/ml) and monoclonal antibody recognize these forms of hCG,had been done.

Immunoreative intact hCG and free hCGß, was elevated (> 5mIU/ml) in the sera of 11 out of 23 patients (48%) with gynecologic cancers (range 22.1 -166 mIU/ml)as shown in Table (2.2). Intact hCG and free hCGß was found to be above normal range in the sera of 4 out of 9 patients (44%) suffering from endometrial cancer range (22.1 -110.5 mIU/ml); mean ± S.E.M (64.5± 41.8 mIU/ml) (Table 2.3). In patients with ovarian cancer (n=14), elevated intact hCG and free hCGß serum values were found in 7 of 14 (50%), range (37.3 -166 mIU/ml).

Intact hCG and free hCGß production occurred exclusively in patients with serous ovarian neoplasm (Table 2. 3).Analysis of the data indicated no significant difference in the occurrence of hCG in patients with endometrial and ovarian cancer (P> 0.05, NS). Furthermore, elevations of intact hCG and free hCGß in the sera from patients with ovarian cancer (100.55± 45.32) were significantly higher than in the sera from patients with endometrial cancer (64.6±41.8).

The mean age of intact hCG and free hCGß positive group (61.5± 12.1) was similar to that of intact hCG and free hCGß negative group (60±11.3) of patients with endometrial and ovarian cancers (Table 2.2).

In benign condition positive findings were made in 7 of 59 cases (12%) including adenomyosis and myomas of the uterus, and various form of benign ovarian tumors (Table 2.3). The elevation level of intact hCG and free hCGß varied between (8.3-14.4) mIU/ml. The mean elevation (11.2± 2.2) mIU/ml was smaller than in the malignant group (87.6± 47.9)mIU/ml (p<0.05), the percentages of elevation intact hCG and free hCGß in benign and malignant conditions are shown in Fig 2.2. In the benign group hCG-positive patients were significantly older (60.6 ±5.8 years) than hCG- negative patients (44.2±8; p<0.05; Table 2.2).

Human chorionic gonadotropin is one of the earliest embryonic gene products appearing in ontogeny. The finding of circulating hCG in association with cancers has been regarded generally as another expression of embryonic antigens in cancer. It's place in fetal survival and placental development has been related to its role in the control of "pseudo malignancy" because trophoblastic cell has angiogenic and invasive properties[201]. Trophoblasts express oncogenes and proto-oncogene[202] [203] and demonstrate cytokine expression[204] and resistance to immune rejection that we associate with malignancy.

The percentage of elevated of intact hCG and free hCGβ was (48%) for patients with gynecologic cancer. This is comparable to that reported by Grossmann et.al.,[107] (37%) for gynecologic cancers based immunoradiometric design. Crawford et. al.,[111] demonstrated the usefulness of free hCGβ and hCG$_{βcf}$ in assessing prognosis of primary cervical cancer. De-Bruijn et.al.,[109] studied the progressive vulvar cancer(n=104) and found elevated free hCGβ in 50% of patients and indicated that the synthesis of free hCGβ can be increased during progression. Ind et.al.,[108] investigated the elevated free hCGβ and other tumor markers in women with primary epithelial ovarian cancer. They found a correlation between markers levels and cancer stage. Elevated levels of intact hCG or free hCGβ were found also by other authors in different nontrophoblastic tumors in addition to gynecologic cancers, such as tumors of the head and neck, lung cancer, gastrointestinal track tumors and melanoma. Hoermann et.al.,[106] found elevated level of free hCGβ in 55% of patients with (breast, pancreas, colon, stomach, ovary, cervix, bladder and hepatocellular cancer), whereas intact hCG was not elevated in all serum malignant samples. Marcillac et.al.,[103] detected elevated free hCGβ level in 47% of bladder, 32% of pancreatic and 30% of cervical cancer, they also found low elevation level of intact hCG in the same patients (1.3%). Alfthan et.al.,[104] observed elevated levels of serum free hCGβ in 72% of pancreatic cancer and 86% of biliary cancer. Iles and colleagues[105] demonstrated that the incidence of free hCGβ was approximately 30% in serum of patients with bladder cancer and 15 out of 36 patients with pancreatic adenocarcinoma had elevated plasma hCGβ[110]. In all these studies, the hCG assay sensitivity was increased for nontrophoblastic cancers by inclusion of free hCGβ subunit measurements. These studies indicated that most hCG-producing malignancies secret hCGβ in addition to the intact hCG molecule and free hCGβ subunit may even be the major form of hCG in nontrophoblatic cancers .This pattern of secretion contrasts with that observed in pregnant women who display a predominant secretion of intact hCG in large excess[79] .The result of an earlier studies based intact hCG determination confirm this observation where the values obtained are low. Donaldson et.al.,[99] in their assessment of markers of gynecologic cancers found low percentage (22%) of patients has total hCG serum level greater than 5 mIU/ml . Rutanen et. al.,[97]. also found an elevation of total serum hCG in only 18% of gynecologic cancers. In early studies[97-101], the overall incidence was in the range of 18-25% of all tumors studies.

Serum intact hCG and free hCGβ levels were elevated in 44% of patients with adenocarcinoma of the uterus and in 64% of patients with serous epithelial ovarian cancers. On considering whether the histological type of tumor or cells give any clue of its ability to secrete hCG, it is obvious, by the spectrum of histological appearance of hCG-producing tumors , that the morphological features of neoplastic are not necessarily linked to hormone producing. Donaldson et.al.,[99] among their analysis of ovarian and cervical tumors found elevated levels of hCG in selected cases of epithelial ovarian cancer (serous 27%, endometriod 100%, mucinous 0%, and Granulosa cell 0%) whereas small-cell, keratinizing, and nonkeratinizing types of squamous carcinomas had elevated levels in 13, 40 and 20% of cases respectively of cervical cancer, while none of the cervical adenocarcinoma had elevated levels of hCG. Mohabeer et.al.,[128] when examining ovarian epithelial tumors, found no relationship between the hCGβ positive and the histological type of tumors, or whether the tumor was of a benign, or malignant type (with no trophoblastic proliferation), nor even the histological grade of malignant tumors .Other authors such as Collins et.al.,. [132]

demonstrated that hCGβ positive was found in adenocarcinoma of the uterine cervix in which the cells of tumor have some histological resemblance to trophoblastic cells. Rutanen et.al.,[97]. did not find bias in the distribution of hCG in patients with poorly and well–differentiated endometrial cancer, while Mc Manus et.al.,[100] have commented that hCG positive was most frequent in poorly differentiated anaplastic tumors. There are also nongestational and extragonadal tumors which both produce hCG and bear a striking resemblance to trophoblastic neoplasms such that many authors use the term "choriocarcinomatous" to describe them. Cases have been reported with either poorly differentiated transitional cell carcinoma of the bladder or adenocarcinoma of the urothelial, stomach colon, and uterus[129-132].

Low elevated intact hCG and free hCGβ were found in 12% of patients with nonmalignant gynecologic tumors, and women with positive intact hCG and free hCGβ were older than women with negative intact hCG and free hCGβ, using cut off level 5mIU/ml. If we take an important consideration that hCG excreted in healthy subjects[80] and there are direct evidence for production of the intact hCG by the pituitary gland[62] and its expression is related to the age[81], then the elevated intact hCG and free hCGβ in benign group could be explained. It is highly likely that the low level of intact hCG and free hCGβ (11.2±2.2) found in the sera of nonmalignant gynecological patients originates from the pituitary gland and not from the tumor. This level is significantly low in comparison with malignant group (87.6±47.9).

Theories to explain the production of ectopic proteins in general involve the possibility of derepression of genes that are expressed in fetal life, or the recruitment of uncommitted cells, or the occurrence of random mutation[132]. These hypotheses assume that hCG are not a normal features of later life, but the presence of very low levels of hCG in both tissues and body fluids during normal metabolism and tissue turnover has been documented[136][137][81][80]. The source of such hCG in healthy subjects is postulated to be the normal stem cells of replicating tissues. If hCG is synthesized in many normal sites, then the neoplastic cells because of their increased number would produce larger quantities of stem cell proteins compared to the quantities produced by normal cells.

Circulating intact hCG and free hCGß could be demonstrated in early as well as in advanced stage of the two kinds of gynecologic cancers (Table 2.4). Of the 9 patients with endometrial cancer, intact hCG and free hCGß elevated in 2 out of 6 and 2 out of 3 of the stage I and II, respectively. The highest intact hCG and free hCGß serum level (110 mIU/ml) was found in women with the stage II, however the analytical data indicated no significant differences in serum intact hCG and free hCGß concentration between two stages of endometrial cancer ($p>0.05$). In ovarian cancer, intact hCG and free hCGß was elevated in 2 out of 5 (40%) of stage I and 5 out of 9 (56%) of stages (II –III (Table 2.4). The highest level of intact hCG and free hCGß was found in patients with stage III (166 mIU/ml), but also no significant different in the serum intact hCG and free hCGß concentration found between different stages. It was found no correlation between the concentration of intact hCG and free hCGß and the spread of the tumor.

Although only a few cases could be examined, women with invasive endometrial and ovarian cancer as assessed by staging had the highest serum concentration of intact hCG and free hCGβ, but difference between stages was not statistically significant .These results confirm previous observation from other investigations, which characterize hCG as a key factor defining the metastasis phenotype specifically of nontrophoblasic malignancies. Grossmann et.al.,[107] obtained that the advanced disease of cervical cancer was associated with elevated intact hCG and free hCGβ serum values (4/13 (31%) of stage II and 11/27(41%) of stages III-IV) , also obtained that women with more aggressive endometrial cancers had higher serum concentrations of intact hCG and free hCGβ. Sheth et.al.,[101] using RIA techniques, found elevated levels of serum hCG in 41% patients with squamous carcinomas of cervix FIGO stages II-IV with an increasing frequency in the higher clinical grades. Marcillic et.al.,[103] found that the production of free hCGβ was associated with tumors of high prognosis such as cancers of the lung, pancreas, and liver. Walker[127] also noted a relation between the presence of hCG and metastases of breast cancer. Syrigos[110] and de-Bruijn[109] also indicated, that the patients with elevated serum hCGβ had a worse progressive tumor compared with the group of normal hCGβ in patients with vulvar and pancreatic cancer.

The sensitivity, specificity and predictive value of the intact hCG and free hCGß in the diagnosis of gynecologic cancers are shown in Table 2.5. When pre-treatment serum intact hCG and free hCGß levels were compared between patients with benign and those with malignant gynecologic tumors, they distinguished the malignant from benign tumor with a predictive value of (73% and 80%) in uterine corpus and ovarian tumor, respectively. In all healthy females sera, intact hCG and free hCGß activity was measurable. All females showed levels within the upper limit of normal range (5mIU/ml) (Table 2.2). In women older than 45 years the mean concentration (4.8 ± 0.5 mIU/ml) was significantly higher than that of the women younger than 45 year(3.8± 0.8 mIU/ml). It is noteworthy that hCG was present at low levels (<5 mIU/ml) in serum of many cancer patients as well as in healthy women's and patients with benign disease (Fig 2.3).

Postoperative Data:

Intact hCG and free hCGß levels were postoperatively measured in 6 hCG –positive cancer patients who had undergone radical surgery including hysterectomy and ovariectomy. The blood was obtained from them in the first postoperative week, while they were still in hospital. The data showed reversion of serum intact hCG and free hCGß levels in 4 patients , while 2 remained hCG–positive (Table2.6). On the basis of clinical examination no tumor was demonstrated in any of the 4 hCG negative patients, while the 2 hCG-positive patients had a metastasis (stage III). One of these 2 women, multiple serum specimens were obtained from her during the period of different weeks. intact hCG and free hCGß concentration paralleled the clinical course of this patient (Fig.2.4).

In several cases elevated intact hCG and free hCGβ levels returned to normal level after operation, and that indicate the hormone has been produced by the tumor itself, but in two ovarian cancer cases with metastasis, intact hCG and free hCGβ levels remained elevated after operation which give an example of usefulness of hCG in the following of patients with ovarian cancer. Crossmann et.al.,[107] reported the importance of intact hCG and free hCGβ in following the ovarian cancer cases and commented that for the optimal management of gynecologic cancer, multiple markers measuring could be helpful. Marcillac et.al.,[103] and Alfthan et.al.,[104] have also reported that free hCGβ is promising marker in diagnosis and following the bladder and pancreatic cancer. Donaldson et.al.,[99] obtained that measuring hCG with other markers could help to follow up patients with gynecologic cancers.

Malignant and Non Malignant Tissues:

Intact hCG and free hCGß immunoreactivity has been determined in tumor tissues of the patients with endometrial and ovarian cancer (n=23), and in tumor tissues of patients with the benign tumor (n=25). For comparison, placenta intact hCG and free hCGß content has also been determined in 2 term pregnancy women. In malignant tissues, intact hCG and free hCGß levels ranged between (14.5 -212 mIU/ml; Table 2.7). Statistical analysis of differences between blood serum and tumor tissue intact hCG and free hCGß content in patient with malignancies has been performed. The elevation of tissue intact hCG and free hCGß level appeared to be statistically significant than in the corresponding serum samples ($P<0.05$). Remarkably, of these 12 women, which showed normal serum values (<5mIU/ml) (Fig 2.5).

In 25 non-malignant tissues, intact hCG and free hCGß levels ranged from (5.3 -9.5 mIu/ml), mean 8.19 ± 1 mIU/ml which were significantly lower than in the specimen from women with ovarian and endometrial cancer ($P<0.05$) (Table 2.7). If the upper limit of normal range for tissue as 10 mIU/ml, then 23 out of 23 (100%) of the malignant tissue samples were considered pathological and 25 out of 25 (100%) of the benign were judged normal.Determination of hCG in placenta had been obtained for comparison. The term pregnancy placenta gives higher level of intact hCG and free hCGß, even more than cancer tissue (371 ± 11 mIU/ml) (Table 2.7).

Effort to improve the diagnosis efficiency in gynecologic tumors has been made by assessing intact hCG and free hCGß in tumor tissues. Malignant cells are assumed to be source of ectopic hormone production, however, the rate at which hCG is actually released into the circulation may vary considerably. Consequently, one might expect that measuring hCG in compartments within the neoplasm increases the sensitivity of marker detection. In our study, determining intact hCG and free hCGß in malignant tissue enhanced the sensitivity of this marker as compared to serum. Concentrations of intact hCG and free hCGß were significantly higher in malignant tumor tissue than in the corresponding sera samples (p<0.05;(Table 2.6),and intact hCG and free hCGß was elevated even in patients with normal serum values (23/23) using 5mIU/ml as reference value (Fig 2.5). In benign tissues, the highest concentration of intact hCG and free hCGß measured in 25 specimens was (9.5 mIU/ml), using 10mIU/ml reference value as suggested by some studies, 23/23 (100%) of the malignant samples were considered pathological and 25/25 (100%) of the benign samples were judged normal.

These data are extended to other findings that intact hCG and free hCGß in tissue might be useful for the diagnosis of the malignancies and it expression in cancer, defines the metastic aggressiveness of the tumors in which it is found. There is several documentation of the localization of hCG, hCG like material, or free hCG subunits in a variety of nontrophoblastic tissues and was extrac Table from virtually all normal and malignant human tissues[126-137]. Using immunohistochemical technique, various investigators found hCG in adenocarcinoma of large bowel[134] and uterine cervix[132], breast cancer[127], hepatoblastoma[133], epithelial ovarian cancers[128], colorectal cancer[130], Urothelial carcinoma[129], and malignant gastric tumors[131]. Yoshimoto et.al.,[163] reported that extracts of cancer tissues from stomach, liver, colon, and lung contained hCG-like material higher than normal tissue. The work of Acevedo and other authors[114-124] has shown that free hCGß express in human fetal tissue and cancer cells of different histological types and origins. Bellet et.al.,[64] demonstrated that activation occurs with the gene cluster responsible for hCGβ in breast, bladder, prostate, and thyroid cancer and suggested that methylation play a role in maintaining the active/inactive status of the gene. The stimulation of growth of several cell lines by hCG has been reported[138-141], also resolution of the three dimensional structure of hCG had shown this hormone to be a growth factor[76][72][77], thus cancer cells are able to regulate independently its own growth. Indeed, hCG has been shown to exert inhibitory effects on the growth of breast and prostate cell lines[199]. In the breast, some patients have increasing hCG levels, and some breast cancer tissue and human breast cell lines contain native hCG.

Several hypotheses have been proposed to explain the origin of hCG producing cells and include the following:[132]

a. Displaced gonadal cell(extragonadol choriocarcinoma).

b. Metastasis from an intrauterine or gonadal lesion

c. An evaluation from a somatic cell that underwent a morphological and functional transformation into a cell functionally similar to the trophoblast.

Table 2.2: The Values of Intact hCG and Free hCGß Levels in Sera of Malignant and Benign Gynecologic Tumors (All details are explained in the text)

Clinical categories	No. of patients	Range of intact hCG and free hCGß (mIU/ml)	Range of age (year)
Malignancy			
Intact hCG and free hCGß positive*	11	22.1-166	40-80
Intact hCG and free hCGß negative*	12	3.8-5	38-70
All patients	23	3.8-166	38-80
Benign			
Intact hCG and free hCGß positive*	7	7.3-14.4	53-67
Intact hCG and free hCGß negative*	52	3.9-5	22-65
	59	3.9-	22-67

All patients		14.4	
Normal	30	2.5-5	20-68

*positive: patients having levels of intact hCG and hCGß above 5mIU/ml

Table 2.3: Preoperative Elevated Serum Levels of Intact hCG and Free hCGß in Malignant and Benign Gynecologic Tumors
(All details are Explained in the text)

Diagnosis	*Elevated/Total >5mIU/ml	%	Range (mIU/ml)	Intact hCG and Free hCGß Conc.mIU/ml (mean±S.E.M.)
Malignancy				
Endometrial Adenocarcinoma	4/9	44	22.1-110.5	64.6±41.8
*Ovarian cancer				
Serous	7/11	64	37.3-166	100.55±45.32
Mucinous	0/3	Zero		
All patients	11/23	48	22.1-166	87.6±47.9

	Elevated/Total	%	Range	Mean±SD
Benign				
Adenomyosis	1/4	25	11.2	11.2
Fibromyoma	3/20	15	8.3-13.1	10.2±2.5
Endometriosis	0/3	0		
Follicular cyst	1/10	14	12.3	12.3
Serous cyst	2/9	33	10.2-14.4	12.3±3.5
Musinous cyst	0/3	0		
Dermoid cyst	0/4	0		
Poly cystic	0/6	0		
All patients	7/59	12	8.3-14.4	11.2±2.2
Normal				
Female <45 year	0/15	0		
Female >45 year	0/15	0		

*Elevated /Total: Number of patients having levels of the intact hCG and free hCGß above 5mIU/ml/total number of patients.

Table 2.4: Elevated Serum Level of Intact hCG and Free hCGß in Relation to Staging of Endometrial and Ovarian Cancers (All details are explained in the text)

	Endometrial cancer			Ovarian cancer		Intact hCG and
	Elevated/total	%	hCG/hCGß	Elevated/total	%	

Clinical Stage*			mean±S.E.M.			free hCGß mIU/ml mean± S.E.M
I	2/6	33.3	62.9±36.34	2/5	40	95.4±82.17
II	2/3	66.7	66.3±62.5	1/3	33.3	66.1
III				4/6	67	112.1±44.4

*FIGO classification

NS: not significant

Elevated/Total: numbers of patients having levels of intact hCG and free hCGß above 5mIU/ml/ total number of patients.

Table 2.5 : Preoperative Serum Intact hCG and Free hCGß Levels: Diagnosis of Gynecologic cancers (All details are Explained in the text)

	Uterine corpus		Ovary	
	Benign	Malignant	Benign	Malignant
No. Of patients *Positive/total	4/24	4/9	3/35	7/14

Sensitivity	100%	100%	100%	100%
Specificity	83%	44%	91%	50%
Predictive value		73%	80%	

* positive: intact hCG and free hCGß >5mIu/ml

Table 2.6: Pre and Postoperative Elevated Serum Intact hCG and Free hCGß in Gynecological Cancers (All details are explained in the text)

Time of study	No. of patient	Intact hCG and free hCGß(mIU/ml) mean±S.E.M.
Endometrial Pre-op* Pos-op*	4 2	64.6± 41.8 4.7±0.14
ovarian Pre-op Pos-op	7 2/4 negative 2/4 positive	87.4±59.6 4.8±0.4 9±1.4

*pre-op. : preoperative

*pos-op. : postoperative

Table 2.7: Comparison of Levels of Intact hCG and Free hCGß in Sera and tissues of Benign and Malignant Gynecologic Cancers (All details are explained in the text)

Clinical categories	Intact hCG and free hCGß in blood		Intact hCG and free hCGß in tissue	
	N	mIu/ml	N	mIu/ml

	o. of cases	mean±S.E.M.(range)	o. of cases	mean±S.E.M.(range)
Endometrial cancer	9	31.5±40.6 (3.8-110.5)	9	52.6±52.8(14.5-168)
Ovarian cancer	14	52.5±59.8(4.2-166)	14	79.7±74(15.2-212)
Benign tumor	25	6.3±3.2(3.9-14.4)	25	8.19±1(5.3-9.5)
Placenta	2		2	371±11(360-382)

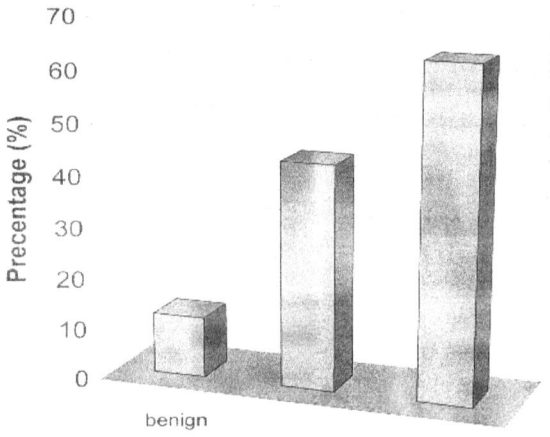

Fig.2.2 The percentage of Elevation Intact hCG and Free hCGβ in Malignant and Benign Endometrial and Ovarian Tumors

(All details are explained in the text)

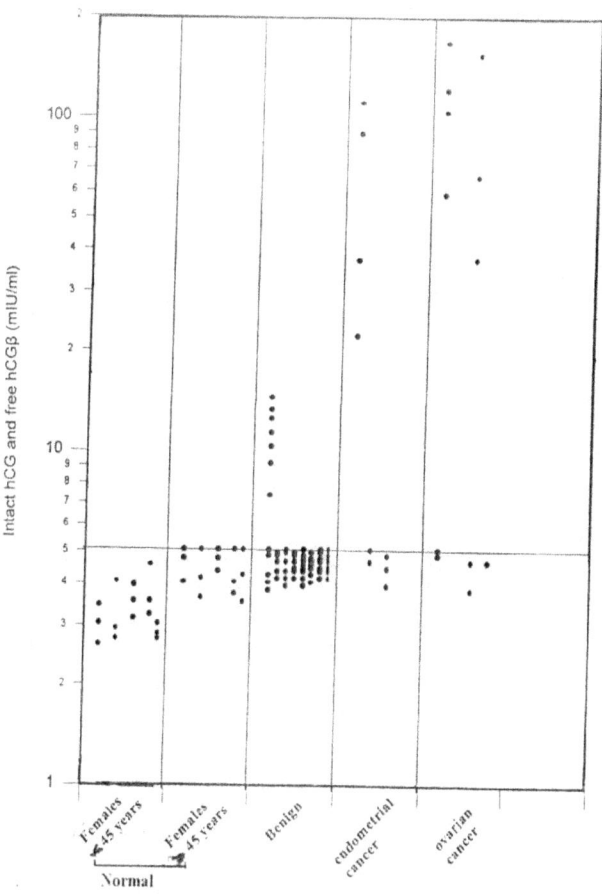

Fig. (2.3) Distribution of the intact hCG and Free hCGβ in healthy Females and in Patients with Gynecologic Tumors
(All details are explained in the text)

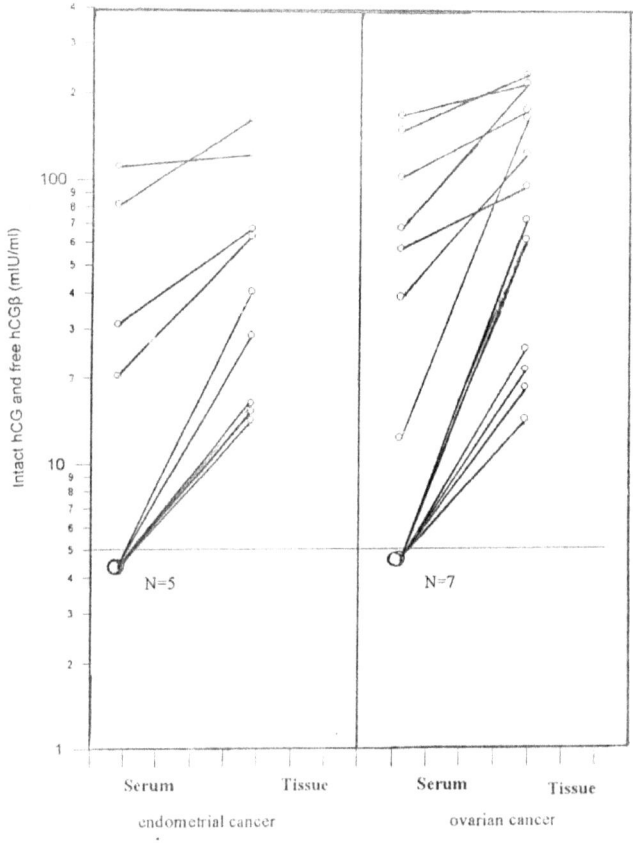

Fig. (2.5) Comparison of Intact hCG and free hCGβ Levels in tissues and corresponding sera of women with ovarian (n=14) and endometrial (n=9) cancer. The large circle give the median of values that were below normal limit (5 mIu/mL) which represented by the dashed lines
(All details are explained in the text)

Chapter Three

Development of a Radio-Receptor Assay for Detection and Analysis of hCG-Receptor in Benign and Malignant Gynecologic Tumors.

Abstract

A quantitative radioreceptor assay for detection and analysis of membrane-associated hCG-Rs of human cancer tissues of gynecologic cancers was developed by binding highly specific, and biological active radiolabeled hormone (^{125}I-hCG) to particulate membrane preparations. The optimum conditions for ^{125}I-hCG-receptor binding were obtained as follows : tracer hormone concentration (17pmol) for ovarian cancer and (4pmol) for endometrial cancer ; receptor concentration (500μg) protein for ovarian cancer and (700μg) protein for endometrial cancer;pH optimum in range between 7-7.4 for ovarian and endometrial cancers.Reaching the equilibrium for ^{125}I-hCG binding activity of the receptor is temperature dependent. The ^{125}I-hCG binding to the receptor was specifically displaced by low concentration of unlabeled hCG, malignant hCG from choriocarcinoma ,and LH.

The radiorecptor assay for 48 tissues from ovarian and uterine tumors, appeared that malignant tumors of ovarian have a specific hCG-binding activity 2 fold higher than that of benign ovarian tumors. In endometrial cancer, the binding activity for ^{125}I-hCG was lower than that in malignant and benign ovarian tumors.

Introduction:

The hCG-R belongs to the family of glycoprotein hormone G protein –coupled receptors, which also comprise TSH-R and FSH-R[149]. All three receptors are glycoprotiens containing a large extracllular leucin rich repeats (LRR), N-terminal domain with several N-linked glycosylation. A plasma membrane domain composed of seven hydrophobic α-helices ending with an intracellular C-terminus[154]. These three receptors have been shown to transduce hormone binding via coupling to the heterotrimeric Gs protein which activates adenylyl cyclase, or other G proteins that activate phosphlipase C (as assessed by measuring inositol phosphates, diacylglycerol, and / or intracellular Ca^{+2}) [151]. The mechanism of the coupling in hCG-R is different from other G protein- coupled receptors whose ligands are much smaller and intercalate among the transmembrane helices[150].

The hCG-R, which is present on gonadal cells, plays a pivotal role in reproductive physiology. Until recently, it was believed that hCG-R were only present in gonadal cells, where they regulate steroidogenesis. However, it is now known that these receptors are also expressed in human and animal uteri[178] [188] [189], human pituitary[185], skin[186] fallopian tubes[179], human placenta, fetal membranes, decidua[178], human gestational trophoblastic neoplasms, human breast[197] and breast cancer[183], endometrial and myometrial blood vessels[187] and brain[182]. These studies have shown that hCG can directly regulate the functions of nongonodal tissues indicating that the receptor are functional in these tissues.

The evidence for hCG receptors has been obtained by ligand binding measurements or immunocytochemistry using receptor-specific antibodies or evaluation the presence of hCG receptor mRNA. The traditional hCG-R binding studies involved binding a radioactive form of either the hormone itself or a biologically active analog (agonist) to membrane preparations (particulate or solubilized)of target tissues, have been used to identify and characterize the hCG-R[154] [189] [205]. In most hCG-R binding studies, ^{125}I-labeled hCG prepared by radio iodination with chloramines or lactoperoxidase has been used. Tritium, has also been used in some studies for labeling.

The binding of LH or hCG to cell surface receptors was thought to be a reversible process, thus saturation binding experiments were performed under steady state conditions[206]. It has now become clear that the hCG-receptor interaction is much more complex and cannot be described as a simple bimolecular reaction[175]. The hormone-receptor complex is not only involved with the stimulation of adenylate cyclase, but also functions in the ligand-induced regulation of the receptors[154]. From different separation techniques of bound from free hormone, the choice was usually between centrifugation and precipitation of the hormone receptor complex[149].

The present investigation was undertaken to develop a radioreceptor assay for hCG-R determination in benign and malignant ovarian and uterine tumors. The nature of the binding was characterized by determination of the equilibrium dissociation constant (k_d), Scatchard plots, inhibition studies, and rate constants.

Materials and Methods

3.1 Buffers and Reagents:

Buffers and reagents of chapter 2 (2.4) were used in the experiments of this chapter. Other solutions used were indicated in each experiment.

3.2 Patients and Specimens:

The benign and malignant tissues of patients described in section (2.3) are included in radio receptor experiments. The crude tissue homogenates were prepared as described in section (2.4), but further centrifugation was carried out on the supernatant at 9000xg for 30min. The pellet containing the receptor, was resuspended in 50 mMTris/HcL, pH 7.4 containing 1mM PMSF and 5 mM EDTA,

20% glycerol, 0.1% BSA in ratio of 1:4 (w:v) and then homogenized and subjected for binding studies.

3.3 Binding Studies of ^{125}I-hCG with Its Receptor in Benign and Malignant Ovarian and Uterine tumors

3.3.1 Preliminary Test of hCG-R Binding:

The binding of ^{125}I-hCG to the tissue homogenates was primarily checked. The experiment was carried out in duplicate, 250µl of ovarian or uterine homogenates (700 µg protein) were incubated with 80µl of I^{125}-hCG (1 µg/ml) and the volumes were completed to 500µl with 50 mM Tris/HCL, pH 7.4 containing 0.1% BSA. Incubation was carried out at 37°C for 1hr or at 4°C for 24h. This was followed by centrifugation for 30 min at 1500xg and then decanted. The radioactivity of the bound hCG(CPM) was counted in the pellet for each tube using gamma counter. Non specific binding(NSB) was also determined by the same method in the presence of 200 fold excess of unlabelled hCG as competitor.

The specific binding (CPM) was calculated by subtraction of the radioactivity (CPM) obtained in the presence of excess of unlabelled hCG (NSB) from total binding (TB).

SB (CPM) = TB (CPM) - NSB (CPM)

Percentage (%SB)= SB(CPM) / Total count (CPM) of the ^{125}I-hCG used in each tube X100

3.3.2 The Effect of ^{125}I-hCG Concentration on the Binding:

Increasing concentration of ^{125}I -hCG ranging from (1-20 pM) were added to 250 µL (700 µg protein) of crude ovarian or uterine homogenates with a final volume of 500 µl (completed with 50 mM Tris/ HCL, pH 7.4 containing 0.1% BSA.

After incubation for 1hr at 37°C, the bound hCG was determined as described in section (3.3.1). The specific binding (SB%) was calculated and plotted against the concentration of ^{125}I-hCG.

3.3.3 The Effect of Receptor Concentration on the Binding:

Forty microlitres of ^{125}I –hCG (Iµg/mL) was added to 250 µl of increasing amount (300,400,500,600,700,800µg) of crude ovarian homogenates in a final volume of 500 µl (completed with Tris -0.1% BSA buffer pH7.4). After incubation for 1hr at 37°C, the bound hCG was determined as described in section (3.3.1) The same experiment was carried out using 250µl of increasing amount (300,400,500,600,700,800,ug protein) of uterine homogenate added to a set of tubes contained 20 µl of ^{125}I -hCG (1µg/ml) in a final volume of 500µl. The SB% was calculated and plotted against the concentration of receptor.

3.3.4 The Effect of pH on the Receptor Binding:

Two hundred fifty microliters of crude ovarian homogenates (500µg protein) were added to 40 µl (1µg/ml) of ^{125}I -hCG. The mixtures were then completed with buffers of different pH (5, 5.5, 6, 6.4, 6.8, 7, 7.2, 7.4, 7.6, 7.8, and 8.0) to 500µl. The experiment was repeated using 250 µl of crude uterine homogenates (700µg protein), added to 20 µl (1µg/mL) of ^{125}I -hCG and completed in the same manner. After incubation for 1hr at 37°C, the bound hCG was determined as mentioned in section (3.3.1).The SB% was calculated and plotted against pH .

3.3.5 Temperature Dependency of the Binding:

Fourty micro liters (1µg/ml) of ^{125}I -hCG and 250 µl (500µg protein) of crude ovarian homogenates were mixed in a final volume of 500µl (completed with Tris -0.1% BSA buffer pH 7.2) .After incubation for 1hr at different temp. (4,25,37,45°C), the bound hCG was determined as described in section (3.3.1). The experiment was repeated using 250µL of crude uterine homogenates (700 µg protein) added to 20 µL of ^{125}I -hCG and completed in the same manner. The SB% was calculated and plotted vs. the temperature of incubation.

3.3.6 Time Course of Receptor Binding

Crude ovarian homogenate (500µg protein in 250µl) was added to 40 µl (1ug/ml) of ^{125}I -hCG in a final volume of 500 µl (completed with Tris -0.1% BSA buffer PH 7.2). The tubes were incubated at 4°C. At certain time intervals (0.5, 1, 2, 6, 8, 16, 24hr), two tubes were taken and the bound hCG was estimated as mentioned in section (3.3.1). The experiment was repeated at 25, 37°C. The same protocol was repeated for the uterine homogenate using 250µl (700µg protein) of this homogenate, added to 20µl (1ug/ml) of ^{125}I -hCG. The SB% was calculated and plotted vs. the time of incubation.

3.3.7 Stability of Receptor Homogenate and Hormone Receptor Complex :

In order to investigate the effect of temperature on receptor properties, the stability of receptor homogenate and receptor – hCG complex have been measured. For measuring receptor stability, 250 µL of crude ovarian homogenates were incubated at 25°C 37°C. At certain time intervals (1, 2, 3, 4hr), two tubes from each set were taken and mixed with 40ul of ^{125}I-hCG in a final volume 0f 500µL (completed with Tris-0.1% BSA buffer PH 7.2). After incubation for 1h at 37°C, the specific activity was determined as described in section (3.3.1) . In a control (with no preincubation) ovarian homogenate , the binding assay was performed and considered 100% binding .The relative binding was plotted against preincubation time,. The same protocol was repeated for the uterine homogenate using 250 µl (700µg protein) of this homogenate.

For measuring receptor-hCG complex stability , the experiment was performed as described in section 3.3.5 at 37°C. After the evaluation of the bound hCG, the receptor-hCG complex was reincubated at three temperatures (4, 25, 37°C). Between 0 and 8hr, the remaining binding of ^{125}I-hCG was measured and the relative binding was calculated and plotted against the time of incubation.

3.3.8 Determination of the Equilibrium Binding Constants of hCG-R

Ovarian homogenate (500µg protein in 250µL were added to 40µL of increasing concentration (1-20pM) of ^{125}I -hCG with or without the addition of a 200 fold excess of unlabelled hCG in a final volume of 500 µL (completed with Tris -0.1% BSA, PH 7.2). After incubation for 1hr at 37°C , the bound hCG was determined as described in section 3.3.1. the same protocol was repeated for the uterine homogenate using 250µl (700µg protein) of this homogenate added to 20µl of increasing concentration (10-2000fM) of ^{125}I -hCG with or without the addition of a 200fold excess of unlabelled hCG.

The concentration of receptors and the association constant (K_a) were determined according to Scatchard equation[205].

$B / F = B_{max} - B / K_d$ and $K_d = 1 / K_a$

Where B= the concentration of bound hCG.

F= the concentration of free hCG.

B_{max}= the maximum number of binding sites (binding capacity).

K_d= equilibrium binding constant (dissociation constant).

K_a= equilibrium binding constant (association constant).

The bound and free hCG were calculated by multiplication of the ratios specific bound (CPM) / total count (CPM) and free (CPM) / total count (CPM) with the concentration of ^{125}I-hCG respectively. The B/F was plotted vs. B. The receptor concentration and the association constant were calculated from the x-axis and the slope of the straight line respectively.

3.3.9 Specificity of Binding and Determination of IC_{50} and K_i

Ovarian homogenates (500µg in 250µl) were added to 40µl (1µg/ml) of the ^{125}I-hCG without or with the addition of increasing concentration (1-1000ng) of unlabelled hCG in a final volume of 500µl. After incubation for 1hr at 37°C, the bound hCG was determined as described in section (3.3.1). For the study of the competitive effect on the binding of uterine receptor, 250µl (700 µg protein) of this homogenate was added to 20 µl (1µg/ml) of ^{125}I-hCG without or with the increasing concentreation (5-500ng/ml) of unlabelled hCG in a final volume of 500µl. After incubation for 1h at 37°C, the bound hCG to uterine receptor was determined and the SB% was calculated as mentioned in section 3.3.1. The experiment was repeated with different competitors [FSH, LH,TSH , Serum from choriocarcinoma] . The relative binding of ^{125}I-hCG was calculated and plotted against competitors conc., competition curves were generated and the concentration of unlabeled ligand displacing 50% of specific radioligand binding (IC$_{50}$) was determined. The binding affinity constant of unlabeled ligand (k$_i$) were also determined according to the following equation:

$$k_i = \frac{IC_{50}}{1+[L]/kd}.$$

Where [L] : the concentration of radioligand .

RESULTS AND DISCUSSION

Membrane Preparation

In a series of preliminary experiments, curde plasma membranes were prepared by centrifugation of ovarian and uterine homogenates at various centrifugal force from 3000 to 9000 xg. As expected, the recovery of the receptor increased with increased rotor speed, membranes were subsequently prepared by centrifugation at 9000xg in order to obtain maximal recovery of the receptor. The stability of the crude membrane receptor was markedly improved by supplementation of the buffer with glycerol; the hormone binding activity of the membrane extract was completely retained after storage at 4°C for at least 3 day. The proteolysis of the receptor (and/or nicking), by endogenous protease, had been inhibited by the inclusion of an inhibitor cocktail containing PMSF, and EDTA in extraction buffer.

Previous reports describing the extraction of the hCG-Rs have been based on the use of either particulate membrane or soluble membrane extract from bovine[157][158] and rat[156][207][208] gonadal tissues. Other non gonadal tissues were also used such as rat kidney and liver[208] and porcine uterus[189]. The membrane fraction was prepared by centrifugation[207][149] at various centrifugal forces. While preparation of soluble hCG-R has been base almost exclusively on the use of triton X-100[157][158][149]. Emulphogene (anonionic detergent) has also been used[156]. Ligand-receptor binding experiments can be done on membrane preparation of tissues as particulate or solubilized, in contrast purification studies have been done only on solubilized receptors.

Human Chorionic Gonadotropin Receptors in Ovarian and Uterine Tumors:

Initial studies were performed by ligand–receptor binding method; involve binding highly specific, purity, and biological active radiolabelled hCG (^{125}I-hCG) to particulate membrane preparations. The presence and binding activity of hCG-Rs in 48 cases of benign and malignant ovarian and uterine tumors were investigated (Table 3.1). The ^{125}I-hCG binding activity of the receptor after 24hr of incubation at 4°C was the same as at 37°C for 1hr, which indicates that the equilibrium of reaction was depended on the temperature. Table (3.1) shows that the malignant tumors of ovarian have a specific hCG-binding activity of 1.6±0.36 pmol/mg of protein, which is approximately 2 fold higher than that of benign tumors. In endometrial cancer, binding activity for hCG (0.2±0.15 pmol/mg of protein) is lower than that in malignant and benign ovarian tumors. Normally hCG-Rs are functional in gonad cells, where both LH and hCG regulate; through this receptor; the ovarian and testicular steroidogenesis[209] ,LH also increases ovarian blood flow in several species. Human chorionic gonadotropin and/or LH may also regulate glandular and Luminal epithelial cell functions in gonadal cells via cAMP mediation or by increasing the local synthesis of steroid hormones[210]. Crude homogenates prepared previously from ovarian of pseudo pregnant rats have a specific hCG-binding activity of 0.8-1.3 pmol/mg of protein (Bruch et.al.,[156], Shao K.[207], Segaloff et.al.,[149]) which is approximately 10 fold higher than that of homogenate prepared from porcine luteal tissue[149], rat testis[211], bovine luteal[157][158], or mouse leydig cell tumors[149].

The high specific hCG-binding activity (1.6±0.36 pmol/mg) in ovarian tumor tissues seems to be compatible with the increasing the number of cell produced by malignant tumor. There was no obvious relation between the specific activity and staging of malignant ovarian tumor. Further studies in a greater number of patients will be necessary to determine whether the concentration of hCG-R also is correlated with aggressiveness of the tumor and the prognosis of the patients. In malignant tumor of uterus, it appeared important to detect a tumor containing hCG-Rs because most hormonal markers (including estrogen, progesterone), EGF receptors, and so-forth, are present in only a subset of the cancer[212]. Specific ^{125}I-hCG binding activity was observed for all uterine preparations (Table 3.1),revealing the presence of hCG-Rs in these tissues. Recent evidence shows that in addition to their classical endocrine effects, hCG and LH may also be ectopically synthesized and exert paracrine effects,regulating the growth of various cell types . This direct action implies the existence of the corresponding receptors in these tissues.

Reshef et.al.,[178] demonstrated for the first time the presence and cellular distribution of hCG-R in nonpregnant human uterus using specific immunostaining procedure. They obtained that the receptors are present in glandular and luminal epithelial cells and stromal cells, of endometrial, and in circular and elongated myometrial smooth muscle and in vascular smooth muscle of myometrium. Ziecik et. al.,[189] reported that all of the porcine myometria and 30% of the endometria examined contained hCG-Rs which increase from the proliferate to the secretary phase of the cycle. The physiological significant of human uterine hCG-Rs is not known, but the presence of receptors in different uterine cell types as well as differences between them suggest that hCG may regulate different functions in different cell type in the uterus. The presence of differences during the menstrual cycle suggests that these receptors are regulated by other hormones and hCG was reported to have a relaxing effect on porcine myometrium[189]. In vitro, hCG was reported to increase cAMP levels and progesterone synthesis in rat uterus[190]. Lei.et. al.,[187] found vascular hCG-Rs mRNA transcript and receptors protein in human endometrial and myometrial, suggested that LH or hCG might directly regulate blood flow in human uterus. Evidence already exists in support of such possibilities, for example LH increases ovarian blood flow in several species[213][214]. Kornyel et.al.,[188] considered human myometrium a direct target of gonadotropin regulation because of the presence of hCG-R in this tissue. They demonstrate that hCG can directly regulate myometrial smooth muscle cells, causing hyperplasia as well as hypertrophy. Logically hCG and LH in uterine cells can make all their functions already exist in gonadal cells[215][216]. The expression of hCG-R has been reported in human endometrial carcinoma by immunocytochemistery[184]. The expression of receptor varied among individual cancers. This receptor has also been reported in breast cancer by two different monoclonal antibodies raised against human hCG-R, which allowed Meduri and his coleagues[183] to detect this receptor in cancer and benign breast lesions. They also reported that the receptor was also present in epithelial cells of normal human and sow breast.

Factors Affecting the Binding of ^{125}I-HCG with Its Receptor

A binding saturation study could be performed at a steady state when criteria of radioligand purity, minimization of nonspecific binding have been satisfied. In addition, linearity of radioligand binding to the tissue must be demonstrated and the appropriate incubation time and temperature to reach a steady state must also be determined. Thus, the investigation of these effects is quite necessary to get reliable measuring for ligand binding in the tissue.

1 <u>The Effect of ^{125}I-hCG Concentration on the Binding</u>

Ligand receptor binding studies usually followed the kinetics of very similar to those of classic enzyme-substrate interactions[206]. For reversible ligand receptor interactions:

$$a[L] + b[R] \frac{K+1}{K\ 1} c[RL]$$

Where: [L]=concentration of free ligand.

[R]= concentration of unoccupied receptor sites.

[LR]= concentration of ligand-receptor complex.

K_{+1}=association rate constant.

k_{-1}=dissociation rate constant.

k_d=equilibrium binding constant ($\frac{k\ 1}{k+1}$)

The classic law of mass action for enzyme-substrate interaction adapted to ligand-receptor interaction which is [LR]= $\frac{B_{max}[L]}{[L]+Kd}$

One of the property of ligand-receptor interactions is saturability, that is, only a finite number of specific receptor sites exist per unit of tissue, usually designated B_{max}. To fulfill this criterion, various concentrations of the ^{125}I-hCG with a fixed concentration of tissue have been incubated[217]. Saturation curves for specific binding of hCG to the receptor homogenates of malignant and benign tumors are shown in Fig. 3.1.

In ovarian tumors the hCG-Rs are saturated with ^{125}I-hCG at 17 pmol and 7.8 pmol for malignant and benign conditioned respectively, while in uterine tumors, the hCG-Rs saturated with ^{125}I-hCG at 4 pmol and 2 pmol for malignant and benign conditions respectively.

2 <u>The Effect of Receptor Concentration on the Binding:</u>

One of methodological problems apparent involves the choice of appropriate tissue (receptor) concentration to utilize in the binding experiment[218]. The receptor concentration influences the K_d value and the linearity of radioligand binding with tissue. In order to demonstrate that, increasing concentration of homogenate were incubated with fixed concentration of ^{125}I-hCG. As shown in Fig.(3.2) the percentage of ^{125}I-hCG bound was increased linearly with increasing of receptor concentration in benign and malignant conditions. This experiment shows the homogeneity of tissue preparation from ovaries and uterus with similar affinities of labeled ligand to these tissues. Jacobs and Cuatrecases[217] pointed out that non linearity may arise if both ligand and receptor preparation are heterogeneous. This situation also arises if the nonspecific binding (to tissue, to tube, etc) is not define or too high[218]. The correct definition of nonspecific (and thus specific) binding is very important.

3 The Effect of PH on The Receptor Binding

The effect of pH on the binding is shown in Fig. 3.3. The binding of ^{125}I-hCG to the human ovary and uterus homogenates exhibited pH optimum between 7.0-7.4. It was not determined whether the decreased binding at suboptimum pH values resulted from a decrease in binding affinity, binding rate, or inactivation of the receptors. It is known that these two homogenates incubated at pH 4.5 for 1h at 37°C loses irreversibly 75% of its binding capacity. Optimum pH binding has been reported for purified hCG-R by Pandian and Bahl[158] in bovine corpus luteal plasma membrane preparation to be (7.2-7.4). Bellisario et.al.,[208] observed that the optimum pH binding of rat testis hCG-Rs was between (6.3-7.3). The binding experiments of rat ovarian receptor in previous reports have been carried out almost exclusively at pH 7.4[156][207][149].

4 Temperature Dependency of The Binding

The temperature dependency of the association of ^{125}I-hCG with it receptor isolated from malignant and benign tumors of ovary and uterus was investigated. Fig.(3.4) shows that the binding of ^{125}I-hCG to the receptor was greater at 37°C than at 4, 25 and 45°C when incubation for 1hr. when binding was carried out for 1h at 37°C, 15-30% of the ^{125}I-hCG was bounded for different tissue preparations. It seems that the hCG has high binding affinity to the receptor and need only 1hr to reach equilibrium, also seems that the equilibrium of binding depend on temperature, at 4 and 25°C need more than 1hr to give maximum binding, while at 45°C thermal inactivation of the receptor may be occur. Thus, time course experiment is important to confirm these observations. In all previous reports, binding experiments have been done either at 37°C for 1hr or at 4°C for 24hr[154].

5 <u>Time Course of Receptor Binding</u>

Association characteristics of ^{125}I-hCG with human homogenates were examined at different temperatures (4, 25, 37°C). As illustrated by Fig.(3.5), maximum binding with both ovarian and uterine homogenates was greater at 37°C than 25°C. Furthermore, the maximum binding that occurred after 24hr of incubation at 4°C was nearly the same as at 37°C for 1hr. When binding was carried out at 25°C, the maximum binding of ^{125}I-hCG was at (6-10hr). As shown in Fig.(3.5) incubation of ^{125}I-hCG with ovarian and uterine homogenates for time periods longer than that required for maximum binding resulted in decreased binding, the loss of binding was more rapid at 37°C than at 25°C. This inactivation process could result from inactivation of ^{125}I-hCG or the human receptor during the binding assay. The results (as shown later) from stability experiment confirmed a time-dependent loss of binding activity of receptor at various temperature. Bellisario et.al.,[208] observed partial degradation of ^{125}I-hCG occurs during the binding assay in addition to the loss of rat testis receptors activity depending on temperature and time incubation. The results from previous reports are in agreement with our results. Bruch et.al.,[156] observed that the maximum binding of ^{125}I-hCG to the rat ovarian receptor at 4°C, was after 24hr of incubation. Pandian et.al.,[158] found that the binding of ^{125}I-hCG to the bovine corpus Luteal receptor was greater at 37°C than at 4 and 25°C and the equilibrium reached in 2 to 10hr depending on the temperature. Huhtaniemi an Catt[211] observed in their work on testicular receptor that the maximum binding with ^{125}I-hCG was at 24°C for 16h.

6 Stability of Receptor Homogenate and Hormone Receptor Complex

To prove the receptor inactivation at various temperatures at time course experiment, preincubation of the ovarian and uterine homogenates at 25°C and 37°C were performed. Time-dependent loss of binding activity was shown in Fig. (3.6) for hCG-Rs at these two temperatures.

The loss of receptor activity was greater at 37°C than at 25°C and was irreversible. Preincubation of the receptor homogenates in benign and malignant conditions at 37°C and 25°C for 2hr resulted in a loss of (45-50) and (9-10) of the initial binding activity respectively. In this regard, the stability was similar to the stability observed previously with ovarian and testicular receptor[156][208].

For further investigation, the effect of temperature on receptor properties, the ^{125}I-hCG-receptor complex was reincubated at three temperature (4, 25, 37°C) and the relative binding was estimated. As shown in Fig.(3.7), hCG-receptor complex were readily dissociated at the higher temperature. Approximately, 80% of dissociation was observed at 37°C after 8hr. This result was consistent with the apparent thermal sensitivity of the crude native receptor homogenate at 37°C when incubated in the absence of hCG. The hCG binding characteristic of the receptor at 4°C was stable, while (40-45%) of performed complex were dissociated at the 25°C after 8hr Thermal sensitivity of the hCG-receptor complex has been previously observed by Bruch et.al.,[156] when they studied the dissociation of hCG receptor complex at different temperature (4, 22, 37°C). Approximately 60% of preformed complex were dissociated at 22°C while complete dissociation was observed at 37°C after 24hr.

7- Determination of the Equilibrium Binding Constants and Maximum Binding Capacity

Ligand - receptor interaction usually follow a reversible binding reactions, thus they require to be performed under steady state or equilibrium conditions. At equilibrium or steady state the rate of the forward reaction equals the rate of the reverse reaction. The equilibrium binding constant may then be defined either as an association binding constant (K_a) or as dissociation binding constant (K_d). Knowing the concentrations of ligand bound (B) and free (F) at equilibrium allow to determine of both the equilibrium binding constant (K_d) and the maximum number of binding sites (B_{max}) by Scatchard equation[206]

$$\frac{B}{F} = \frac{B_{max}}{Kd} B$$

the Scatchard equation allows a linear transformation of binding data so that K_d and B_{max} may readily be determined from radioligand saturation experiments, by incubating the concentrations of the radioligand with a fixed concentration of tissue receptor. A binding saturation study can then be performed over a range of radioligand concentration, from 10-20% of the estimated K_d to four to five time this value[217].

Scatchard analysis of saturation curves in our study (Fig. 3.8) indicated a single class of binding sites for hCG-R in benign and malignant tissues and, also indicated high affinity and low capacity for receptor in uterus tissue. The values of K_a of malignant and benign ovarian and uterine receptor did not alter significantly and were of the order of 10^{10} M^{-1} (Table 3.2). This result was consistent with the previous results of the determination of K_a. Bruch et.al.,[156] demonstrated that the value of K_a of the crude receptor from rat ovarian was 6.88($\pm$0.33) x 10^{10} M^{-1}. Pandian and Bahl[158] reported the value of K_a of the receptor in bovine corpus Luteal membrane, which was 4.9x10^{10} M^{-1}, while Dattatreyamurty et.al.,[157] pointed out that the value of K_a of the crude receptor did not alter from that of purified. Dufau et.al.,[219] showed the presence of a single order of binding sites with high affinity (10^{10} M^{-1}) for hCG of gonadotropin receptor of the rat testis, Bellisorio and Bahl[208] confirmed this result in their study. Keutmann[229] indicated that there was only modest increase in the association constant of the purified receptor over that was determined in crude receptor from bovine corpora lutea homogenate. (3.28($\pm$0.21)x10^{10}M^{-1}, 8.88($\pm$ 1.08) x10^{10} M^{-1}, respectively). Huhtaniemi and Catt[211] in their study found a differential binding affinities of rat testis hCG-R for hLH and hCG obtained that the value of K_a for hCG was 6.5x10^{10} M^{-1} where for LH was 2.25 x 10^{10} M^{-1}.

8 <u>Specificity of Binding</u>

The specificity of hCG binding was evaluated in competition experiments with unlabeled glycoprotein hormones[218]. A fixed concentration of ^{125}I-hCG was incubated with increasing concentration of unlabeled hormone, competition curves were then generated, and the concentration of unlabeled hormone displacing 50% of specific ^{125}I-hCG (IC$_{50}$), and affinity constant of unlabeled ligand (K_i) were determined.

The ^{125}I-hCG binding to the receptor was inhibited in a dose-dependent manner by the hCG (native and from serum of choriocarcinoma (chCG)) and LH (Fig.3.9). as shown in the Figure, the binding of ^{125}I-hCG by homogenates is specifically displaced by low concentration of hCG,unlabeled hCG, and LH. The curves of hCG and LH yield similar slopes.

The doses of hCG and LH that resulted in 50% displacement of the ^{125}I-hCG (IC_{50}) were shown in (Table 3.3). The ChCG was about (1.5-2.5) times more effective in competition for binding than native hCG. Other glycoprotein hormones such as FSH, TSH are inactive in competition for binding, with relative potency 0.014, 0.008, respectively. The binding inhibition observed with TSH and FSH may result from low level contamination with LH in these preparations. The results of competition experiments revealed a high affinity receptor sites for hCG and LH in malignant and benign homogenates, but hCG was more effective in competing with ^{125}I-hCG for binding sites than LH, this results was expected for hCG receptor, where hCG is higher affinity for these receptor as compared to LH[149]. The same binding affinities for the hCG, LH have been previously observed[156][208][158]. In addition, hCG from serum of choriocarcinoma was more effective in competing with ^{125}I-hCG than native hCG (Fig. 3.9),this may results from asialo hCG produced by malignant cells[221] . Choriocarcinoma produce tumor-specific glycosylation variant of hCG. Numerous studies have demonstrated that asilo hCG exhibited greater binding to the hCG-R than native hCG. Bellisario and Bahl[158], in their study, demonstrated that asialo hCG exhibited greater binding to the rat testis and liver homogenates than native hCG, they indicated to the presence of additional binding sites for the asialo, hormone. Bruch et.al.,[156] also obtained that asialo hCG had higher binding affinities than intact hCG to ovarian homogenate. Keutmann et.al.,[222] in their characterization of deglycosylated hCG obtained a high affinity of asialo hCG to receptor than native hCG. However, it has been believed that hCG by its carbohydrate chains bind to a lectin-like membrane component to give the biological response and remove of the terminal sialic acid enhance binding to these lectine component by decreased electrostatic repulsion[19][223].

Crude membrane preparation		protien mg	^{125}I-hCG Binding activity pmol(±S.E.M)	specific activity pmol/mg protein
ovary	Malignant	0.7	1.1±0.25	1.6±0.36

Table 3.1 Binding activity of ^{125}I-hCG with hCG-R in particulate membrane
preparations of benign and malignant ovarian and uterine tumors

(All details are explained in the text)

			Ovarian homogenates	Uterine homogenates
uterus	Benign	0.7	0.55±0.28	0.8±0.4
	Malignant	0.7	0.2±0.15	0.3±0.2
	Benign	0.7	0.08±0.014	0.11±0.2

Table 3.2 Scatchard analysis of saturation binding curves for hCG-R in ovarian and uterine homogenate.

(All details are explained in the text)

	Benign	Malignant	Benign	Malignant
$K_d \times 10^{-10}$ (M^{-1})	0.15 (±0.06)	0.13 (±0.09)	0.13 (±0.02)	0.12 (±0.07)
$K_a \times 10^{10}$ (M^{-1})	6.6 (±1.2)	7.7 (±1.7)	7.7 (±1.0)	8.3 (±1.5)
B_{max} (pM/mg)	3.2 (±0.49)	3.6 (±0.28)	0.93 (±0.33)	1.3 (±0.3)

Table 3.3 Specificity hCG binding to the receptor.
(All details are explained in the text)

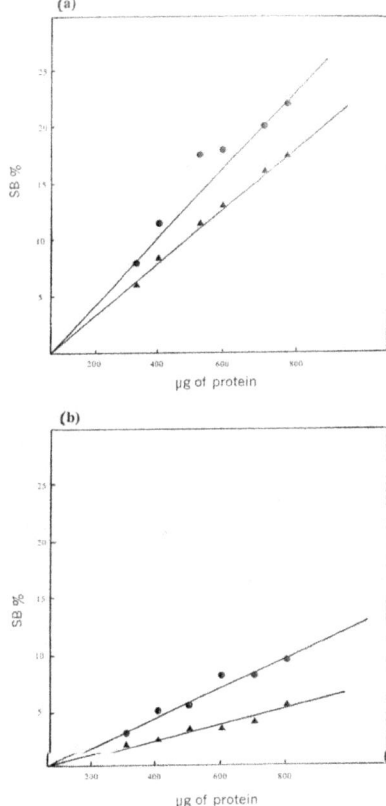

Fig 3.2 Effect of receptor concentration on the binding of ^{125}I-hCG to hCG receptor in crude homogenates.
(a) ovarian homogenate : (●—●) malignant ,(▲—▲)benign .
(b) uterin homogenate : (●—●) malignant ,(▲—▲)benign .
(All details are explained in the text)

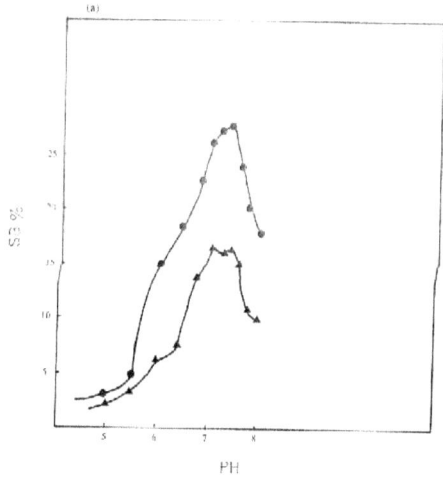

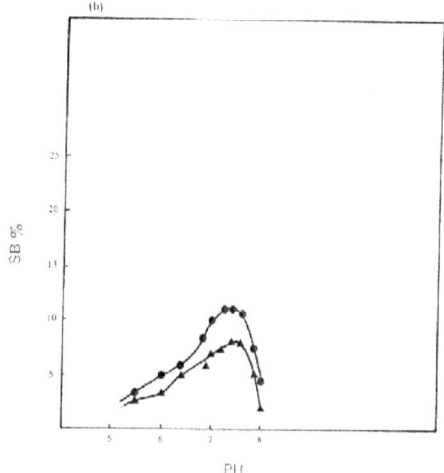

Fig 3.3 Determination of optimal pH of the binding of ^{125}I-hCG to hCG receptor in crude human homogenate
(a) ovarian homogenate : (●—●)malignant, (▲—▲)benign
(b) Uterin homogenate : (●—●)malignant, (▲—▲)benign
(All details are explained in the text.)

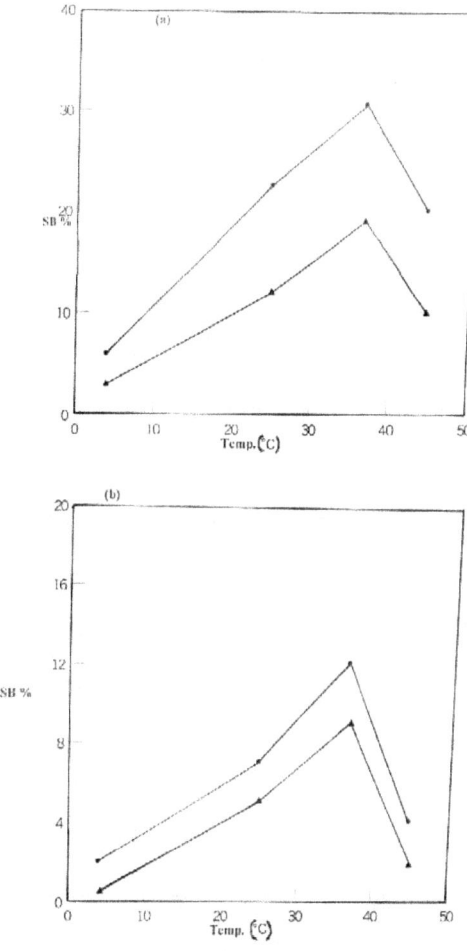

Fig. 3.4: Determination of optimal temperature of the binding of ^{125}I-hCG to hCG receptor in crude human homogenates.
(a) Ovarian homogenate : (•–•) malignant, (▲–▲) benign.
(b) Uterine homogenate : (•–•) malignant, (▲–▲) benign.
(All details are explained in the text)

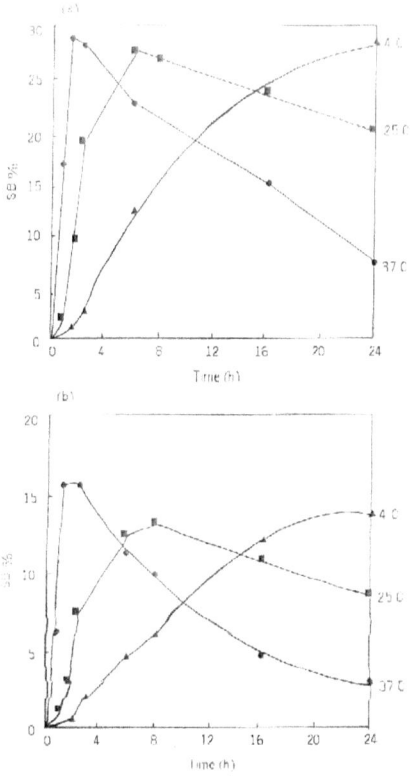

FIG. 5.5 Time course of the binding of ^{125}I-hCG to hCG receptor in crude human homogenates
a) Ovarian homogenate: ●—● mlignant at 37 °C.
■—■ mlignant at 25 °C.
▲—▲ mlignant at 4 °C.
b) Ovarian homogenate ●—● benign at 37 °C.
■—■ benign at 25 °C.
▲—▲ benign at 4 °C.

(All details are explained in the text)

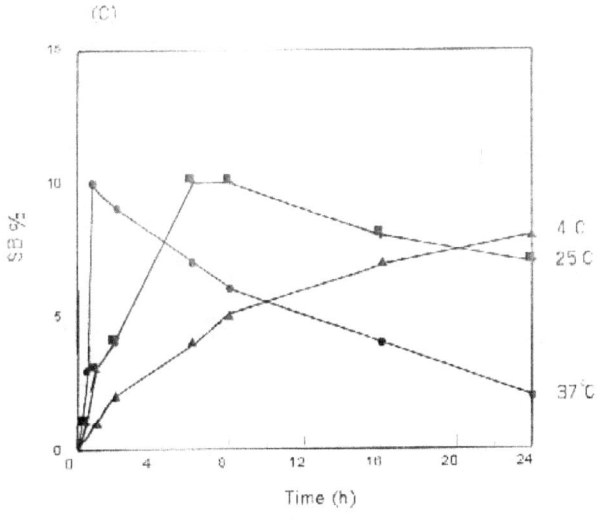

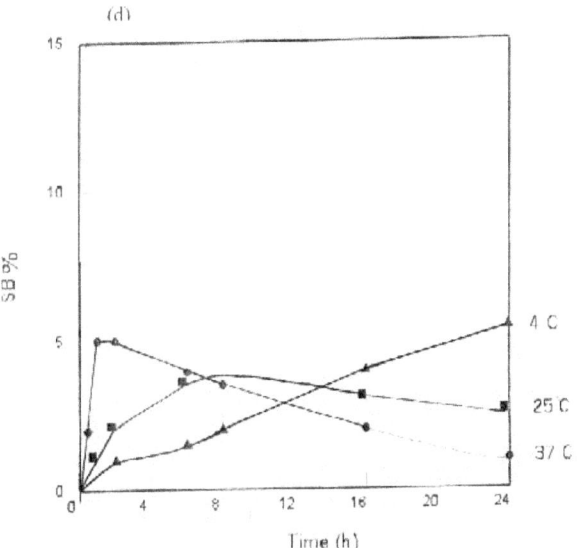

FIG. 3.5 Time course of the binding of ^{125}I-hCG to hCG receptor in crude human homogenates

c) Uterine homogenate: •–• mlignant at 37 °C.
■–■ mlignant at 25 °C.
▲–▲ mlignant at 4 °C.

d) Uterine homogenate •–• benign at 37 °C.
■–■ benign at 25 °C.
▲–▲ benign at 4 °C.

(All details are explained in the text)

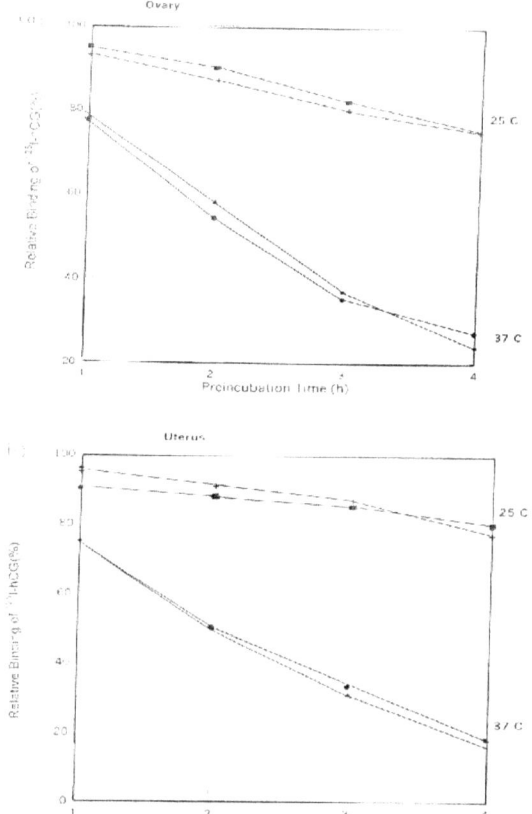

Fig. 3.6: Time dependent Loss of ^{125}I-hCG-binding to hCG receptor in human homogenates
  a) ovarian homogenates
   ■—■ malignant at 25 °C   ■ — ■ benign at 25 °C
   ▲—▲ malignant at 37 °C   ●—● benign at 37 °C

  b) uterine homogenates
   ●—● malignant at 25 °C   ●—● benign at 25 °C
   ▲—▲ malignant at 37 °C   ●—● benign at 37 °C
  (All details are explained in the text)

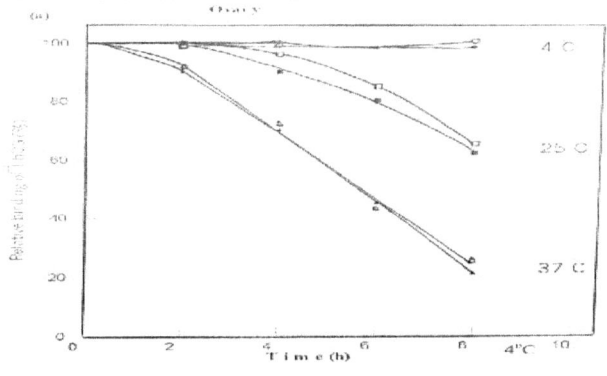
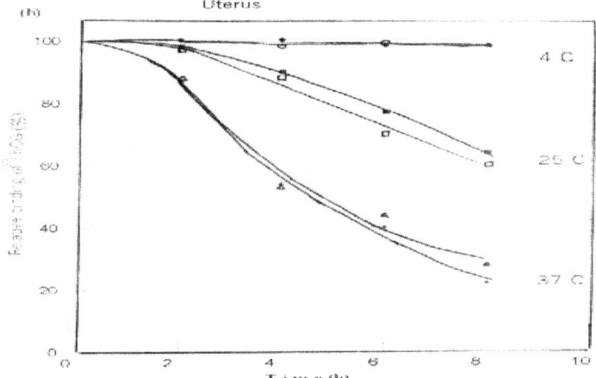

Fig. 3.7: Stability of ^{125}I-hCG receptor complex.
   a) receptor in ovarian homogenates
      •–• malignant at 4°C   o–o benign at 4°C
      ■–■ malignant at 25 °C   □–□ benign at 25 °C
      ▲–▲ malignant at 37 °C   △–△ benign at 37 °C
   b) receptor in uterine homogenates
      •–• malignant at 4°C   o–o benign at 4°C
      ■–■ malignant at 25 °C   □–□ benign at 25 °C
      ▲–▲ malignant at 37 °C   △–△ benign at 37 °C

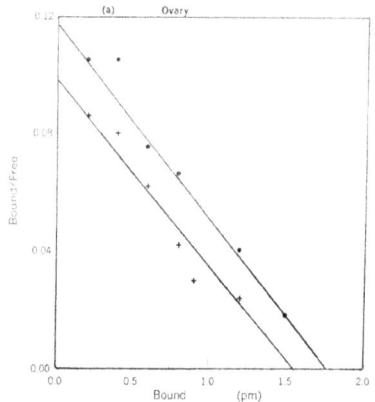

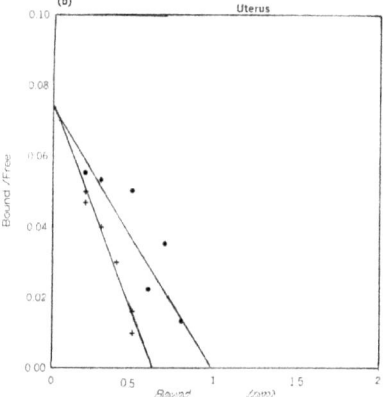

Fig. 3.8. Scatchard plots analysis of ^{125}I-hCG binding to hCG receptor in crude human ovarian and uterine homogenates.
a) Ovarian homogenate: (•–•) malignant, (+ – +) benign.
b) Uterine homogenate : (•–•) malignant, (+ – +) benign.
(All details are explained in the text)

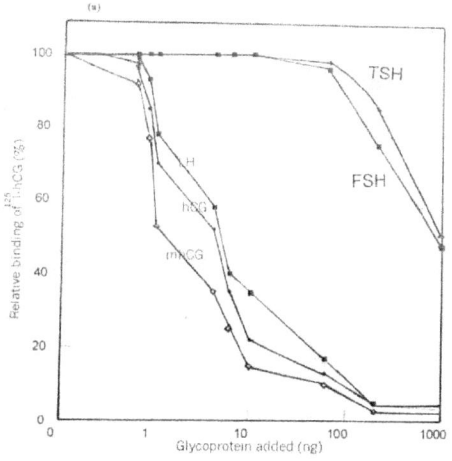

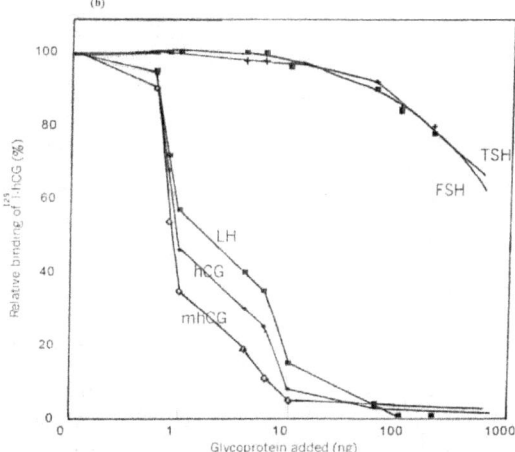

Fig 3.9. Competitive binding curve for ^{125}I-hCG binding to hCG receptor in crude human homogenates.
    a) **Ovarian homogenate**
    b) **Uterine homogenate**
    (All details are explained in the text)

Chapter Four

Purification and Characterization of hCG-R from Ovarian Cancer

Abstract

Human chorionic gonadotropin receptors of ovarian cancer from detergent soluble membranes extract were purified 8,538 fold. The purification scheme based on gel filtration and hCG-Sepharose affinity column. The pure receptor showed three protein components with molecular weight 80,350 D; 53,703D;42,657D as indicated by polyacrylamide gel electophoresis under reducing conditions.Analysis of the purified receptor under nonreducing conditions showed the molecular weight 204,000D and stock's radius 54^0A. The pI of the purified receptor was found to be 6.4, and the carbohydrate content was found to be 8%.

Human chorionic gonadotropin binding studies demonstrated that the isolated receptor retained all the specific binding characteristics expected for the hCG-R. The stability of the receptor during solubilization was markedly improved by using 20% glycerol. Further characterization of purified receptor was obtained through the estimation of the kinetic parameters.The Hill plot data indicated that there was no cooperative between the hCG-R binding sites.The UV.studies on hCG-receptor complex indicated that conformation changes of the hCG or receptor may occurr.

Introduction

The evidences that have been accumulated show that the receptor for hCG is located on the plasma membranes of the target cells. One of the major problems in hCG-R isolation and characterization had been its presence in extremely low amounts. Nonetheless, the receptor from some animal gonads has been purified.

Receptor from leydig cells of rat testis[208][224][225], rat ovaries [225], bovine corpus luteum [157][158][226] and porcine lutea [227] tissues have been purified and characterized by binding studies with radio-iodinated gonadotropins. Some studies used the photo or radioactive labeling for receptor to facilitate its isolation and characterization [158][228]. Human chorionic gonadotropin receptor is an integral membranes glycoprotien, which requires detergent for solubilization. The detergent solubilized forms of the receptor of testis and ovary exhibit the properties of asymmetric molecules.

Different techniques have been used to purify this receptor such as sucrose density gradient ultracentrifugation[211], gel filtration, lectin affinity chromatography on wheat germ agglutinin(WGA), agarose column or on immobilized concanavalinA, hCG affinity chromatography which was performed with hCG coupled directly to cyanogen bromide-activated Sepharose[156-158][219]. The reverse immunoaffinity chromatography (using polyclonal antibodies against nonreceptor proteins by Metsikko and Rajaniemi, or by Bruch and et. al., [156]) provided a better yield of hCG-R from pseudopregnant rat ovaries, however, the final product has been biologically inactive .In general, low yield , poor stability and lack of large quantities of starting material have not permitted the isolation of the hCG-R in quantities sufficient for detailed characterization and structural studies. Metsikko and Rajaniemi (reviewed in 149), isolated a protein of Mw 100,00 by indirect immunoaffinity chromatography which was biologically inactive. In contrast, Dufau and co-workers [219] described the isolation of five species by lectin affinity chromatography ranging from Mw 12,000 to 165,000 which retained hCG binding activity. Dattatreyamurty[157] and Saxena[229] purified the hCG receptor in large quantities from bovine ovary by a complex scheme in 13% yield.They demonstrated that the bovine receptor in an oligomeric protein of Mw 280,000 composed of two disulfide linked subunits. In contrast, immunoprecipitation of the labeled receptor and chemical Crosse linking experiments with ovarian membranes[154] and isolation of cDNA for hCG receptor[149],suggest that this receptor is a single polypeptide .

In this chapter, hCG-R was purified from detergent -solubilized preparations of ovarian cancer patients by a scheme based on gel filtration and affinity chromatography on immobilized hCG. Then the purified receptor obtained has been characterized by different techniques.

Materials and Methods
4.1 Buffers and Reagents

Buffers and reagents mentioned of this chapter were mentioned in section (2.4) of this thesis. Other solutions used were indicated in each experiment.

4.2 Patients and Specimens

The ovarian cancer patients described in section (2.3) were included for hCG receptor analysis. The crude membranes pellet described in section 3.2 were pooled and used for the purification with further solubilization with 0.1% Triton X-100 (provided by BDH) in 50 mM Tris buffer pH 7.4 containing 5 mM EDTA, 1mM PMSF 20% glycerol in a volume equivalent to 1:4 for 1hr at 4°C. The resulting mixture was subjected to centrifugation at 9000 xg for 30 min The pellet was extracted again with 0.1% Triton X-100 in a volume equal to half that used in first extraction for 1hr to ensure the complete extraction of the receptor. The supernatants were combined and diluted 2-fold with the same buffer not containing detergent. The receptor activity and protein concentration were measured on this supernatant.

4.3 Protein Determination

Routine protein determination was carried out by the method of lowry et.al.,[200] The method was slightly modified for its application to solutions containing Triton X-100. Turbidity produced during the assay due to the detergent was removed by centrifugation at 3000xg and the absorbance of the clear blue-colored supernatant was read at 750 nm. The calibration curve was prepared under similar conditions using bovine serum albumin containing the same amount of Triton X-100.

4.4 Binding Assay

The soluble receptor was assayed by mixing 250ul of the receptor solution with 40µl of ^{125}I.hCG (1 µg/ml)and completed to 500µl with 50mM Tris buffer 0.1%BSA, pH 7.2. Incubation was carried out at 37°C for 1hr. This was followed by addition of 750µl of 10% polyethylene glycol in Tris buffer. After shaking well and allowing to stand at 4°C for 10min, the reaction mixture was centrifuged at 1500xg for 30 min and the radioactivity was determined in the pellet. The non specific binding was determined in the presence of 200 fold of unlabelled hCG.

4.5 Purification of hCG-R

4.5.1 Gel Filtration Chromatography on Sepharose CL-6B

Preparation of the Column

The Sepharose CL-6B (provided by pharmacia)gel was prepared by swelling it in 50 mM Tris/HCL buffer, pH 7.4, containing 1mM $MgCl_2$, 0.01% Triton X-100 (v/v). The swollen gel was packed in the column (1.5x 87cm) which was equilibrated with the same buffer.

Void Volume Determination:

The void volume (V_0) of the column was estimated by using blue dextran 2000 with concentration of 1mg/ml dissolved in Tris buffer used for equilibration .One ml of blue dextrane solution was applied to the column, and then elution was carried out with the same Tris buffer. Fractions of 2ml each were collected at a flow rate of 18ml/hr and their absorbance were measured at 600nm. The volume of the buffer that was required to elute the blue dextrane represents the void volume (44ml) of the column.

Purification Procedure

The supernatant of crude solubilized membranes (30mg of protein) was applied to the column which was eluted with the same Tris buffer used for equilibration of column.Fractions of 2ml each were collected at a flow rate of 10ml/hr. The protein localization was followed in each tube at low ultraviolet absorption of Triton x-100 (230nm), also specific binding of each fraction was calculated as described in section (4.4) and plotted against the elution volume. The fractions that contained the binding activity were pooled and protein was determined as described in section (4.3) while binding of I^{125}– hCG to the receptor with increasing amount of I^{125} –hCG (1-200pM) was determined as described in (4.4). The pooled fractions were concentrated by dialysis against polyethylene glycol for a subsequent purification steps.

4.5.2 Affinity Chromatography on Sepharose –hCG

Preparation of Sepharose –hCG

Purified hCG (1500mIU/ml) was attached directly without interposing hydrophobic extension to Sepharose 4B[158]. The hormone (0.7 mg protein) in 0.1M phosphate buffer pH 9.5 was incubated with Sepharose 4B activated with CNBr (1gm), purchased from Pharamacia for 24hr at 4°C. The amount of the unreacted protein was determined by measuring the absorbance of the supernatant at 280nm. Under these conditions, approximately 0.5 mg of protein was attached per 1gm of CNBr activated Sepharose 4B.

Fractionation of the Receptor on Sepharose –hCG

The concentrated pooled fractions obtained from gel chromatography on Sepharose Cl-6B column were applied to the Sepharose–hCG column (2x8cm) previously equilibrated with 50 mM Tris /HCL-0.01% Triton X-100, pH 7.4 at flow rate of 20ml/hr. The effluent was recycled 4 times for maximal binding. The column was washed with Tris buffer used for equilibration to remove any nonspecifically bound proteins. Subsequently the flow was stopped and the receptor was eluted by adding 8ml of 50 mM glycine buffer pH 4 containing 5 mM EDTA 1mM PMSF, 20%glycerol and suspending the column in situ. After standing 10 min, the column was drained and then eluted continuously with the glycine buffer.Fractions of 3ml were collected into tubes containing neutralization buffer (1M NH_4HCO_3, pH 8 containing 20% glycerol).The pH of the collected fractions after mixing with the neutralization buffer was 6.8-7. The individual neutralized fractions were immediately dialyzed at 4°C for 6hr against 50mM Tris/HCL pH 7.4 containing 1mM PMSF, 20% glycerol. The eluted fractions were tested for the binding activity (section 4.4), and plotted against the fraction number. The purified receptor was dialyzed against polyethylene glycol. This purification receptor preparation was used for analysis and characterization studies.

4.6 Analytical and Characterization of Purified hCG Receptor
4.6.1 Determination of the Association Constant

Fifty micro liters of purified receptor solution (0.2 µg protein) were incubated with increasing amount of I^{125} –hCG (1000-30000CpM) with or without 200 fold of hCG. The volume of incubation was completed to 500 µl by adding 50 mM of Tris buffer - 0.1% BSA, pH 7.2. The binding of I^{125} –hCG to the receptor was determined as described in (4.4). The association constant (k_a) was determined as described in section (3.3.8).

4.6.2 Analysis of Purified Receptor by Conventional PAGE

The polyacrylamide gel (7.5%) was casted by mixing 7.5 ml distilled H_2O, 33 ml solution A, 22.2 ml solution C, 3.2 ml solution D and 0.1ml TEMED. The gel slab was applied on the LKB 2117 Multiphor electrophoresis system. After pre-electrophoresis at 50mA for 30 min according to the LKB instructions[230], the samples (15µl) were loaded and concentrated for 5-10 min with a current of 20 mA. The field strength was increased to 15V/cm after the concentration step. The run continued until the band of bromophenol blue reached the gel edge. The gel was then divided into two parts. Part one was sliced to 2mm segments and the radioactivity of each segment was measured by mixing them with 100µl of 50 mM Tris buffer- 0.1% BSA pH 7.2, then sliced and cracked softly with glass rod and left for 12hr at 4°C. The mixture was centrifuged for 20 min at 1500 xg and 4° C, then the binding activity was determined as described in section (4.4). Part two of the gel slab was stained for protein by leaving the gel in the staining solution (G) for 2hr after fixing for 1h by solution (F). Distaining has been done by several changes of distaining solution (H).

Solutions:

Solution A (Tris-glycine buffer stock solution, pH 8.9): 75.1g Glycine and 2.5 g sodium azide were dissolved in 3L distilled water. The solution was titrated with Tris to pH 8.9 and made up to 5L with distilled water.

Solution B (Electrode buffer):1 part buffer stock solution plus 1 part distilled water.

Solution C (Acryl amide solution): 22.2 g Acrylamide and 0.6g Bis dissolved in 100ml of distilled water.

Solution D (Ammonium persulphate solution):150 mg Ammonium persulphate in 10 ml of distilled water.

Solution E(Bromophenol blue 0.25%): 25 mg Bromophenol blue in 10ml of Tris glycine buffer.

Solution F (Fixing solution): 57gm Tirchloroacetic acid and 17g sulphosalicylic acid were added to the mixture of 150ml methanol, and 350 ml distilled water.

Solution G (staining solution):1.25g Coomassie brilliant blue (G-250) in the solution of 227ml methanol, 227 ml distilled water and 46 ml glacial acetic acid.

Solution H (Distaining solution):1500 ml Ethanol with 500 ml acetic acid and 3000 ml of distilled water.

4.6.3 Determination of the Molecular Weight by SDS-PAGE

The polyacrylamide gel (10%) was prepared by mixing 33 ml of solution A, 29.7ml of solution C, 3.2 ml of solution D and 0.1 ml of TEMED. The samples were prepared by adding SDS to the receptor preparation in final concentration (1%) and incubated in boiling water bath for 5 min, then dialyzed against solution A. Ten micro liters of bromophenol blue and 10µL of 2-mercaptoethanol were added to 500 µl sample solution. Standard proteins were prepared according to the manufactures. The electrophoresis procedure was followed according to LKB instructions[230].

Fifteen micro liters of sample were applied to the gel slab and the run was performed using the following conditions, current 190mA, field strength 6 v/cm and total time 5h. After electrophoresis, the gel was divided into three part. Part one was sliced to 2mm Segments and treatment as mentioned in conventional PAGE (4.6.2) for binding activity. Part two of the gel slab was processed for protein staining as in section (4.6.2). Part three of the gel slab was stained for carbohydrates according to the method of segrest & Jackson [231], by leaving the gel in solution I for 2hr, then treatment with solution J for (2-3)hr, followed treatment with solution K for (2-3) hr with several changes. When the gel become colorless, it had been soaked in solution L for 1hr at dark place, followed by washing with solution K for 40 min with several change, then the gel was washed with distilled water several times until the appearing of pink bands. The relative mobility (R_m) of each standard protein and sample was measured as follows:

$$Rm = \frac{\text{Distance moved by the protien}}{\text{Distance moved by the bromophenol blue}}$$

The log M.W of the standard proteins was plotted against the R_m values and the molecular weight of hCG-R was calculated from the straight line obtained.

Solutions:

Solution A: 0.2M Tris -glycine buffer stock solution pH 7.1 containing 0.2% SDS.

Solutions (B, C, D, E, F, G, H) : As described in conventional PAGE (4.6.2).

Solution I (Fixing solution for carbohydrate) : 40ml Ethanol with 5ml glacial acetic
  acid and 55 ml distilled water.

Solution J (0.7% periodic acid): 1.4 gm of periodic acid in 200 ml of 5%acatic acid.

  Solution K (0.2% sodium meta bisulfite): 0.4 gm of sodium meta bisulfite in 200ml
  of 5% acetic acid.

Solution L (Schiff reagent): 10gm of basic fuchsine was dissolved in 2L of distilled
  water with heating. After cooling, 200 ml of 1N HCL and 17 gm of sodium meta
  bisulfite were added then, followed by charcoal was added and the mixture was
  filtrated to take the clear solution was taken.

Standard Protein: Pharmacia electrophoresis calibration Kit for low M.W was used. The kit contains a mixture of five lyophilized purified proteins. The mixture was dissolved in 100 µL Tris -glycine buffer(A) containing 1% SDS and treated exactly as described for receptor preparation. The protein mixture consists of:

Protein	M.W (D)
Phosphorylase b	94,000
Albumin	67,000
Ovalbumin	43,000
Carbonic Anhydrase	30,000
Trypsin Inhibitor	20,000
Lactalbumin	14,000

4.6.4 Determination of the Molecular Weight and Stocks Radius by Gel Filtration Chromatography

The same Sepharose CL-6B column used in section 4.5.1 was calibrated for molecular weight determination. Calibration kit for high M.W (purchased from Pharmacia Fine Chemical) was used. The Kit contains six highly purified proteins individually packed.

Protein	M.W (D)
Thyroglobulin	669,000
Ferritin	440,000
Catalase	232,000
Aldolase	158,000
BSA	67,000
Ovalbumin	43,000

Standard protein were dissolved in equilibration buffer (50mM Tris /HCL pH 7.4 containing 0.1 M NaCl, and 1 mM $MgCl_2$, then applied to the column through three portions (1ml each portion). Portion 1 contained protein 1 and 3, portion 2 contained protein 2 and 5, and portion 3 contained protein 4 and 6. Elution was carried out with the same equilibration buffer at a flow rate of 22 mL/hr. The absorbance's of the fractions collected (2ml each fraction)were measured at 280nm to evaluate the elution volume. Receptor preparation was applied to column and eluted at the same conditions. The partition coefficients (K_{av}) of the standard proteins were determined using the following formula:

$$k_{av} = \frac{V_e - V_0}{V_t - V_0}$$

Where

V_0 : Void volume, V_e : Elution volume, V_t : The volume of the bed gel.

The values of Kav were plotted vs. the values of log M.wt of standard proteins. The M.wt of hCG-R was calculated from the standard curve obtained.

The stocks radius of purified receptor was calculated from the straight line obtained from plotting $(K_{av})^{1/3}$ against the stocks radius of standard protein.

Carbohydrate determination for hCG-R fractions was carried out by the method of phenol-H_2SO_4 of Dubois and et.al.,[232].

4.6.5 Determination of the Isoelectric Point (pI) of Purified Receptor.

The polyacryl amide gel (5%) containing ampholine pH range (3.5-9.5) was prepared according to Garfin[233] by mixing 31.57 distilled water,14.8 Acrylamide solution, 12.7ml glycerol solution (25%w/v) 3.3 ml ampholyte solution (40%w/v),3.2 ml Ammonium persulfate, 0.33ml riboflavin 5-phosphate (0.1%w/v), 0.1ml TEMED. LKB 2117 Multiphor electrophoresis system was used for this run. After soaking with electrode buffer solutions, the electrode strips were applied on their place and (20-30 µL) of samples were loaded on small filter paper equipped from LKB for this purpose. The running conditions were 1500 volts and a current of 50mA for 2hr according to the LKB instruction. After electrophoresis, the gel was divided into two parts. Part one was processed for protein staining , where the gel immersed in fixing solution for 30min then in distaining solution for 15-30 min to remove excess ampholine followed by soaking in staining solution for 1hr at room temperature .The gel was then soaked in distaining solution with several changes until a clear background was obtained. Part two was treated as mentioned in conventional PAGE (4.6.2) for binding activity.pH gradient was determined by slicing a part of the gel to 0.5cm segments. Individually each piece was soaked in 2ml distilled water and left for 24hr at 4^0C, and then the pH was measured in each solution.

Solutions:

Acrylamide solution, Ammonium persulfate solution, Fixing solution ,and

Distaining solution: As described in section (4.6.2)

Anode buffer:1M H_3PO_4

Cathode buffer: 1M NaOH

Staining solution: 0.04% Coomassie Brilliant blue G-250 in destining solution

4.6.6 Specificity of Binding for Purified Receptor

The binding of ^{125}I-hCG to purified receptor was performed as described in section (4.4) with or without the addition of increasing concentration (1-1000ng) of unlabelled hCG or LH .The effect of 1 µg quantities of (FSH, TSH, Prolactin (PRO), 5-Fluorouracil(5-Fu), Haloxan and Cisplatin) on binding activity were also performed as in section (4.4) and relative ^{125}I-hCG binding was calculated and plotted against competitors conc.

Solution

FSH,TSH,PRO:Standard hormones were provided by (DPC, USA).

5-Fu drug:250 mg/5ml ampoule provided by (Edewe , Austria,Europe).

Haloxan drug: 2g/100ml preparation provided by ASTA Medica AG,Germany

Cisplatin:50 mg/ 100ml preparation provided by ASTA Medica AG, Germany

4.6.7 U.V Spectrum of hCG, and hCG-Receptor Complex

Forty microlitres of hCG (1 µg/ml) was diluted by Tris buffer, pH 7.4 to 1ml and placed in cuvette in sample beam. The absorption spectrum was measured in the range (200-350nm). To study the hCG receptor complex, 40µL of hCG (1 µg/ml) was added to (200µL) of purified receptor and the volume was competed to 1mL with Tris buffer. The binding was performed as described in section 4.4, and the complex precipitate was redissolved in 1ml Tris buffer, pH 7.4 and placed in cuvette.The absorption spectrum was measured in the range (200-350 nm).

4.6.8 Stability of Crude and Purified Solubilized Receptor

Crude and Purified receptors were stored at -20°C for 3 month and the binding activity was measured at different time of storage as mentioned in section (4.4). The binding activity was calculated (expressed as a percentage of the binding activity of the preparation that had not been stored) and plotted against the storage periods. The effects of glycerol on the stability of the crude solubilized receptor preparation was also investigated, where the ^{125}I-hCG binding activity was determined for the storage preparation at 4°C within 24h at different concentration of glycerol (0-50%). The binding activity was calculated (% of initial) and plotted against the different glycerol concentrations.

4.6.9 Binding Kinetics of Purified Receptor

The association rate constant (k_{+1}) was estimated for purified receptor preparation by performing the experiment as explained in section (4.4) at various time intervals at 4^0C up to the steady state level (B_{eq}) under pseudo first –order condition. The bound ^{125}I-hCG was plotted against the time and the data obtained was used to estimate k_{+1} by plotting $\ln[B_{eq}/(B_{eq}-B_t)]$ against versus time, where B_t:the amount of bound hormone at time t and B_{eq}:the amount of bound hormone at equilibrium. The slope of this line (K_{obs}) is related to the association (k_{+1}) rate constant, hormone concentration $[H]_T$, and receptor concentration $[R]_T$.

$$K_{obs} = K_{+1} \frac{[H]_T [R]_T}{[HR]_e}$$

The dissociation rate constant(k_{-1}) was estimated by incubating ^{125}I-hCG and purified receptor to steady state and then a 1000 fold excess of unlabeled hCG was added. Binding activity was assayed at various time in travels thereafter as in section (4.4). The $\ln B_t$ was plotted against the time where the slope of this plot is k_{-1}.

4.6.10 Estimation of Hill coefficient (n) of purified receptor

The experiment was performed as described in section (4.6.1). The equation of Hill used was

$$Log \frac{B}{(B_{max} - B)} = n \log[H] - \log K'$$

Where

n: the theoretical number of hCG binding sites per receptor molecule.

B_{max}: the maximum number of binding sites.

K': a composite constant composed of the intrinsic dissociation constant K_d and

interaction factors that determine the degree to which K_d is altered.

Results and Discussion

Purification of the Receptor

In this chapter, the hCG-Rs were isolated and purification from detergent –solubilized preparation of ovarian cancer tissues. The crude homogenate (150mg of protein) obtained from (11) specimens of cancer ovarian tissues (5.2 gm) had 0.87 pmol/mg specific binding activity to ^{125}I-hCG (Table4.1). On centrifugation of crude homogenate at 9000xg, the plasma membranes were obtained at sediment, and the radio receptor assay indicated that approximately 89%(Table 4.1) of receptor was recovered in membrane pellet which was provided a significant initial step in the purification of the receptor. The ovarian hCG-R like other receptors of polypeptide hormones, such as insulin, growth hormone, prolactin, thyrotropin, follitropin, exist in particulate form as an integral part of the cell membranes which requires nonionic detergent for solubilization[226], such as Triton, deoxycholate, and Emulphogene. We used low concentration (0.1%) of Triton X-100 for solubilization. The soluble receptor obtained after two extraction was in a good yield. The insoluble material remaining after two extraction, contained (5%) of the initial activity of the membranes. Previous reports describing the extraction and solubilization of the hCG-R have been based exclusively on the use of Triton X-100 [156-158]. The high concentration of nonionic detergents were obtained to inhibit binding of ^{125}I-hCG to the receptor[157][208], which may be due to the formation of Triton x-100 micelles; but low concentration (0.1-0.25%) of the detergents, were effective in solubilizing. In an earlier study, the extracting of the lipids by treatment with ethanol , methanol, and CCl_4 had been found to destroy the activity of the receptor, which may be due to the loss of essential phosphatidylserine and phosphatidylethanolamine which may be the integral components of the hCG-R, whereas the neutral lipids may provide the matrix for the hydrophobic lipid bilayer ,where the receptors located[225][226].

The receptor could not be extracted with nonionic detergent in agood yield. The major loss of binding activity during the step of solubilization, probably due to denaturizing and not to incomplete solubilization .The inactivation of the receptor may result from conformational change during solubilization. The stability of the crude soluble receptor was markedly improved by supplementation of the extraction buffer with glycerol.

The soluble receptor was assayed by the same binding assay used for particulate receptor with difference in separation of the bound hormone from free. Usage polyethylene glycol in a a final concentration of 10% was more effective in separation ^{125}I-hCG-recptor by precipitation than centrifugation.

The solubilized hCG receptors was purified sequentially by Sepharose CL-6B gel filtration and hCG-affinity chromatography. The separate fractionation of the soluble ovarian membrane extract on Sepharose CL-6B yield identical patterns of hormone binding activity and protein separation as shown in Fig (4.1), the receptor preparation shows a single peak with hormone binding activities. The fractions with binding activity contained 27% of the initial crude receptors and the receptor was purified only about 7.4 fold (Table 4.1). Although a significant fraction of the extraneous proteins was removed only a modest increase in specific activity was obtained. Gel filtration of the solubilized hCG-R on a column of Sepharose 6B provided the major step in the separation of adenylate cyclase from the receptor molecules as confirmed by Dattatreyamurty and et.al., [157] and this indicates that the hCG binding sites (the receptor molecules)and the adenylate cyclase are physically separable entities. Dufan and et. al. [225] also found this result. The hCG binding sites are functionally independent of the adenylate cyclase and therefore represent the actual hCG-R. After the formation of the hormone receptor complex, the coupling of a adenylate cycles may occur at a specific site on the enzyme via ATP-ADP dependent G-protein[234][152]. Dattatreyamurty[157] also mentioned that, 5-nucleotidase associated with the plasma membrane proteins was effectively separated on gel filtration.

The receptor fractions from the gel filtration were further purified by affinity chromatography on Sepharose - hCG. This step yielded a single active receptor peak (Fig 4.2), indicated the homogeneity of the receptor preparation . The direct affinity chromatography was quite efficient with an 8,538-fold purification; however, the recovery of the receptor was 12%, which is consistent with other studies on the receptor. The receptor was eluted from the affinity adsorbent with 50mM glycine buffer pH 4 containing 20% glycerol, followed by immediate neutralization (Fig 4.2). Alternative method of elution was also evaluated, including 1M guanine hydrochloride [158], but this elution did not improve the recovery of the receptor from the hCG affinity column (the final recovery of purified receptor was 8%). The methods used to elute the receptor from the affinity resin are the critical point in the purification of the hCG-R. It is desirable that the elution may done quantitatively, specifically and the hCG –binding activity of the eluted receptor is preserved. Elution of the affinity column as described by different studies with brief treatment of acidic buffer at pH 4 was completely dissociated the hCG-receptor complex in ovarian membrane and did not significantly alter either the hormone binding properties of the receptor or the biological [157,225,156] activity. While elution of the bound receptor from Sepharose hCG with increasing concentration of hCG was not successful. The guanidine hydrochloride has also been used for deblocking, but the biological activity

of the receptor was destroyed[158]. Thus, the elution of affinity chromatography columns with acidic buffer has been successfully employed to purify high affinity membranes receptor, such as those for insulin and prolactin in addition to hCG-R[149].

Several authors with different schemes have reported the purification of hCG-R. Bruch et.al.,[156] used reverse immunoaffinity chromatography (using immobilized antibodies to membranes proteins from receptor down–regulated ovary), and subsequent affinity purification on hCG-spharose to isolated homogeneous receptor from rat ovaries in 19% yield. Another scheme involving lectine affinity chromatography followed by hCG affinity chromatography was used by Pandian and Bahl[158] to purify the receptor from crude plasma membranes of bovine. corpus luteal .They obtained a 2.6% yield of purified hCG-R. Dufau et.al[219] purified hCG-R of the rat testis 15,000 fold after detergent extraction by a single affinity chromatography step on agarose coupled to (hCG coupled directly to CNBr –activated Sepharose, or hCG coupled to Sepharose –concanavalinA). Saxena and co-worker[229] have purified the hCG-R in large quantities from bovine ovary by a scheme involved a WGA and hCG affinity chromatography. While Dattatreyamurty et.al.,[157] isolated the hCG-R from bovine corpora homogenate by Ultracentrifugation in a stepwise sucrose density gradient, then the hCG-R fraction was solubilized in Triton X-100 and purified by gel filtration on Sephrose 6B to separate the adenylate cyclase , on sepharose 4B to separate the 5-nucleatidase activity , and on Ultragel AcA34 to remove excess Triton X-100 and to concentrated the receptor.

Analytical and Characterization of Purified Receptor

Human Chorionic Gonadotropin Binding Properties of Purified Receptor

The functional identity of the proteins isolated by affinity chromatography as hCG receptor was established by analysis of hCG binding properties. The equilibrium associated constant (Ka) was calculated from a dose-response curve using ^{125}I-hCG and a fixed amount of the purified receptor (Fig 4.3a). When data analyzed by a Scatchard plot, they indicated a single class of binding sites with an association constant of the value 9.1×10^{10} M^{-1} (Fig 4.3b). Thus, purification of the receptor was accompanied by only a modest increase in the association constant over that determined with crude receptor. This is consistent with findings of Bruch et.al., [157] pandian and Bahl[158], where the value of K_a of the receptor at various stages of purification did not alter significantly and was of the order of 10^{10} M^{-1}.

It should be pointed out that, in contrast to the crude membrane receptor, as one would expect, there is no non specific binding of ^{125}I-hCG when the purified receptor is used in the binding assay. Any radioactivity obtained in the presence of hCG is not due to nonspecific binding but is due to the nonspecific precipitation of radioactivity by polyethylene glycol.

Polyacrylamide Gel Electrophoresis

The homogeneity of the protein isolated by purification methods was evaluated by conventional gel electrophoresis using 7.5% polyacrylamide gel. Electrophoresis of fractions obtained by the purification scheme based on, gel filtration and hCG-Sepharose affinity chromatography is shown in (Fig 4.4). The electrophoresis pattern of the crude soluble membranes (lane 2) in Fig 4.4 indicates that some components of crude homogenate were separated by precipitation. Lane 3 of electrophoresis pattern (Fig4.4) represents purification of the receptor by gel filtration chromatography, this pattern is heterogeneous and indicates that this step only marginally improved the purity of receptor. The purity of the hCG receptor obtained from affinity chromatography was supported by the major protein band obtained in gel electrophoresis (Fig 4.4 Lane 4), which was migrated with Rm of 0.38. The receptor activity of protein band was confirmed by specific binding with ^{125}I-hCG. This band was observed in the crude soluble receptor preparation.

Molecular Weight of the Purified Receptor

The purified hCG receptor was analyzed by SDS-polyacrylamide gel electrophoresis and gel filtration chromatography to determine it's molecular weight.

<u>1-SDS –polyacrylamide gel electrophoresis</u>

The purified preparation (treated with 1% SDS, 2% mercaptoethanol) was separated into three protein components (Fig 4.5) with molecular weights (80,352D), (53,703D), (42,657D) obtained from a plot of Rm versus the log of the known molecular weight of protein markers (Fig 4.6). These three band suggested the presence of possible oligomers of the hCG-R, but only one protein band with Mw of (80,352D) defied as receptor by binding activity with ^{125}I-hCG. Therefore other bands may be explained by proteolysis (and or nicking) of the receptor during the experimental procedures, one must keep in mind that because the receptor binds hormone, it does not necessarily mean that it is structurally intact. Another explanation bear in mind is that the hCG receptor identified and purified could represent an hCG-binding component of the receptor and that an additional, protein(s)co purified not identified by hCG binding might be required for coupling to the Gs protein.

In view of the analysis of the purified receptor by SDS-polyacrylamide gel electrophoresis, the earlier studies indicated to the presence of multiple subunits (joined covalently or non covalently) of purified receptor of rat testis[219] (consist of a dimmer of subunits of Mwt 90,000D), bovine corpora lutea[157] (consist of 3 disulfid linked oligomers of M.wt 160,000, 57,000 and 44,000D), and rat ovaries[156] (consist of 3 subunits of M.wt 94,000, 66,000, and 55,300D). More recently, the studies of chemical cross linking of hCG to the hCG receptor (in MA-1- murine leydig tumor cells and in porcine granulose cells[149]), the studies of isolation and characterization of the cDNA for the rat luteal[235][236], porcine testicular[237] and human hCG receptors[238], the development of antibodies for this receptor[239] and the transfection studies[240] confirm that this receptor is single polypeptide. Segaloff et. al.,[149] and Rosemblit et.al.,[236] in their purification hCG receptor from rat Luteal tissue, used binding of ^{125}I-hCG to probe blots prepared from SDS gel of the WGA - and affinity purified material from rat luteal detergent extract. They observed the presence of specific ^{125}I-hCG binding to the 93kDa protein. The blots were incubated with ^{125}I-hCG alone or together with an excess of unlabeled hCG and the runs were performed in the absence or presence of reducing agents.

The cloning and expression of the rat hCG receptor cDNA with the aid of the polymerase chain reaction using oligonucleotides based on amino acid sequences obtained from the purified rat luteal hCG–R [149] have shown that the mature receptor protein would be 674 amino acids and would be predicted to have a molecular mass of 75KD. The difference between this size and that observed for purified receptor (93KD) is presumably due to the glycoprotein nature of the receptor.

<u>2-Gel Filtration Chromatography on Sepharose Cl-6B</u>

The molecular weight and molecular size of the purified receptor were determined on the Sepharose CL-6B column. The elution of receptor from the column was compared with protein marker of known molecular weight. From the relationship of the logarithm of the molecular weights versus the elution volume (k_{av}) of the marker proteins (Fig 4.7), the molecular weight of the purified LH/CG-R was estimated to be (204,000)D.

The molecular size of purified receptor was also determined on a Sepharose CL-6B column. The stock's radius of the purified receptor was calculated to be ($54^0 A$) as shown in Fig 4.8

The different Mwt of purified receptor obtained by gel filtration and SDS electrophoresis may reflect a different tertiary structure of the receptor under denaturing versus non-denaturing conditions, or whether the receptor exits in the plasma membranes as a noncovalenthy associated oligmer of (80,325D). Our study and the previous studies for determination of Mw of hCG-R design on the binding of hCG to the receptor thus identified hCG-R could represent an hCG binding component of the receptor only.

In agreement with previous observation, almost all of Mw determined by gel filtration was approximately 200 KDa.Dufau et.al., and Ascoli and segaloff [241]reported the molecular weight of the nondenatured and most highly purified testicular and ovarian receptors to be in the range of 194-280 KDa. Pandian and Bahl determined the molecular size of the purified bovine luteal receptor on the Sepharose 6B column. The stock's radius of the receptor- hCG complex was $64^0 A$. Dattatreymaurty et.al.,[157] estimated the bovine corpus Luteal receptor to be 280KDa by Sepharose 4B column. Bellisario and Bahl [208] estimated the molecular size of soluble receptor by Sepharose 6B chromatography. The distribution coefficients (Kav) of the receptor obtained from rat testis, ovaries and interstitial were 0.38 ± 0.02, 0.39, and 0.36 ,respectively and from bovine corpus luteum and testis were 0.41± 0.02 and 0.39 respectively.

Isoelectric Point (PI) of Purified Receptor

Isoelectric focusing technique was used for the estimation of isoelectric point (pI) of purified receptor, using polyacrylamide gel containing ampholine to give a pH gradient range from 3.5-9.6. The result shown protein band consistent with binding activity to ^{125}I-hCG (Fig 4.9), which confirm the purity of the hCG-R and that the receptor compose of a single polypeptide. The pI of the purified receptor was found to be 6.4(Fig 4.10) by sliced the gel and measurement the pH of each segment.

Carbohydrate Composition of the Purified Receptor

The direct evidence for the association of carbohydrates with the receptor was provided by electrophoresis analysis, where the band identified as receptor gave positive stained for Schiff reagent, and by carbohydrate determination was 8%.

All groups of investigators reported a fact that the hCG-R is a glycoprotein[158]. Indirect evidence for the association of carbohydrates with receptor have been provided by binding of the receptor to concanavalin A during affinity purifications[158][219]. Dattateryamurty et.al.,[157] shown the total carbohydrate content of the receptor fraction was 10% , mannose and galactose were the predominant hexoses and the carbohydrate moiety of the receptor contained both N-acetylglucosamine as well as sialic acid. The six potential sites for N-linked glycosylation on extracellular domine of the receptor were proposed from experiments of cloning of cDNA using oligonucleocides based on amino acid sequences obtained from the purified rat Luteal hCG-R[2]. The possible involvement of the carbohydrate moieties of the receptor in the recognition and high affinity binding of hCG have been examined using different approaches. All groups of investigators reported that neuraminidase treatment reduced the molecular mass of the hCG-R without affecting its ability to bind hormone[242][243]. In more recent studies, introduced mutation in each of the six N-linked glycosylation and tested for the effects of these mutations on hCG binding [244][245].

The conclusion of these studies, shows that at least one of the carbohydrate chains of the hCG receptor is needed for hCG binding and carbohydrate may be important for proper trafficking of receptor to the plasma membranes.

Human Chorionic Gonadotropin Binding Specificity of Purified Receptor

The specificity of hCG binding of purified receptor was evaluated in competition experiments with unlabeled glycoprotein hormones, prolactin, and drugs used in chemotherapy (5-Fu, haloxa,and cisplatin.The ^{125}I-hCG binding of the purified receptor was inhibited in a dose-dependent manner by the unlabeled hCG and LH(Fig 4.11). The purified soluble receptor has specificity similar to those of the crude membranes receptor and the expected rank order of binding affinities for the LH and hCG was retained in the purified material.

At the level up to 1mg, no inhibition was observed with FSH, TSH, Prolactin and cytotoxic antineoplastic agents. These results indicate that the purified receptor retains it hormonal specificity and has no significant cross-reaction with other peptide , glycoprotein, and the cytotoxic agents which used for first treatment of cancer or for second line treatment as single or combination to kill cancer cell in the body after surgery to control tumor growth[13][16] .Also showed no effect on hCG binding activity of the purified receptor.

The UV-Spectrum of hCG and hCG-Receptor Complex

The UV spectrum of hCG and hCG-receptor complex were measured to determine their maximum wavelengths, and the alteration in the UV spectra as a result of their interaction.

As shown in Fig 4.12, the spectrum of hCG consisted of two maximum wavelength, λ_{max1} at 275.0 nm with an absorbance 2.159 and λ_{max2} at 231.2 nm with an absorbance 2.359. The ultraviolet absorption spectrum of protein solution in the region (250-310) nm are contributed mainly from phenylalanyle, trosyl, and tryptophanyl residues. At shorter wave lengths contribution comes from other group such as histidyl residues and the peptide [246][247] bonds. Changes in the environment of these chromophores (e.g. pH, temp, ionic strength, solvent etc...) can lead to alteration in the absorption spectrum, also configurational changes of the protein may be also involved environmental changes. The UV spectrum of hCG-receptor complex is illustrated in Fig (4.13), The spectrum consisted of one maximum wavelength at 219nm with an absorbance 1.81 and the λ_{max1} 275 for the hCG was disappeared . Protein shows a strong absorption in the range 180-220nm, which arises from electronic transition in the polypeptide backbone itself and is therefore sensitive to backbone conformation. When the hCG and receptor associated, chromphores may be buried at the newly formed interface or conformational changes of the hormone or receptor may bury previously exposed chromophores [246] .These proposals agree with studies on hCG receptor interaction which indicate, to the prescience of multiple contact sites between the exodomain of the receptor and both subunits of hCG [164][166] ,and conformational changes for both hCG and receptor[158][175-177].

Stability of Purified Receptor

The crude and purified receptor were stored at -20^0C in glycerol-containing media for about 3 month to study the stability of this receptor and to check it efficiencies of binding throughout the storage period every one weak. The results in Fig 4.14 demonstrate that the crude receptor was more stable than purified receptor. The purified receptor was stable for about 2 months while the binding activity of the crude receptor was remained without significant changes for more than two months.

To obtain the stabilization of the crude solubilized receptor by glycerol, receptor (extracted with 0.1% Triton X-100) containing (0-50%) glycerol were tested. The data in Fig 4.15 show that the binding activity of the solubilized receptor is maximal when concentration of glycerol of 20% or greater are utilized. As shown in Fig 4.15 there is 70% loss in binding activity of the receptor during storage at 4^0C within 24h in the absence of glycerol. The instability of the soluble receptor was not dependent on prior storage of the tissue or the concentration of Triton X-100. In contrast if glycerol (20% or greater) is included in the same incubation, there is only a 5-10% loss in binding activity. Therefore the receptor must be maintained in glycerol (of at least 20%) to preserve its binding activity actually 20% glycerol was included in all buffer used to solubilize and store hCG-recptor preparation. Previous studies also used glycerol to obtain stable preparation of the crude soluble receptor from bovine corpus luteal[157], rat testis and rat ovary[156][208] .Glycerol may stabilized the receptor by favoring a thermodynamically stable conformation in solution by sequestration of hydrophobic amino acids from the solvent[248]. Furthermore. It had been shown that the hCG receptor complex is considerably more stable than the native receptor even in crude preparation[248].

In addition, Bruch et.al.,[156] have found that the regulatory and catalytic component of adenylate cyclase system also appears to be stable in the presence of glycerol

Kinetic of Purified Receptor Binding and the Determination of Hill Coefficient

The kinetics of ^{125}I-hCG binding to the purified receptor were examined at 4^{0}C under the condition of radioreceptor assay. Accurate binding parameters could not be obtained at higher temperatures, apparently as a consequence of thermal inactivation of the receptor. However, at 4^{0}C, bound hCG was assayed at various time intervals up to the steady state level. The binding of ^{125}I-hCG to the receptor reached a stable maximum after 24hr of incubation (Fig 4.16a). Addition of excess unlabeled hCG to preformed ^{125}I-hCG receptor complexes partially reversed the binding of the labeled hCG resulting in the dissociation of approximately 80% of the preformed hCG receptor complexes (Fig 4.16 a) .The time course data for the association and dissociation process were analyzed by the pseudo first orders and first order kinetic models, used previously to characterize the hCG binding kinetics of purified hCG receptor preparation [156].

The rate of formation of hCG receptor complex was linear over the first 8h of the time course (Fig 4.16b). In contrast, the kinetics of dissociation were biphasic and were characterized by an initial rapid phase followed by a much slower rate of dissociation (Fig 4.16c). The biphasic nature of the dissociation kinetics has been previously observed with crude and purified preparation of the hCG receptor[156][218]. The rate constants for association and dissociation determined with the purified receptor and the equilibrium binding constant calculated from the kinetic parameters are shown in Table (4.2). The equilibrium binding constants calculated from the kinetic parameters were in good agreement with those determined by Scatchard analysis.

The dissociation rate constant (k_{-1}) was estimated by incubating ^{125}I-hCG and receptor for a given period of time and then infinite dilution for ^{125}I-hCG concentration was performed by adding excess of unlabeled hCG. Negative cooperatively may be a part of the binding process when unlabeled hormone accelerates the dissociation rate, but when the receptor site concentration approaches or exceeds K_d of the ^{125}I-hCG, diffusion of dissociated ^{125}I-hCG away from free receptor sites is progressively hampered probably because of the very high local concentrations of unoccupied binding sites.

The Hill plot analysis was carried out in order to determine the Hill coefficient. The result revealed that this coefficient was equal to 1 (Fig 4.17) suggesting that there was no cooperatively during the binding reaction.

Table 4.1 Purification of the hCG receptor from ovarian cancer(All details are explained in the text)

Step	Protein (mg)	^{125}I-hCG binding(pmol)	Specific activity (pmol/mg)	Purification fold	Recovery %
Homogenate	150	130.2	0.87	1	100
Crude membrane	90	115.6	1.28	1.72	89
Soluble extract	30	51.2	1.71	1.95	39
Sepharose Cl6B	5.6	35.7	6.38	7.36	27
hCG-Sepharose	2.1 (µg)	15.6	7.43	8,538	12

Table 4.2 Kinetic and equilibrium hCG binding constants for the purified receptor.

*k_a (x10^7 m^{-1}·min^{-1})	7.6
*k_d (x10^{-3} (min^{-1}))	2.78
*K_a (x10^9 m^{-1})	
Kinetics	27.33
equilibrium	91.0

k_a: pseudo first –order constant for hCG receptor complex formation

k_d: first –order constant for hCG –receptor complex dissociation

Ka: equilibrium association constant estimated from k_a/k_d (kinetics) or Scatchard analysis (equilibrium)

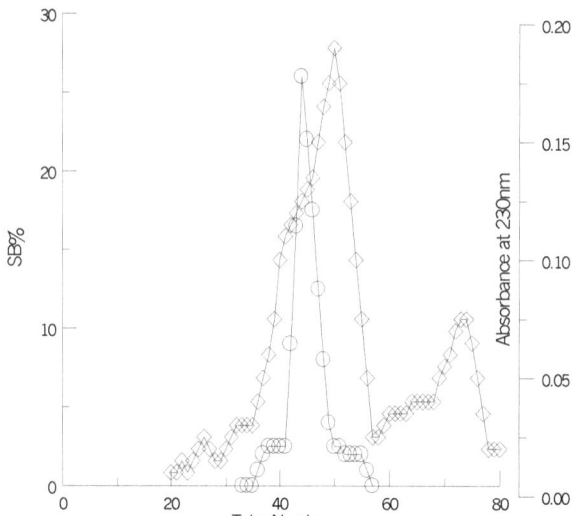

Fig 4.1 Fractionation of soluble receptor on Sepharose CL6B

(0---0) Absorbance at 230 nm
(   ) SB%
(All details are explained in the text)

(All details are explained in the text)

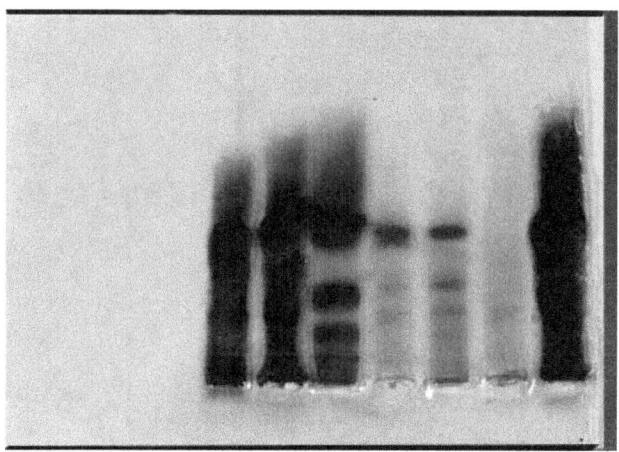

Fig.4.4 conventional- PAGE (7.5%) of Crude and purified hCG receptor in different steps of purification.

Lane 1 Crude hCG receptors homogenate
Lane 2 Crude membranes homogenate
Lane 3 Fractions obtained by gel filtration chromatography
Lane 4 Fractions obtained by affinity chromatography
(All details are explained in the text)

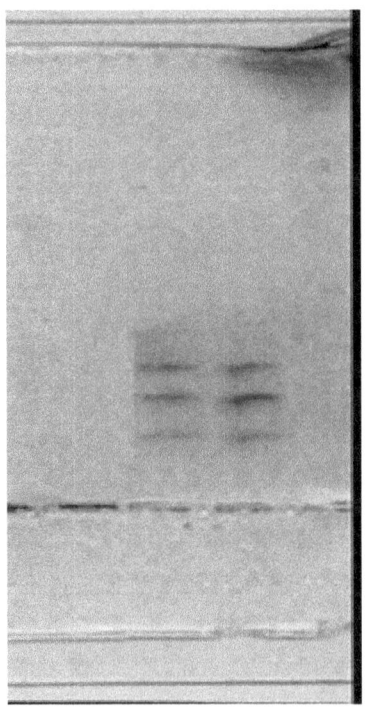

Fig 4.5 SDS –PAGE of purified hCG receptor.
(All details are explained in the text)

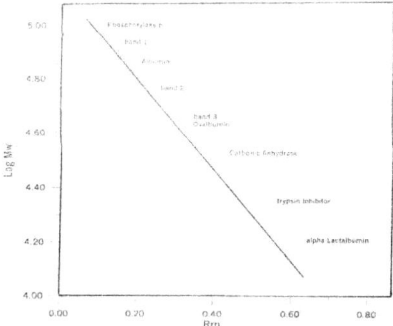

Fig   Calibration Curve for determination of M.wt of the purified hCG receptor by SDS-PAGE(10%)

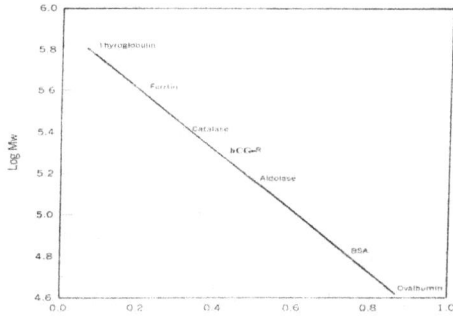

Fig   Calibration Curve for determination of M.wt of the purified hCG receptor By gel filtration chromatography .

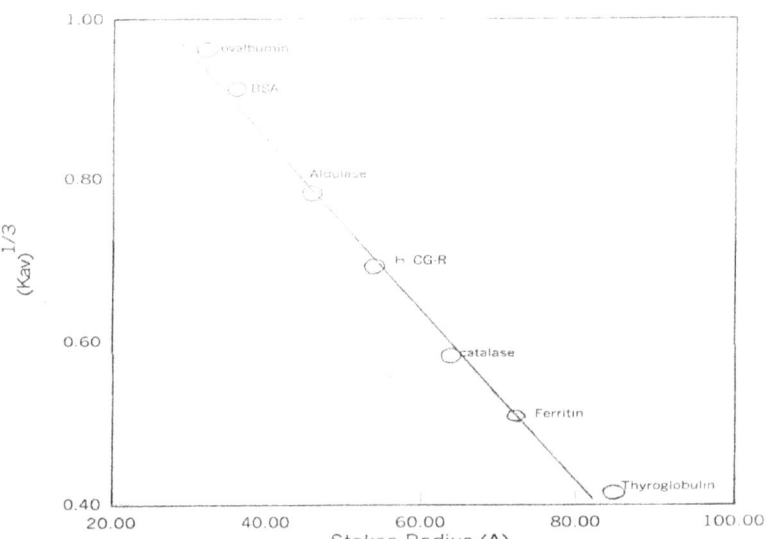

Fig   Calibration curve for determination of molecular siz of the purified receptor by gel filtration chomatography

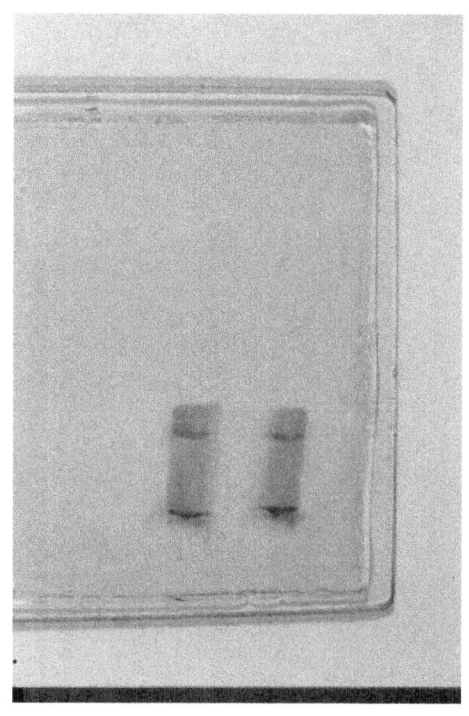

Fig 4.9 Isoelectric focusing of purified hCG receptor
(All details are explained in the text)

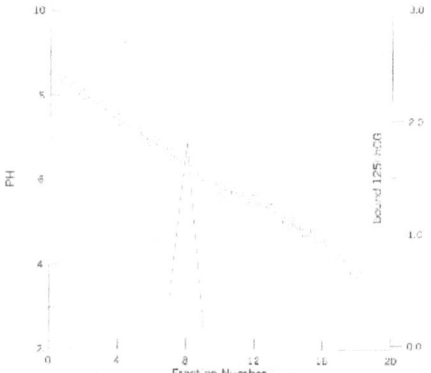

Fig 4.10 PH gradient in isoelectric focusing of purified receptor
(All details are explained in the text)

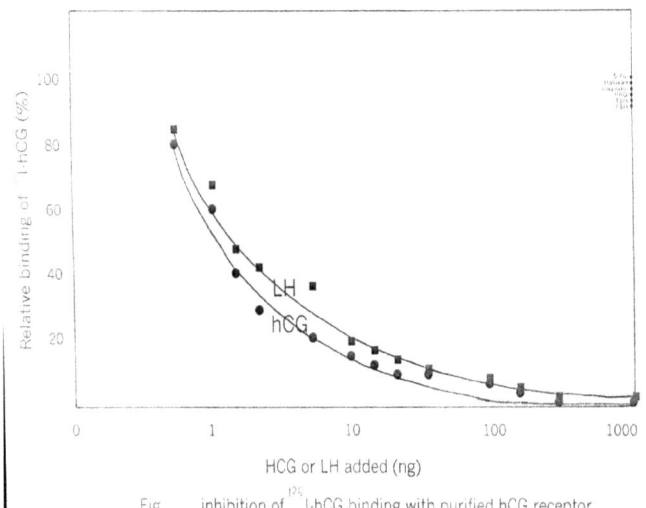

Fig inhibition of ^{125}I-hCG binding with purified hCG receptor by addition of increasing concentration of unlabeled hCG and LH. The effect of 1μg of FSH,TSH,PRO,5-FU,Haloxan,and Cispltin are also shown

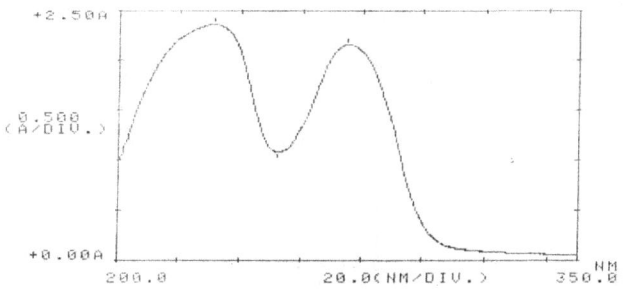

Fig   The UV Spectrum of hCG

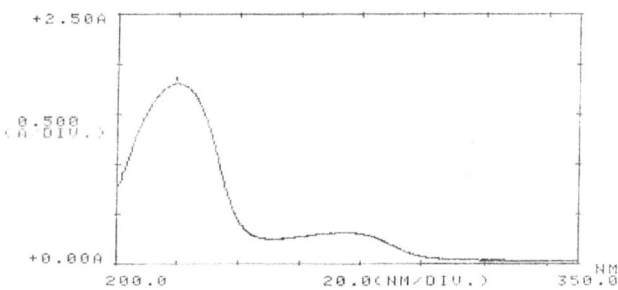

Fig   The UV Spectrum of hCG-Receptor complex

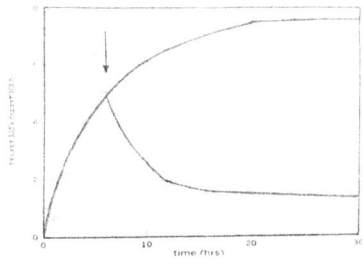

a- Time course of ^{125}I- hCG binding to the purified receptor. At the point indicated by the arrow, excess unlabeled hCG (1 µg) was added to obtain the dissociation time course

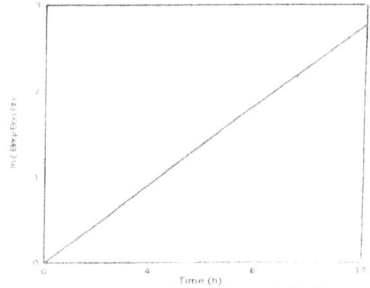

b- pseudo first order kinetic analysis of the association time course

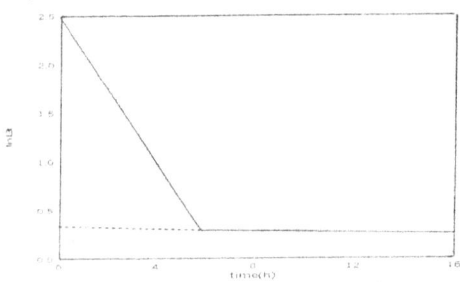

c-first order kinetic analysis of the dissociation time course. For calculation of the dissociation rate constant, the initial more rapid component was corrected for the contribution due to the slower second component as indicated by the dashed line

Fig   Kinetics of ^{125}I-hCG binding to the purified hCG - receptor

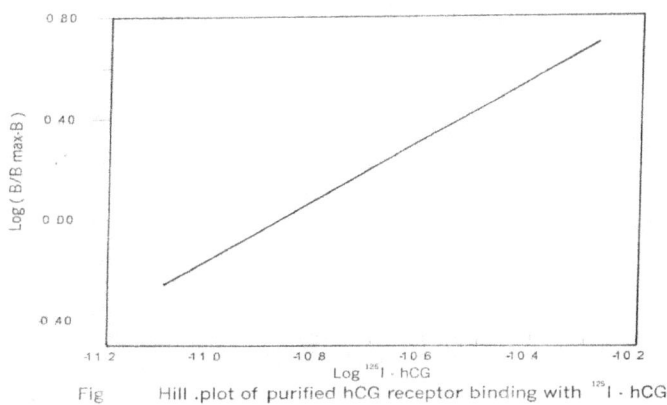

Fig   Hill plot of purified hCG receptor binding with ^{125}I - hCG

Evaluation of Some Biochemical Constituents in Gynecologic Cancers

Chapter Five

Evaluation of some Biochemical Constituents in Gynecologic

Abstract

The levels of CEA were slightly elevated above the reference value(5ng/ml) in the sera of 33% of patients with endometrial cancer and 43% of patients with ovarian cancer..CEA immunoreactivity of malignant ovarian tissue extracts were significantly elevated than in corresponding serum samples, while it was not significantly elevated in malignant uterine tissue extracts. CEA was found in the sera of 7% of control and 11-13% of patients with benign tumors, which indicate the non specify of this tumor marker for gynecologic cancers.

The LDH activity was higher in patients with ovarian and endometrial cancers than those of healthy individuals but not statistically significant. The highest elevation of LDH activity was observed in the invasive stages. The electrophoresis analysis of LDH isoenzymes for serum of ovarian cancer showed extra band.

The total AIP activity and HSAP activity in the sera of cancer patients were significantly higher than in control group. The differences in HSAP activity between malignant and control were clear.

Copper, Zinc, and Calcium concentration were measured in the sera of the patients with ovarian and uterine cancers by flame atomic absorption spectrophotometry. Zinc levels were significantly higher in cancer patients than in control group. Copper levels did not alter in cancer patients than in control. The patients with ovarian and endometrial cancer (without metastasis) showed no significant differences in their sera calcium levels from those of the control group.

Introduction:

Increased levels of several antigens, oncofetal proteins, ectopic hormones and enzymes have been considered as potential tumor markers in gynecologic cancers for helping in screening, diagnosis, staging, prognosis and monitoring of cancer therapy.

The elevation of oncofetal antigen, carcinoembryonicantigen(CEA), have been demonstrated in a large variety of cancers [249,250]. The frequency of elevated concentration varies with the tumor type and the stage of disease, generally, being higher in patients with metastasis. Serial CEA determination has been effective in predicting occult recurrence in patients whose tumor contain high antigen concentration. There are a number of reports on the clinical values of circulating CEA in patients with uterine tumor, elevated plasma level was found in 14-100% of cases of endometrial cancer[251-253]. Elevation plasma concentrations of CEA have also been found in ovarian cancer and often in mucinous histotypes[254][255][26]. Immunohistochemical studies demonstrated that CEA was higher in cervical squamous cell carcinoma tissue than in endometrial cancer[256-258], therefore it might be useful for differentiating endocervical from endometrial.

Lactate dehydrogenase (LDH) is an enzyme in the glycolytic pathway and is released as the results of cell damage. The elevation of LDH in malignancy is nonspecific. LDH has been demonstrated in a variety of cancer such as breast, colon, stomach, liver, germ cell testicular, lung cancer and Hodgkin's lymphoma[44]. The significance of increased serum LDH levels in gynecological cancers have been documented[259][43][45-47]. The serum LDH level has been shown to be correlated with tumor mass in solid tumors and provided a prognostic indicator of disease progression. The changes in activity of serum LDH isoenzymes provided some specificity for organ involvement[44][45].

Alkaline Phosphatase (ALP)enzyme may arise from liver, bone or placenta. Elevated levels of ALP are seen in primary or secondary liver cancer and prostate cancer. Its level may be helpful in evaluating metastasis cancer with bone or liver involvement. Other malignancies such as leukemia, breast, and lung cancer with bone or liver metastases also manifest elevated ALP. The heat stable alkaline phospatase (HSAP) has been designated as carcinoplacental protein following the observation that an isoenzyme of ALP biochemically and immunologically is vary similar to placental alkaline phosphatase which ectopically expressed by some cancer cells[48][260]. This isoenzyme has been measured in the sera of patients with varying degree of tumors of the cervix, corpus uteri and ovaries[49][261][262] and with Hodgkins disease[263] and breast cancer[50].

Different elements are constituents for, or interact with enzymes and hormones that regulate the metabolism of much larger amounts of biochemical substrates, affecting the permeability of cell membranes, and other mechanism[264]. Therefore it is attracted attention in cancer research. Many of the trace elements such as copper, zinc, and essential elements such as calcium have been studied in different cancers. Zinc is needed as cofactor in nucleic acid polymerases and is therefore connected to high mitotic activity of tumor cells. Thus, an elevated zinc serum levels in cancers patients might be a consequence of tumor development[265]. Serum copper levels have also be studied in different type of malignancies such as breast, lung, Kidney tumors, it appeared to be increase in some cancers and frequently return to normal levels after resection[266].Cancer has been considered to be the commonest cause of hypercalcaemia which occurs in approximately 10-20% of patients suffering from cancer[267] . Hypercalcaemia in cancer without the presence of bone metastasis was described by the finding of the parathyroid hormone (PTH)-like peptides[268], or by a mechanism involving elevated levels of 1,25-dihydroxy vit.D3[269], or by releasing locally osteoclast activating factors which stimulate osteoclastic resorption of bone[270]. Hypercalcaemia is almost associated with squamous cell tumors and parathyroid hormone-like material was extracted from a sequamous cell lung tumors[271]

In this chapter CEA levels in sera and tissues of patients with ovarian and uterine tumors were determined by immunoradiometric assay (IRMA). The activity of LDH and ALP were measured by colorimetric methods. LDH isonenzymes were analyzed by electrophoresis. HSAP was also determined. In addition copper, zinc, and calcium concentration were determined in the sera of the ovarian and uterine tumors patients. All biochemical constituents were evaluated for their usefulness in gynecologic cancers.

Material and Methods

5.1 CEA,LDH, and AIP Kits

1- Carcinoemryonic antigen (CEA) IRMA Kit was provided by (OCPA.France) and contains one vial of ^{125}I-rabbit-antiCEA antibody, one hundred test tubes coated with anti CEA antibodies, seven vials of human CEA standard with concentrations ranging of (0-200 ng/ml), two vials of control serum, one vial of assay buffer, and tube of wash reagent.

2- Lactate dehydrogenase (LDH) Kit for colorimetric determination was provided by (Randox, England) and contains one vial of 0.8 mmol/l sodium pyruvate in phosphate buffer PH 7.45, 10 Vial of NADH (1mg/ml)(each vial content reconstituted with 10ml of buffer before used), and one vial of color reagent (1mmol/L 2,4-dinitrophenydrazine in hydrochloric acid).

3- Alkaline phosphates Kit for colorimetric determination was provided by (bio Merieux-Frane) and contains substrate buffer vial (5mmol/l Dinatriumphenylphosphate in 50 mmol/l carbonate bicarbonate buffer pH 10), standard phenol reagent equal to 20 Kind and King U, inhibitor reagent (60mmol/l amino -4antipyrine with 75 g/l sodium arsenate), color reagent (150mmol/l potassium ferricyanide).

5.2 Apparatus

The apparatus, mentioned in section 2.2 were used in the experiments of this chapter.

5.3 Patients and Specimens

The study group consisted of benign and malignant patients, and control group described in section 2.3. The sera of these groups were collected as described in section 2.4.and tissues homogenates and were prepared as described in section 2.4.

5.4 CEA Quantitative Measurement

The level of CEA was measured in the sera and tissues of patients by the immunoradiometric assay (IRMA). The assay protocol was described in Table 5.1., according to manufacturer instructions.

5.5 Determination of LDH Activity

Colorimetric method was used to determine LDH activity. The method is based on the reduction of pyruvate to lactate in the presence of NADH by the action of lactate dehydrogenase.

$$Pyruvate + NADH + H^+ \xrightarrow{LDH} Lactate + NAD^+$$

The pyruvate that remains unchanged reacts with 2,4-dinitrophenylhydrazine to give the corresponding phenylhydrazone which is determined calorimetrically in alkaline medium at 520 nm. The assay protocol was described in Table 5.2, according to manufacture instructions

Table (5.1): Assay protocol of CEA

Tube number	CEA standards ng/ml						Control serum	Samples
	0	2.3	4.8	9.6	19.6	60.0	210.0		
	1,2	3,4	5,6	7,8	9,10	11,12	13,14	15,16	17,18
Standard Control Patient	←-------------------------------------- 100uL ----------- --------------------→							

samples (µL)	
Assay buffer (µL)	←---------------------------------- 100 µL -------→
	Shaking for 2h at 25 °C
Wash buffer (ml)	←----------------------------1 ml ----------------------→
	Aspirate, wash once with wash buffer
^{125}I-anti CEA (µL)	←----------------------------100uL --------------→
	Shaking for 1h at 25°C
Wash buffer (ml)	←----------------------------1 ml ----------------------→
	Aspirate, wash once with wash buffer, measurment of the radioactivity bound in gamma counter

Calculation

1- The radioactivity of all standards measured in (CPM) bound was plotted vs. their concentration (Fig 5.1).

2- The samples and control values of CEA were read from standard curve.

Table 5.2 : Assay protocol of LDH

Tube number	Sodium pyruvate with NADH(U/L)	Serum

	0	111.4	223.1	334.9	390.6	446.3	sample
	1,2	3,4	5,6	7,8	9,10	11,12	13,14
Sodium pyruvate with NADH(ml)	0.9	0.7	0.5	0.3	0.2	0.1	0.9
Distilled water(ml)	0.1	0.3	0.5	0.7	0.8	0.9	
Serum sample(ml)							0.1
	Mixing, incubation in a water bath at 37^{0}C for exactly 30 min							
Color reagent	←---------------------------------1 ml ---------------------------------→						
	Mixing, let stand at room temperature for 20 min							
1.6% NaOH(ml)	←---------------------------------10 ml ---------------------------------→						
	Mixing, let stand at room temperature for 10 min. Reading absorbance of sample or standards against distilled water at 520 nm							

Calculations:

1- The absorbance of samples (1-12) are plotted vs. the corresponding values in U/L (Fig 5.2).

2- The serum LDH activity was determined from calibration curve.

5.6 Analysis of LDH by Electrophoresis

LDH isoenzyme was separated by poly acrylamide gel electrophoresis as described in section (4.6.2). The enzyme activity bands were developed according to following reactions[272]:

Lactate + NAD $\xrightarrow{LDH}$ Pyryvate + NADH + H

NADH + Phenazain methosulfate $\longrightarrow$ NAD$^+$ + Phenazin methosulfate

Phenaz methosulfate (PMS) + nitroblue tetrazolium (MBT) $\longrightarrow$ Nitroblue tetrazolium formazan

(Violt ppt) + phenazain methosulfate

After electrophoresis, the working solution was added on the gel, then incubation for 60 min at 37^{0}C, The violet bands appeared and the gel was placed in fixing solution for 10 min to fix the bands.

Solutions:

Working solution : was prepared by mixing 1ml of 2 mol/l sodium lactate, 3ml of 1mg/ml NBT, 0.3ml of 1mg/ml/PMS and 1ml of 0.056 mol/l Tris buffer pH 7.4 then 15 mg of NAD$^+$ was added.

Fixing solution: was prepared by mixing methanol, distilled water, glacial acetic acid in ratio of 5:4:1

5.7 Determination of Alkaline Phosphates Activity

Colorimetric method was used to determine AIP activity according to the following reaction

Alkaline phosphatase, pH10

Phenylphosphate → phenol + phosphate

The liberated phenol was measured in the presence of amino 4-antipyrine and potassium ferricyanide. The presence of sodium arsenate in the reagent stops the enzymatic reaction. The assay protocol was described in Table 5.3: according to manufacturer instructions.

Table 5.3 Assay protocol of ALP

Tube number	Serum sample 1,2	Serum blank 3,4	Standard 5,6	Reagent 7,8
Substrate buffer (ml)	←------------------2 ml------------------→			
	Incubation for 5 min at 37^{0}C			
Serum sample (µL)	←--50 µl--→			
Standard (µL)				←--50 µl--→
	Incubation exactly 15 min at 37^{0}C			
Inhibitor	←------------------0.5 ml------------------→			
	Mixing			
Color reagent	←------------------0.5 ml------------------→			
Serum sample			←--50 µl--→	
Distilled water	←--50 µl--→			
Mixing, let stand for 10 min in the dark. Reading absorbance at 510				

nm against reagent blank

Calculation

$A_{\text{serum sample}} - A_{\text{serum blank}} / A_{\text{standard}} \times 142$

Kind and king U/L=Enzyme activity unit=One kind and king unit is that an mount of enzyme which in the given conditions liberates 1mg of phenol in 15 min at 37^0C

5.8 Determination of Heat Stable Alkaline Phosphatase Activity (HSAP):

Heat stable alkaline phosphatase (HSAP) activity was measured after the complete inactivation of all AIP isoenzymes by heating the sample for 15 min at 55^0C in a water bath[50]. Fifty microliter of serum sample was incubated for 15 min at 55^0C in a water bath. Immediately after incubation the inactivated serum was cooled at ice temperature. The remaining activity was measured by the same methods of AIP (section 5.7)

5.9 Determination of the Elements (Cu^{++}, Zn^{++}, Ca^{++}) Concentrations in the Sera of Tumor Patients

Copper, zinc,Ca were determined by flame atomic absorption spector- photometry[73].Serum trace element levels were measured using serum diluted (1:2) with deionized distilled water, while Ca^{++} level was measured using serum diluted (1:10) with deionized distills water. The concentrations of the elements were determined by comparing the signal from diluted serum with that from appropriate aqueous calibration standards, which were prepared. Patients with metastases were excluded when calcium was determined.

Results and Discussion

Biochemical Constituents in Ovarian and Endometrial Cancer Patients

The levels of the biochemical constituents in the sera of patients studies were as summarized in Tables 5.4, 5.5, and 5.6. As can be seen from the Tables, all the biochemical constituents had higher mean values in patients as compared to the controls.

CEA levels of sera of ovarian and endometrial cancer patients were slightly elevated and most modestly elevated in endometrial cancer. CEA was elevated in 43% of ovarian cancer and 33% of endometrial cancer patients as compared to 7% in the control population Table (5.4). The incidence of antigen elevation in patients with invasive cancer was 25% higher than that in patients with stage I cancer. The higher levels of CEA were found in patients with mucinous ovarian neoplasms. CEA immunoreactivity also has been determined in the patient's tissues. The elevation of tissue CEA levels appeared to be statistically significant than in the corresponding sera samples ($p<0.05$) in ovarian malignant condition (Fig 5.3), while the difference was not significant in endometrial cancer. The results of determination of the CEA in the sera and tissues of gynecologic tumor patients indicated that CEA was not tumor specific and was found in some benign conditions. These results confirms previous observation from other investigator, where CEA activity in tissues can be demonstrated in most gynecological cancers while serum levels are often too modestly elevated to be useful for diagnosis or for monitoring. Stall et.al.,[255] showed slightly elevated plasma CEA level in approximately 50% of ovarian cancer patients , and mucinous tumors tend to have highest levels. Tuxen et.al., [26] also found elevated serum CEA levels in patients with mucinous histotypes ovarian cancer. Bast et.al., [39], and Tuxen et.al., [26][254] obtained that the combination of CEA and CA125 would have a higher sensitivity and specificity than CA125 alone, in monitoring human ovarian cancer. In the uterus, adenocarcinoma that arise in the endocervix express a greater amount of CEA than adenocarcinomas of the endometriums [251-253]. It is reflected in the reported frequency of elevated plasma CEA level:68% in cervical adenocarcinoma and 34% in endometrial adenocarcinoma, therefore, have therapeutic consequences when doubt exists as to the origin of a uterine tumor[40]. Squamous cell carcinoma of cervical cancers also express CEA and high levels indicated in advanced disease[41]. In immunohistochemical studies, CEA was detected in 60-63% of cervical squamous cell carcinoma , and in 77% of cases of cervical adenocarcinoma [256-258]. Generally, in gynecologic cancers a high pretreatment values of serum CEA indicate awide spread of tumor[99] and poor prognosis[251-253]. CEA have also been demonstrated in a large variety of malignancies included, cancers of the colon and pancreas and elevated concentration, varies with the tumor type and the stage of disease [249][250]. CEA also elevated in several non cancerous condition, such as smoking or hepatic dysfunction.

It is evident from the Table 5.5 that the mean levels of LDH activity were higher in cancer patients then those of healthy individuals but the differences were not statistically significant. The highest elevation of LDH activity was observed the invasive stage of ovarian cancer (stage I FIGO.133±24.343 U/L, and satageII+III FIGO.185.6±47.54U/L). The electrophoresis pattern of serum LDH isoenzyme in ovarian cancers (Fig 5.4) showed an additional band (extra band). This extra isoenzyme moved faster than LDH_1 which means that its charge is more negative than LDH_1.LDH, is elevated in the sera of patients with a variety of cancers[259]. Evaluation of serum levels of this enzyme was documented to be significantly advantageous in diagnosis and treatment monitoring of patients with gynecological cancers[43-47], and its level reflect tumor burden as well as tumor growth or regression .However LDH as single marker has not proved sufficient to meet the full requirements of clinical application. Increased serum LDH value in malignancies is multifunctional and no single explanation can be put forward of account for whole spectrum of malignancies. LDH is often elevated in epithelial ovarian cancer[259], cervical cancer[47] and many solid malignancies[17], making it a non-specific tumor marker. This isoenzyme provide marginal specificity for organ involvement[44]. The elevation of LDH_5 was associated with liver metastases and central nervous system metastases[17]. In dysgerminom, the two fast fraction(s)(LDH_1 and LDH_2) were increased and their activity appeared to parallel the response to therapy[45], LDH_3 was found to be elevated in multiple myeloma. Add ional band has been found in hepato cellular cancer patients[17].

The results of total AIP determination indicated that the mean values in cancer patients were higher than control and significantly different (Table 5.6). Different stages of ovarian and endometrial cancers appeared different levels of total AIP. The results of HSAP estimation showed that mean value in patients with ovarian and endometrial cancers were significantly increase compared to the control group (Table 5.6).

The HSAP levels did not correlate with tumor stage. This isoenzyme has been found to be more useful as tumor marker in gynecologic cancers than total AIP because the differences between cancer and control groups were clear. It has been reported that HSAP appeared in 21-22% of cases of cervical cancer and in 25% of cases of endometrial cancer[260][261]. Elevated serum HSAP has been found in 64% of ovarian cancer patients[49], but by using a radioimmunoassay and a polyclonal antibody for placental AIP determination, only 18% of cases showed elevated levels .HSAP has previously been reported to be useful in combination with lipid bound sialic acid (LSA), in diagnosis, and treatment monitoring for cancer patients [50][274]. The changes in the HSAP level were also reported in benign tumor of ovary[49][261].

Copper, Zinc, Calcium in the Sera of Ovarian and Endometrial Cancer Patients

The results of determination copper, zinc, and calcium ions concentration are illustrated in Table (5-7). These results showed that sera zinc levels were significantly higher ($p<0.05$) in ovarian and endometrial cancers groups than in the control group. In benign group, zinc levels remained without alteration. Serum copper levels were not significantly higher in the two cancer groups when compared with those of controls, although the mean values were higher in cancers groups. No alteration in the level of copper ion was observed in the sera of benign tumor as compared to healthy individual. The patients with ovarian and endometrial cancers showed no significant difference in their sera calcium levels from those of the control group. However, two patients of ovarian cancer and one of endometrial cancers were hypercalcemic (10.5, 11.1 and 10.6 mg/dl respectively).

High levels of zinc in the ovarian and endometrial cancer are in agreement with studies that indicate, the zinc level was increased in different cancer patients[275][265][261]. It has been suggested that the zinc might play a possible role in tumor growth promotion.

The copper in serum exists largely in the form of ceruloplasmin (more than 90%) a metalloprotein of Mwt (132 kDa) contain 6 atoms of copper/ molecule[275].

The copper levels of cancer sera in our study did not significantly change but some studies reported elevation in ceruloplasmin level in some cancer such as breast and lung cancers[256]. It is known that copper plays some role in angiogenesis, where it is required for development of new tissue as well as tumor growth possibly through the mediation of copper –dependent amine oxidases. The increase of zinc to copper ratio in the sera of cancer patients reflects the implication of the zinc and copper in the cancerous process of the tissue, so it may be useful as a tumor marker for the evaluation of the status of gynecologic cancers.

Although the mean values in the present study for ovarian and endometrial cancers patients showed no significant difference in their sera calcium levels, from those of the control group, but some cancer patients were hypercalcaemic. Hypercalcemia is a common complication of cancer and usually associated with bone metastasis[270]. Calcium is present in serum in three distinct fractions, free or ionized calcium account for about 50% of total calcium, about 5% complexed with variety of anions particularly phosphate and citrate, and the remaining 45% of Ca is bound to plasma protein especially to albumin[276]. Calcium homeostasis involves three major organs, the small intestine, the kidney and the skeleton. Hypercalcemia has been reported in breast, lung, kidney cancers without the presence of bone metastases[268]. PTH like peptides were found in about 49% of patients with malignant hypercalcaemia and 16% of patients with malignant normocalcaemic state. It have also been reported that cancer tissue in the human being was capable of direct osteolysis[277]. Rieke et.al.,[269] reported a hypercalcemic state (with the absence of bone destruction) in Hodgkins disease. They suggested elevated levels of 1,25 dihydroxy Vit D3. Hypercalcamia on malignancy was thought to be due to progression of the tumor tissue[277]. Hypercalcacmia has also referred to osteoclast activating factors which are released locally by tumor. These include prostaglandin E2[278], interlukin I, tumor necrosis factor[279] such as Beta lymphotoxin, epidermal growth factor [280], and transforming growth factor beta[281].

Table 5.4 The values of CEA levels in sera of patients with gynecologic tumors

(All details are explained in the text)

Tumor	Group	%of elevated CEA >5ng/ml *(Elevated /total)	Mean ± S.E.M of all patients ng/ml	Statistical *significant
Ovarian	*C	7(2/30)	3.96±1.25	
	*B	11(4/35)	4.11±1.79	*NS
	*M	43(6/14)	12.25±12.8	P<0.05
Uterine	C	7(2/30)	3.96±1.25	
	B	13(3/24)	3.97±1.76	NS
	M	33(3/9)	6.33±3.64	P<0.05

*C: control

*B: benign

*M: malignant

NS: not significant

Elevated/Total: number of patients with CEA above 5mg/ml/number of patients with CEA below 5mg/ml.

Table 5.5 Lactate dehydrogenase (LDH) activity in sera of patients with gynecologic tumors (All details are explained in the text)

Tumor	Group	*N	mean±S.E.M(U/ml)	Statistical significance
Ovarian	*C	30	144.78±32.5	
	*B	35	145.6±24.7	NS
	*M	14	156.57±28.8	NS
Uterine	C	30	144.78±30.5	
	B	24	143.93±26.5	NS
	M	9	146.51±67.34	NS

C:control
B:benign
M:malignant
NS:not significant
N:number of patients.

Table 5.6 AIP activity and HSAP activity in sera of patients with gynecologic tumors (All details are explained in the text)

Tumor	Group	*N	mean ±S.E.M of AIP U/L	Statistical significance	mean ±S.E.M. of HSAP U/L	Statistical significance
Ovarian	C	30	58.5±19.4		2.3±0.3	
	B			NS		NS
	M	3	60.3±1	P<0.05	2.2±0	P<0.05

Tumor	Type	N	mean±S.E.		mean±S.E.	
		5	2.3		.6	
		1	81.5±26.5		5.5±0.4	
		4				
Uterine	C	30	58.5±19.4	NS P<0.05	2.3±0.3	NS P<0.05
	B	24	65.9±14.6		2.4±0.9	
	M	9	78.5±32.3		4.6±0.5	

*M: Malignant

*B: Benign

* C control

*N: Number of patients

*NS: Not Significant

Table 5.7 Copper (Cu^{++}), Zinc (Zn^{++}) and Calcium (Ca^{++}) concentration in the sera of ovarian and endometrial cancers patients

Tumor	No. of patients	Copper mean±S.E.M µmol/l	Zinc mean± µmol/l	Zn^{++}/Cu^{++}	Calcium men±S.E.mg/100ml
Ovarian *M *B *C	8 10 25	17.3±1.5 16.6±1.6 16.7±0.71	13.3±3.2 7.4±1.1 7.1 ± 0.45	0.75 0.44 0.42	10.1±1.2 9.5±0.08 9.7 ± 0.7
Uterine	9	17.4±1.8	10.7	0.61	9.9±0.3

*M *B	10	17.1± 0.85	± 2.15 6.8 ± 1.9	0.4	9.6±0.16
*C	25	16.7± 0.71	7.1 ± 0.45	0.42	9.7±0.7

*M: Malignant
*B: Benign
*C: Control

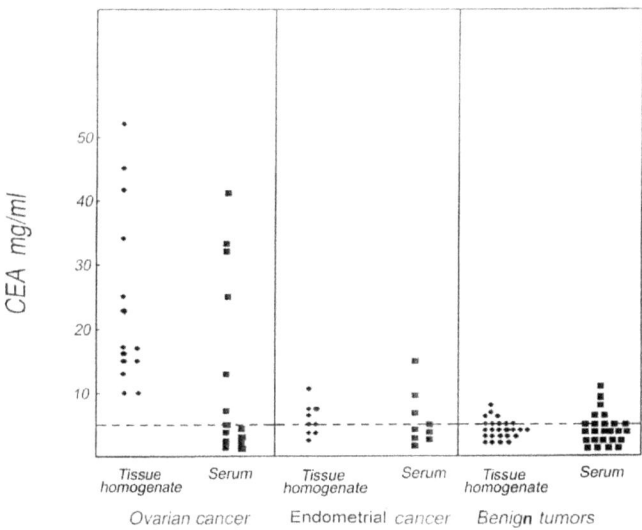

Fig   Distribution of the individual levels of the CEA (ng/ml) in the sera and tissues homogentes of benign and malignant gynecologic tumors

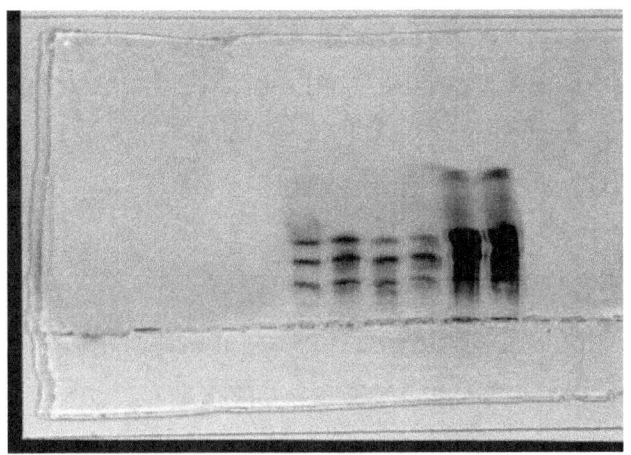

Fig. Electrophoresis pattern of serum LDH isonzymes of ovarian cancer patient
1, 2: control
3, 4: benign
5, 6: malignancy

Conclusions :

1- Determination of total serum intact hCG and free hCGβ or free hCGβ is of great clinical diagnostic value for ovarian and endometrial cancers than determination of only intact hCG.

2- Determination of intact hCG and free hCG in malignant tissues found to be of diagnostic value for gynecologic cancers.

3- The developed protocol for the assay of hCG-Rs is capable to analyze these receptor and the procedure is suitable for the assessment of such receptors in benign and malignant gynecologic tumors.

4- Purification of hCG receptor from ovarian cancer indicated that this receptor is functional in malignant conditions.

5- Determination of CEA immunoreactivities in cancer tissues may be useful for differentiating between gynecologic cancer.

6- LDH enzyme activity can be used for the assessment of the aggressiveness gynecologic cancers. Also LDH isoenzymes analysis may be useful for diagnostic these cancers , where it had been showed extra band different from normal serum.

7- HSAP isoenzyme could be used for the diagnostic of gynecologic cancer better than total ALP activity .

8- Serum Zinc to Copper ratio may be introduced as a biochemical marker in the diagnosis of gynecologic cancers .

The Future Work

According to the results obtained in this thesis, the flowing works are suggested for the future :

1- Investigation the free hCGβ subunit in more cases, and more type of gynecologic cancers such as cancer of cervix, vagina, and vulva .

2- Studies the expression of hCG and its subunits on the cell membrane of cultured human cancer cells using flow cytometric technique.

3- Preparation of monoclonal antibodies for free hCGβ subunit and for hCG-Rs to use them for hCG and its receptor identity as diagnostic tools for tumor .

4- Isolation of the m-RNAs encoding hCG-Rs and cloning the cDNA of these receptors , then comparison the purified receptors obtained in this study with those produced by the cDNA technology to elucidate the actual native species of hCG-R within the ovarian tumor.

5- Evaluation of the anti-hCG vaccine, which used for the prevention of pregnancy to prevent cancer. The studies must be including the adjuvant application of anti-hCG vaccines for the diagnosis and treatment of patients with cancer.

References:

1- Berek J.S.,Hacker N.F. eds. Practical Gynecologic Oncology. 3rd ed. Philadelphia: Lippincott Williams & Wilkins, 2000.

2- Results of IRAQI Cancer Registry. (a):1992-1994,(b)1995 - 1997. Cancer board Minsitry of health .IRAQI Baghdad; 1996 (a) , 1999 (b).

3- Eisner R. P. and Berek J.S. Gynecologic Cancers. In Dennis A., Casiato M.D. , Barry B. and Lowirz M.D. (eds) Manual of Clinical Oncology 4th ed. Philadelphia:lippincott Williams & Wilkins, 2000, P238-268.

4- Parazzini F. , Vecchia C.L. , Bocciolone L. and Franceschi S. 1991. The Epidemiology of Endometrial Cancer (Review). Gynecol . Oncol. 410:1-16.

5- Seiden M.V. 2001 Ovarian Cancer. The Oncologist 6:327-332.

6- Burke T.W., Luna G. T., Malpica A. Baker V.V. Whittaker L. and Mitchell M.F. 1996 Endometrial Hyperplasia and Endometrial Cancer. Gynecol Cancer prevention 23:411-456.

7- Ambros R.A., Ballouk F. Malfetano J.H. et.al. 1994 Significance of Papillary (Villoglandular) Differentiation in Endometrioid Carcinoma Am. J.Surg. Pathol .18:569-574.

8- Brinton L.A., Berman M.L., Mortel R., et.al. 1992 Reproductive , Menstrual and Medical Risk Factors for Endometrial Cancer Am. J.

Obstet. Gynecol .167:1317-1322.

9- Allan J. Ovarian Cancer. Clinical Symposia (Vol 48) New Jersey:Ciba- Geigy Corporation 1996.

10- Alper O., DeSantis M.L. Stambery K. et.al. 2000 Antisense Suppression of Epidermal Growth Factor Receptor Expression Alters Cellular Proliferation, Cell-adhesion , and Tumorigenicity in Ovarian Cancer Cell. Int. J. Cancer. 88:566-574.

11- Rosal J. ed. Ackermans Surgical pathology 8th ed. St. Louis (U.S.A): Mosby year book, 1996, p1461.

12- Tait Dl, Thigpen JT. Advances in Gynecologic Cancer Program and Abstracts of the 37th Annual Meting of the American Society of Clinical Oncology, San Francesico, California, May 12-15, 2001.

13- Vasey P.A. Atkinson R., Coleman R. et. al. 2001 Docetaxel-carboplatin as first line chemotherapy for Epithtial Ovarian Cancer. Br. J. Cancer 84:170-178.

14- Markman M. Kennedy A., Webster K.,et al. 2001 Combination Chemotherapy with Carboplatin and Docetaxelin in the Treatment of Cancer of the Ovary and Fallopian Tube and Primary Carcinoma of the Peritoneum. J. Clin Oncol. 19:1901-1905.

15- Williams C.J. 1998 Tamoxifen in Relapsed Ovarian Cancer: a systematic review .Int. J. Gyncol Cancer 8:89-94.

16- Markman M., Bookman M.A., 2000 Second- line Treatment of Ovarian Cancer. The Oncologist 5:26-35.

17- Onsrud M. 1991 Tumor Markers in Gynecologic Oncology. Scand J Clin Lab Invest. Suppl 206:60-70.

18- Daunter B. 1990 Tumor Markers in Gynecologic Oncology (Review) Gynecol. Oncol. 39:1-15.

19- Rademacher T.W., Parekh R.B., and Dwek R.A. 1988 Glycobiology Ann. Rev .Biochem. 57:785-838.

20- Kato H.,TorigoeT. 1997 Radioimmunoassay for Tumor Antigen of Human Cervical Squamous Cell Carcinoma. Cancer 40:1621-1624.

21- Senikjian E.K.,Young J.M., Weiser P.A. 1987 An Evaluation of Squamous Cell Carcinoma Antigen in Patient with Cervical Squamous Cell Carcinoma Am J Obstet Gynecol. 157:433-437.

22- Scambia G., Panici P.B., Baiocchi G., et.al. 1991 The Value of Sequamous cell Carcinoma Antigen in Patients with locally Advanced Cervical Cancer Undergoing Neadjuvant Chemotherapy. Am J obsret Gynecol 164:631-634.

23- Bast B. J., Klug T. L., St John E., et al. 1983 A Radioimmunoassay Using Monoclonal Antibody to Monitor the Course of Epithelial Ovarian Cancer. N Engl J Med. 309:883-887.

24- Bon G., Kenemans P., Verstraeten R. , Van Kamp G., and Hilgers J. 1996 Serum Tumor Marker Immunoassays in Gynecologic Oncology: Establishment of references values. Am J Obstet Gynecol 174:107-114.

25- Tuxen M.K., Soletormos G., Dombernowsky G. 1995 Tumor Markers in the Management of Patients with Ovarian Cancer. Cancer Treat. Rev. 21:215-245.

26- Tuxen M.K., Soletormos G., Petersen P., Schioler V., and Dombernowsky G., 1999 Assessment of Biological Variation and Analytical Imprecision of CA.125, CEA, and TPA in Relation to Monitoring of Ovarian Cancer. Gynecol Oncol. 74:12-22.

27- Niloff J.M., Klug A., Schaetzl E., et.al. 1984 Elevation of Serum CA.125 in Carcinoma of the Fallopian Tube, Endometrium, and Endocervix. Am J Obsret Gynecol 148:1057-1059.

28- Abrao M.S., Podgaec S., Filho B.M., Ramos L.O., Pinotti J.A., and Oliveira R.M. 1997 The use of Biochemical Markers in the Diagnosis of Pelvic Endometriosis. Human Reproduction 12:2523-2527.

29- Buamah P.K.,,,,, Skillen A.W. 1994 Serum CA.125 Concentration in Patients with Benign Ovarian Tumors. J. Surg Oncol 56:71-74.

30- Jacobs I.J., Fay T.W., Lower A.M. 1989. The London Hospital Ovarian Cancer Screening Project. Proc 2nd meeting Int. Gynecol Cancer Soc, Toronto :167.

31- Onsurd M., Vergoto I., Moen M. 1988 CA-125 in Peritoneal Fluids of Healthy Women and of Patients with Ovarian Carcinoma .Proc 24th meeting Scand Soc Obstet Gynecol, Trondheim:49.

32- Koprowski H., Steplewski Z., Mitchell K., et.al. 1979 Colorectal Carcinoma Antigens Detected by Hybridoma Antibodies. Somatic Cell Genet 5:957-961.

33- Hilkens J., Bulis F., Hilgers J, et.al. 1984 Monoclonal Antibodies Against Human Milk–fat Globule Membranes Detection Deferential Antigens of the Mammary Gland and its tumor. Int J. Cancer 34:197-200.

34- Scambia G., Benedetti P., Baiocchi G., et.al. 1988 CA 15-3 as aTumor Marker in Gynecological Malignancies. Gynecol Oncol 30:265-273.

35- Bon G., Keneman P., Vankamp G.J. Yedema C.A., and Hilgers J 1990 Review on the Clinical Value of Polymorphic Epithelial Mucin Tumor Marker for the Management of Carcinoma Patients. Nucl Med Allied Sci. 34:151-162.

36- Yedema C., Kenemans P., Wobbes T., 1991 Carcinoma – Associated Mucin Serum Marker CA M26 and CA M29: efficacy in detecting and monitoring patients with caner of the breast,colon,ovary, endometrium and cervix. Int J. Cancer 47:170-179.

37- Knauf S. 1988 Clinical Evaluation of Ovarian Tumor NB/70K: monoclonal antibody assays for distinguishing ovarian cancer from other gynecologic disease. Am J Obstet. Gynecol. 158:1067-1071.

38- Petru E., Sevin B., Averette H.E., 1990 Comparison of Three Tumor Markers –CA-125, Lipid associated Scialic Acid (LSA), and NB /70K. in Monitoring Ovarian Cancer. Gynecol Oncol.38:181-185.

39- Bast R.C., Klug T.L., Schaetzl E., et. al. 1984 Monitor Human Ovarian Carcinoma with a Combination of CA-125, CA 19-9 and a CEA. Am J. Obstet Gynecol 149:553-558.

40- KjQrstad K.E., Rjaster H. 1977 Studies on Carcinoembryonic Antigen level in Patients with Adenocarcinoma of the Uterus. Cancer 40:2953-55.

41- KjQrstad K.E., Rjaster H. 1977 Carcinoembryonic Antigen levels in Patients with Squamous cell Carcinoma. Obstet Gynocol 51:293-296.

42- Genest D., Sheets E., Lage J., et.al. 1994 Flow Cytometric Analysis of Nuclear DNA Content in Endometrial Adenocarcinoma .A Typical Mitoses are Associated with DNA Aneuploidy. Am J. Clin pathol 102:341-346.

43- Bishop A.J. , and Schiestl R.H.,2001. Homologous Recombination as a Mechanism of Carcinogenesis. Biochimica et. Biophysical Acta 1471:109-121.

44- Jones M.H, Koi S. and Fujimota I 1994 Allelotype of the uterine Cancer by Analysis of RFLP and Microsatellite Polymorphisms :frequent loss of heterozygosis on chromosome 3P, 9q, 10q, and 17P, Gene Chromosome cancer 9: 119-125.

45- Kacinski B.M., Carter D., Mittal K. et al. 1988 High level of Expression of fms Proto-oncogene mRNA is Observed in Clinically Aggressive Human Endometrial AdenoCarcinoma Int J. Radiat Oncol Biol Phys. 15:823-826.

46- Baiocchi G., Kavanagh J., Talpaz M, et.al. 1991 Expression of the Macrophage Colon- Stimulating factor and its Receptor in Gynecologic Malignancies. Cancer 67:990-996.

47- Urban J.L. 1999 Oncogenes, Cancer and Imaging J. Nuclear M. 40:498-504.

48- Takai N., Miyazaki T., Fuijsawa K., et.al.2001 Expression of Polo-like Kinase in Ovarian Caner is Associated with Histological Grade and clinical Stage. Caner letters 164:41-49.

49- Warburg O. 1956. On the Origin of Cancer Cells Science 123:309-314.

50- Gerorge M.A. 1978 Carcinoma of the Ovary and Serum Lactic Dehydrogenase levels Surg Gynecol Obstet 146:893-896.

51- Wood D.C. , Verla V., Plamqist M., and Weber F., 1973 Serum Lactic Dehydrogenase and Isoenzyme Change in Clinical Cancer J. Surg Oncol. 5:251-257.

52- Fujii S., Koniski I., Suzuki A.1985 Analysis of Serum Lactic Dehydrogenase Levels and Its Isoenzymes in Ovarian Dysgerminoma .Gynecol Oncol .22:65-68.

53- Das H.K.,Chakravorty M., and Sanyal B. 1985 Serum Enzymes as Biochemical Markers in the Diagnosis and Prognosis of Carcinoma of the Cervix Uteri. Ind J. Cancer 22:121-131.

54- Patel P.S. , Rawal G.N., Balar B.D. 1993 Importance of serum Sialic Acid and lactate Dehydrogenase in Diagnosis and Treatment Monitoring of Cervical Cancer Patients. Gynecol. Oncol 50:294-299.

55- Fishman W.H. Inglis N.R. , Green Setal . 1968 Immunology and Biochemical of Regan Isoenzyme of Alkaline Phoshatase in Human Cancer. Nature 219:697-700.

56- Vergote I. Onsrud M., and Nustad K. 1987 Placental Alkaline Phoshatase as a Tumor Marker in Ovarian Cancer . Obstet Gynecol 69:228-231.

57- Patel P.S. , Baxi B. R., Adhvaryn S. G. , and Balar D.B. 1990. Evaluation of Serum Sialic Acid , Heat stable Alkaline phosphatase and Fucose as Markers of Breast carcinoma . Anti Cancer Res. 10:1071-1074.

58- Lappohn R. , Burger H., Krans M. 1991. Inhibin as a Marker for Gronulosa Cell Tumor, Proc Symposium Scand Soc Gyncol Oncol, Gothenburg.

59- Hirose T. 1920 Exogenous Stimulation of Corpus Luteum Formation in the Rabbit; Influence of Extracts of Human Placenta, Decidua, Fetus Hydaditimole and Corpus Luteum on the Rabbit Gonad.J. Jpn Gynecol Soc 16:1055.

60- Ashheim S. and Zondek B. 1928 Die Schwangerschafsdiagnose ausdem Harn durch Nachwies des Hyprophysenvorderlappen –hormone. Praktishe und theoretische Ergebnisse aus den harnuntersuchungen .klin wochenschr 7:1453-1457.

61- Pierce J.C., and Parsons T.F. 1981 Glycoprotein hormones:Structure and Function. Annu. Rev. BiO chem. 50:465-495.

62- Hoermann R, Spoettl G, Moncay O.R, Mann k. 1990 Evidence for the presence of human Chorionic Gonadotropin (hCG) and Free ß- subunit of hCG in the Human Pituitary. J.Clin Endocrinol Metab 71:179-186.

63- Jameson J.L. and Hollenberg A.N. 1993 Regulation of Chorionic Goandotropine Gene Expression. Endocr Rev.14:203-221.

64- Bellet D., Lazer V., Bieche I., Paradis V. , et. al 1997 Malignant Transformation of Nontrophoblastic Cells is Associated with the Expression of Chorionic Gonadotropin ßGenes Normally Transcribed in Trophoblastic Cells .Cancer Res. 57:516-523.

65- Ryan R.J., Charlesworth C., Mc Cormick D.J. et.al. 1988 The Glycoprotein Hormones Recent Studies of Structure Functions Relationship FASEB J. 2:2661-2669.

66- O'conner J. F., Birken S., Lustbader J.W., et al 1994 Recent Advances in the Chemistry and Immunochemistry of Human Chorionic Gonadotropin. Impact on Clinical Measurement. Endocrine Reviews 15:650-683.

67- Carlsen R.B., Bahl O.P. Swaminathan N., 1973 Human Chorionic Gonadotropin liner Amino Acid Sequence of the ß Subunit J. Biol Chem.. 248:6810-6827.

68- Bellisario R., Carlsen R.B., and Bahl O.P. 1973 Human Chorionic Gonodotropin linear Amino Acid Sequence of the Alfa subunit. J. Biol Chem. 248:6796-6809.

69- Nisula B., Connor J. 1993 Standardization of Protein Immunoprocedures choriogonadotropin (CG). Scand J. Clin Lab Invest .53 (Suppl 216):42-78.

70- Thotakura N.R. Weintraub B.D., and Bahl O.P. 1990 The Role of Carbohydrates in Human Choriogonadotropin (hCG) Action Effects of N-Linked Carbohydrate chains from hCG and other Glycoproteins on Hormonal Activity. Molecular and Cellular Endocrinology 70:263-272.

71- Shao K. Balasubramanian S.V., Pope C.M. Bahl O.P. 1998 Effect of Individual N-Glycosyl Chains in the ß-Subunit on the Conformation of Human Choriogonadotropin. Molecular and Cellular Endocrinology .146:39-48.

72- Lustbader J. W., Yarmush D.L. Birken S., et.al. 1993 The Application of Chemical Studies of Human Chorionic Gonadotropin to Visualize its Three Dimensional Structure. Encoder.Rev 14:291-311.

73- Moyle W.R., Matzuk. M. M., Campbell P.K. et.al.1990 Localization of Residues that confer Antibody Specificity Using Human Chorionic Gonadotropin Luteinizing Hormone ß-subunit Chimera and Mutants J. Biol Chem. 265:8511-8518.

74- Berger P. Bidart J.M., Delves P.S., Dirnhofer S., et.al. 1996 Immunochemical Mapping of Gonadotropins. Molecular and cellular Endocrinology 125:33-43.

75- Shao K., Purohil S., Bahl O.P 1996 Effect of Modification of all loop Regions in α and ß- subunits of human Choriogonadotropin on its signal Transduction Activity. Molecular and Cellular Endocrinology 122:173-182.

76- Wu H., Lustbader J.W., Liu Y., Canfield R.E., and Hendrickson W.A., 1994 Structure of Human Chorionic Gonadtropin at 2.6 A° Resolution from MAD Analysis of the Selenomethionly Protein. Structure 2:545-558.

77- Lapthorn A.J. Harris D.C. Littejohn A., Lustbader J.W., et.al. 1994 Crystal Structure of Human Chorionic Gonadotropin. Nature 369:455-461.

78- Lobel L. Pollak S., Wang S. Chancy M., and Lustbader J.W., 1999 Expression and Characterization of recombinant beta subunit hCG Homodimer. Endocrinology 10:261-70.

79- Bidart J. M.,and Bellet D. 1993 Human Chorionic Gonadotropin Molecular Forms, Detection, and Clinical Implications Trends. Endocrinol Metab. 4:285-291.

80- Odell W.D., Griffin J., Sawitzke A., 1990 Chorionic Gonadotropin Secretion in Normal Nonpregnent Humans. Trends Endocrinol Metab. 1:418-421.

81- Alfthan H., Haglund C., Dabek J. ,Stenman U.H. 1992 Concentration of Human Choriogondotropin, Its ß-subunit, and the Core fragment of the ß-Subunit in Serum and Urine of men and nonpregnant women. Clin chem 38:1981-1987.

82- Cole L.H. 1997 Immunoassay of hCG, Its Free Subunits and Metabolites. Clin chem. 43:2233-2243.

83- Skarulis M., Wehmann R.E. Nisula B.C., and Blithe D.L. 1992 Glyecosylation Changes in Human Chorionic Gonadotropin and Free α Subunit as Gestation Progresses. J. clin Endocrinol Metab.75:91-96.

84- Cole L.A., Tanaka A., K.m G.S., Park S-Y, et.al.1996 Beta Core Fragment (ß -core/UGF/UGP) a Tumor Marker :seven year report. Gynecol Oncol. 60:264-270.

85- Cole L.A. , Kadana A., Park S-Y., and Braunseten 1993 The Deactivation of the hCG by Nicking and Dissociation. J. Clin Endocrinol Metao 76:704-710.

86- Kardana A., Cole L.A., 1994 Human Chorionic Gonadotropin ß-Subunit Nicking Enzymes in Pregnancy and Cancer patients serum J. Clin Endocrinol Metab 79:761-767.

87- Birken S., Krichevsky A., O'Connor., SchlattererJ., et.al. 1999 Development and characterization of Antibodies to a Nicked and Hyperglycosylated Form of hCG from a Choriocarcinoma patient. Endocrine J. 7:15-24.

88- Ichikawa N.Zhai Y.L.,Shio zawa T.,Koki T.,Nogchi H., Nikaido T., Fujii S.2001 Immunohistochemical Analysis of Cell

Cycle Regulation Gene Production in Normal Trophoblast and Trophoblastic Tumor. Int. J.Gynecol. Pathol 17:235-240.

89- Nilson J.H., Bokar J.A.,Clay C.M., Farmeric T.A., et.al. 1991 Different Combination of Regulatory Elements may Explain Why Placenta –Specific Expression of the Glycoprotein Hormone α-Subunit Gene Occurs Only in Primates and Horses .Biol Reprod 44:231-237.

90- Spencer K., Coombes E.J., Mallard A.S., and Ward A.M. 1992 Free Beta Human Choriogonadotropin in Down's Syndrome Screening :a Multicentre Study of its Markers .Ann Clin Bio Chem. 29:506-518.

91- Cole L.A., Shahabi., Rinne K.M., et.al. 1999 Urinary Screening Tests for Fetal Down Syndrome Hyperglycosylated hCG. Prenat Diagn 15:24-30.

92- Nishimura R., Koizume T., Yokotani T., et.al.1998 Molecular Heterogeneity of hCG-Related Glycoprotiens and the Clinical Relevance in Trophoblastic and Nontrophoblastic Tumors Int., J. Gynecol Obstel., 60 (Suppl 1):529-532.

93- Zon R.T., Nichol C.,Einhorn L.H. 1998 Management Strategies and Outcomes of Germ Cell Tumor Patients with Very High Human Chorionic Gonadotrophin Levels J.Clin Oncol 16:1294-297.

94- Elliott M., Kardana A., Lustbader J., Cole L.1997 Carbohydrate and Peptide Structure of the Alpha –and Beta –subunits of hCG from Normal and Aberrant Pregnancy and Choriocarcinoma. Endocrine 7:15-32.

95- Gregory J.J, and Finlay J.L 1999 Alpha –Fetoprotein and Beta –Human Chorionic Gonadotropin: their clinical significance as tumor markers. Drugs 57: 463-467.

96-Suzuki K., Nakazalo H., Kurokava K. Suzaki T., and Yamanaka H.1999 Treatment of stage I Seminoma:should beta hCG Positive Seminoma be Treated Aggressively. In. urol. Nephrol 30:593-598.

97-Rutanen E.M.And Seppala M.1978 The hCG-Beta Subunit Radioimmunoassay In Nontrophoblastic Gynecologic Tumors. Cancer 41:692-696.

98-Tormey D.G. Waalkes T.P., and Simon R.M. 1977 Biological Markers in Breast Carcinoma II Clinical Correlation with Human Chorionic gonadotrophin. Cancer 39:2391-2396.

99- Donaldson E.S., Van Nagell J.R. , Pursell S., Gay E.C., Meeker W.R, Kashmiri R. and Voorde J. 1980 Multiple Biochemical Markers in Patients with Gynecologic Malignancies. Cancer 45:948-953.

100- Mc Manus L.M., Naughton M.A. and Martinez H.A. 1976 Human Chorionic Gonadotrophin in Human Neoplastic Cells. Cancer Res. 36:3476-3481.

101- Sheth N.A., Adil M.A., Nadkarni J.J., Rajpal R. M., and Sheth A.R. 1980 Inappropriate Secretion of Placental Lactogen and β Subunit of Human Chorionic Gonadotrophin by Cancer of the Uterine Cervix. Gynecol oncol. 11:321-329.

102- Birkenfeld S., Noiman G., Krispin M., Schwarz S. and Zakut H. 1989 The Incidence and Significance of Serum hCG and CEA in Patients with Gastrointestinal Malignant Disease. Eur .J. surg. On col. 15:103-108.

103- Marcilliac I., Troalen F., Bidart J.M.,Ghillami P.,Ribrag V., Escudier B., Malassagne B. et al 1992 Free Human Chorionic Gonadotropin βSubunit in Gonadal and Nongonadal Neoplasms. Cancer Res. 52:3901 – 3907.

104- Alfthan H., Haglund C. Roberts P., and Stenman U.H. 1992 b. Elevation of Free β subunit of Human Choriogonadotropin and Core β

Fragment of Human choriogonadtropin in the serum and urine of Patients with Malignant Pancreatic and Biliary Disease. cancer Res. 52:4628-4633.

105-Iles R.K., Jenkins B.J. Oliver R.T. Blandy J.P., and Chard T. 1989 ß Human Chorionicgonadatrophin in Serum and Urine. A Marker for Metastatic Urothelial Cancer. Br J urol 64:241-244.

106- Hoermann R.,Gerbes A.L., Spoettl G., Jungst D., and Mann K. 1992 Immunoreactive Human Chorionic Gonadotropin and Its Free β Subunit in Serum and Ascits of Patients with Tumors. cancer Res. 52:1520-1524.

107- Grossmann M., Hoermann R.,Gocze P.M., Ott M.,Berger P., and Mann K. 1995 Measurement of Human Chorionic Gonadotropin Related Immunoreactivity in Serum, Acites ,and Tumor Cysts of Patients with Gynecologic Malignancies .Eur J. clin Inv. 25:867-873.

108- Ind T., Iles R.K., Shepherd J., and Chard T., 1998 Serum Concentration of Cancer Antigen 125, Placental Alkaline Phophatase, Cancer Associated Serum Antigen and Free Beta Human Chorionic Gonadotrophin as Prognostic Markers for Epithelial Ovarian Cancer Br. Obstet. Gynaecol 104:1024-1092.

109- De-Bruijn H.W., Ten-Hoor K.A., Krans M., and Vander-Zee A.C. 1997 Risinf Serum Values of β - Subunit Human Chorionic Gonadotropin (hCG)in Patients with Progressive vulvar Carcinomas Br.J.cancer 75:1217-1218.

110- Syrigos K.N., Fyssas I.,Konstandoulakeis M.M. Harvington K.J.,Popadapouls S. et al. 2001 Beta Human Chorionic Gonadotropin Concentration in Serum of Patients with Pancreatic Adenocarcinoma.Gut 42:88-91.

111- Grow ford R.A., Iles R.K. Carter P.G., Caldwell C.J., Shepherd J.H. ,and Chard T. 1998 The Prognostic Significance of

Beta Human Chorionic Gonadotropin and Its Metabolites in Women with Cervical Carcinoma. J. clin pathal 51:685-688.

112-Schulter E.M. Mulder C., Van-Kamp G.J., and Keneman S.P., 1997 The Beta–Core Fragment Chorionic Gonadotropin :Biochemical background and Clinical Application . Anticancer Res., 17:1255-1272.

113- Schwarz R.U., Petzoldt B., Waldschmidt. et.al., 1997 UGP A Tumor Marker of Gynecologic and Breast Malignancies Specificity and Sensitivity in Pretherapeutic Patients and the Influence of Hormonal Substitutions on the Expression of UGP .Anti Cancer Res 17:3041-3045.

114- Acevedo H.F., Krichevsky A., Campbell -Acevedo E.A., Galyon J.C. ,Buffo M.J., and Hartsock R.J. 1992 Flow Cytometry Methods for the Analysis of Membranes Associated Human Chorionic Gonadotropin , Its subunits, and Fragments on Human cancer cells, Cancer 69:1818-1828.

115- Acevodo H.F. ,Tong J.Y. and Hartsock R.J. 1995 Human Chorionic Gonadotropin-beta Subunit Gene Expression in Cultured Human Fetal and Cancer Cells of Different Types and Origins. cancer 76:1467-1475.

116- Sztumowicz M.SlodKowsta J. Zych J., Rudzinski P., Sakowicz A., Rowinskas 1999 Frequency and Clinical Significance of Beta Subunit Human Chronic Gonadotropin Expression in Non-small lung Cancer Patients tumor Biol 20:99-104.

117-Doi F. , Chi D.J., Charuwork B.B., Conrad A.J. Russel J. Morton D.L. 1996 Detection of ß- Human Chorionicgonadotropin mRNA as a Marker for Coetaneous Malignant Melanoma. Imt. J. cancer 65:454-459.

118- Lazar V., Diez S.G., Laurent A., Giovanrandi I., Radvanyi F., Chopin D., Bidaart J.M. Bellet D. and Vidaud M. 1995 Expression of Human Chorionic Gonadotropin ß Subunit Genes in Superficial and Invasive Bladder Carcinoma. cancer res. 55:3735-3738.

119- Regelson W. 1995 Have We Found the (Definitive Cancer Biomarker)? The Diagnostic and Therapeutic Implication of Human Chorionic Gonadotropin Beta Expression as a Key to Malignancy. Cancer 76:1299-1301.

120- Bhalang K., Kafrawy A.H., and Miles D.A. 1999 Immunohistochemical Study of the Expression of Human Chorionic Gonadotropin Beta in Oral of Sequamous Cell Carcinoma cancer 85:757-762.

121- Yokotan T. Koizumi T. , Tamiguchi R., Nakagawa T., Isobet., et.al. 1997 Expression of Alpha and Beta Genes of Human Chorionic Gonadotropin in Lung Cancer . Int.J. cancer 71:539-544.

122- Bieche I,, Lazer V., Nogues C., Poynand I., Giovangrand Y. ,Bellet D. et.al. 1998 Prognostic Value of Chorionic Gonadotropin Beta Gene Transcripts in Human Breast Carcinoma Clin. Cancer Res 4:671-676.

123-Dirhnofer S., Koessler P. , Ensinger C., Feichtingerl T., Madersbacher S., and Berger P., 1999 Production of Trophoblastic Hormones by Transitional Cell Carcinoma of the Bladder Association to Tumor Stage and Grade. Hum Pathol 29:377-382.

124- Piecioli M., Gamberi B., Pasquinelli G., Poggi S., Ascani S. et.al. 2000. Beta hCG Aberrant Expression in Primary Mediastinal Large ß-Cell Lymphoma. Am J. Surg pathol 23:717-721.

125- Acevedo H.F., Hartsockk R.J. 1996 Metastic Phenotype Correlation with High Expression of Membranes Chorionic Gonadotropin in vivo. Cancer 78:2388-2399.

126- Gebauer G., Muller-Ruchholtz W. 1997 Tumor Marker Concentration in Normal and Malignant Tissue of Colorectal Cancer

Patients and their Prognostic Relevance. Anti cancer Res. 17:2731-2734.

127- Walker R.A. 1978 Significance of ß- subunit hCG Demonstrated in Breast Carcinoma by the Immunoperoxidase Technique J. clin pathol. 31:245-249.

128- Mohabeer J. Buckly C.H., and fox H. 1983 An Immunohistochemical Study of the Incidence and Significance of Human Chorionic Gonadotrophin Synthesis by Epithelial Ovarian Neoplasms. Gynecol oncol. 16:78-84.

129- Oyasu R. Nan L., SmithD.P., and Kawanata H. 1994 Human Chorionic Gonadotropin β-Subunit Synthesis by Undifferentiation Urothelial carcinoma with syncytiophoblastic differentiation. Arch. Pathol. Lab. Med 118:715-717.

130- Campo E., Palacin A., Benasco C. Quesada E., and Cardesa A. 1987 Human Chorionic Gonadotrophin in Colorectal Carcinoma. Cancer 59:1611-1617.

131- Yakeishi Y., Mori M., and Enjoji M. 1990 Distribution of β-human Chorionic Gonadotropin –Positive Cells, in Non Cancerous Gastric Mucosa and in Malignant Gastric Tumors cancer 66:695-701.

132- Collins R.J. and wong L.C. 1989 Adenocarcinoma of the Uterine Cervix with βhCG Production: A case report and review of the literature. Gynecol. Oncol 33:99-107

133- Morinaga S., Yamaguchi M., Watanabe I., Kasai M., Ojima M., and Sasano N. 1983 An immunocytochemical Study of Hepatablastoma Producing Human Chorionic Gonadoropin.

134- Buckly C.H. and Fox H. 1979 An Immunohistochemical Study of the Significance of hCG Secretion by Large Bowel Adenocarinoma .J.clin pathol 32:368-372.

135- Iles R.K. and Chard T. 1989 Immunochemical Analysis of the Human Chorionic Gondotropin Like Material Secreted by Normal and Neoplastic Urothlial Cells. J. mol Endocrinol 2:107-112.

136- Yoshimoto Y., Wolfsen A.R., and Odell W.D. 1979 Human Chorionic Gonadotropin like Material :presence in normal human tissue Am.J.Obstet .Gynecol 134:729-733.

137- Braunstein G.D., Kamdar V.,and Rosor J., Swomina than N., and Wade M.E. 1979 Widespread Distribution of a Chorionic Gonadotropin –like Substance in Normal Human Tissue. J. clin Endocrial Metab 49:917-925.

138- Gillott D.J., Iles R.K, and chard T 1996 The Effects of Beta Human Chorionic Gonadotrophin on the Invitro Growth of Bladder Cancer Cell lines Br. J. cancer 73:323-326.

139- Muckherjee K., and Das S.A. 1984 Placental Glycoprotein , Human Chorionic Gonadotropin as a Growth Stimulant of Murine Tumor Cells. IRCS med Sci :12 1101-1102.

140- Melmed S. . and Braunstien G.D. 1983 Human Chorionic Gonadotropin Stimulated Proliferation of Nb2 rat Lymphoma cell J. clin. Endocrinol. Metab. 1068-1070.

141-Zygmunt M. ,Hahn D., Munstedt K., Bischof P. Lana U. 2000 Invasion of Cytotrophoblastic JEG.3 Cells is stimulated by hCG in Vitro. Placenta 19:587-593.

142-Acevedo H.F. Raikow R.B., Powell J.E., and Stevens V. 1987 Effects of Immunization Against Human Choriogondotropin on the Growth of Transplanted Lewis Lung Carcinoma and Spontaneous Mammary Adenocarcinoma in Mice: Cancer Detect Prev (Suppl 1): 477-486.

143- Triozzi P.L. , Gochnour D., Martin E.W., and Aldrich W. et. al. 1994 Clinical and Immunological Effects of Synthetic Beta Human Chorionics Vaccine. Int J. Oncol 5:1447-1453.

144- Syner L., Woo D., Triozzi P.L. ,and Stevens V., Synthetic Hormone /Growth Factor Subunit Vaccine with Application to Anti-fertility and Cancer . In:Powell M.F., and Newman M.J., editors

Vaccine Design :the Subunit Approch. New York: plenum Publishing Corp 1995, P907-930.

145- Dufau M.L., Pock R., Neubarer A., Catt K.J. 1976 Vitro Bioassay of LH in Human Serum: Assay .J Clin Endocrinol Metab. 42:958-969.

146- Van Damme M.P., Robertson D.M., and Dicz Falusy E. 1974 An improved in Vitro Bioassay Methods for Measuring Luteinzing Hormone (LH) Activity Using Mouse Leydig Cell Proportions Action. Endocrinol 77:655-671.

147- Jia X.C., Olkawa M., Bo M., Tanaka T., et.al. 1991 Expression of Human Luteinizing Hormone (LH) Receptor Interaction with LH and Chorionic Gonadotropin from Human but not Equine, rate and Ovine Species Mol. Endocrinol 5:759-768.

148- Vaitukaitis J.L.,Braunstien G.D., and Ross G.T. (1972) Radioimmunoassay Which Specifically Measures Human chorionic Gonadotropin in the presence of Human Luteinizing Hormone Am J. Obstet Gynecol 113:751-758.

149- Segaloff D.L. ,Sprengel R., Nikolics K., and Accoli M. Structure of the Lutropin Choriogondotropin receptor. In Clark J.H(e.) Recent Progress in Hormone Research Vol. 46, A academic Press, INC. San Diego, 1990 , P261-303.

150- Ji T.H., Grossmann M., and Ji J.G. 1998 G Protein -Coupled Receptors 1.Diversity of Receptor Ligand Interaction. J. Biol. Chem, 273:17299-17302.

151- Herrlick A., Kuhn B., Grosse R.et.al. 1996 Involvement of Gs and Gi Proteins in Dual Coupling of the LH Hormone Receptor to Adenyl Cyclase and Phospholipase C.J. Biol .chem 271:16764-16772.

152- Gudermann T., Birnhaumer M., and Birnbaumer L. 1992 Evidence for Dual Coupling of the Murine Luteinizing Hormone

Receptor to Adenyl Cyclase and Phosphinositide Break Down and Ca^{+2} Mobilization J. Biol Chem 267:4479-4488.

153- Ekstrom R.C. , and Hunzicker D.M. 1989 Homologous Desension of Ovarian LH/hCG Receptor Responsive Adenyl Cyclase is Dependent Upon GTP. Endocrinology 124:956-963.

154- Segaloff D.L., and Ascoli M. 1993 The Lutropin / Choriogonadotropin Receptor 4 Years Later Endocr Rev. 14:324-347.

155-Dias , J.A. 1992 Recent Progress in Structure Function and Molecular Analysis of the Pituitaryl Placental Glycoprotein Hormone Receptors .Biochem Biophys Acta. 1135:278-294.

156-Bruch R.C., Thotakura N.R., and Bahl O.P. 1986 The Rat Ovarian Lutropin Receptor: Purification, Hormone Binding Properties, and Subunit Composition. J. Biol. Chem. 261:9450-9460.

157- Dattatreyamurty B., Rathnam P., and Saxena B.B., 1983 Isolation of the Luteinzing Hormone -Chorionic Gonadotropin Receptor in High Yield from Bovine Corpora Lutea Molecular Assembly and Oligomeric Nature. J., Biol Chem. 258:3140-3158.

158- Pandian M.R., and BaHl O.P. 1977 Labeling of Bovine Corpus Luteal Membranes Human Chorionic Gonadotropin or Luteinzing Hormone (hCG/LH) Receptor and its Purification and Properties Arch. Bioch. Bioph 182:420-436.

159- Koo Y.B., Ji I., Slaughter R.G., and Ji T.H., 1991 Structure of the LH Receptor Gene and Multiple Exone of the Coding Sequence Endocrin 128:2297-2310.

160- Loosfelt H. Misrahi M., Atger M. et.al. 1989 Cloning and Sequencing of Porcine LH-hCG receptor CDNA: variants lacking transmembrance domine. Science 245:525-528.

161- Thomas D.M., and Segaloff D.L. 1994 Hormone Binding Properties and Glycoslation Pattern of a Recombinnant form of the Extracellular Domain of the LH/CG-R Expressed in Mammalian Cells .Endocrin 135:1902-1912.

162- Hong S., Phang T., Ji I., and Ji T.H. 1998 The Amino Terminal Regions of the Lutaining Hormonal Choriogonadotrophin Receptor Contacts Both Subunits of Human Choriogonadotropin I-Mutational Analysis J. Biol .chem 273:13835-13840.

163- Kanada M., Jablonka S.A., Sato A. et.al 1999 Genetic Fusion of an Alpha -Subunit Gene to the Follicle Stimulating Hormone and Chorionic Gonadotropin Beta Subunit Genes :Production of a bifunctional Protein .Mol Endocrinol 13:1873-1881.

164- Bhowmick N., Huang J., Puett D., Isaacs N. W., and Laphthorn A.J., 1996 Determination of Residues Important in Hormone Binding to the Extracellular Domain of the Luteinizing Hormone /Chorionic Gonadotropin Receptor by site- Directed Mutagenesis and Modeling. Mol Endocrine .10:1147-1159.

165- Nagayama Y., Wadsworth H.L. Russo D., Seto P., and Rapoprt B., 1991 Thyrotropin Luteinizing Hormone/ Chorionic Gonadotropin Receptor Extracellular Domain Chimera as Probes for Thyrotrophic receptor Function. Proc Natl Aced Sci USA 88:902-905.

166- Roche P.C. Ryan R.J. and McCormick D.J. 1992 Identification of Hormone Binding Regions of the Luteinizing Hormone Human Chorionic Gonadotropin Receptor Using Synthetic Peptides. Endocrinol 131:268-274.

167- Davis D.P. , Rozell T.G., Liu X., Segaloff D.L.1997 The Six N-Linked Carbohydrates of the lutropin/ Choriogonadotropin Receptor are not Absolutely Required for Correct Folding, Cell Surface Expression, Hormone Binding or Signal Transduction. Mol. Endocriol 11:550-562.

168- Braun T., Schofield P-R., and Sprengel R. 1991 Amino-Terminal Leucine -Rich Repeats in Gonadotropin Receptors Determine Hormone Selectivity EMBO J. 10:1885-1890.

169- Kobe B., and Deisenhofer J. 1994 The Lucien Rich Repeat :A Versatile Binding Motif. Trends Biochem Sci. 19:415-421.

170- Kobe B., and Deiesenhofer J 1995 A Structural Basis of the Interaction Between Lucien -Rich Repeats and Protein Ligands. Nature 374:183-186.

171- Ji T.H., Murdoch W.J. and Ji I 1995 Activation of Membranes Receptors Endocrine 3:187-194.

172- Moyle W.R., Campbell R.K., Raos N.V. et.al. 1995 Model of Human chorionic Gonadotropin and Lutropin Receptor Interaction that Explains Signal Transduction of the Glycoprotein Hormones J. Biol. Chem .270:20020-20031.

173- Phang T., Kunda G., Hong S., Ji I., and J. T., 1998 The Amino Terminal Region of the Luteinizing Hormonal Choriogondotropin Receptor Contacts Both Subunits of Human Choriogonadotropin II Photo affinity Labeling J., Biol Chem 273:13841-13847.

174- Mcfarland K.C., Sprengel R., Phillips H.S., et.al. 1989 Lutropin -Choriogonadotropin Receptor: An Unusual Member of the G Protein-Coupled Receptor Family. Science 245:494-499.

175- Jiang X., Dreano M., Buckler D.R.,et.al 1995 Structural Predictions for the Ligand -Binding Region of Glycoprotein Hormone Receptors and the Nature of Hormone Receptor Interactions. Structure 3:1341-1353.

176- Hong S., Ji I., Ji T. 1999 . The β-Subunit of Human Choriogonadotropin Interacts with the Exodomine of the Luteinizing Hormone /Choriogondotropin Receptor and Changes Its Interaction with the α -Subunit. Mol Endocrinol 13:1285-1294.

177- Ji I., Pan Y., Lee Y., Phang T., and Ji T.H 1995 Receptor Binding Dependent Structural Changes in Human Choriogonadotropin :photochemical inter subunit cross linking .Endocrine 3:907-911.

178- Reshef E., Lei Z.M., Rao C., Pridham D.D. et.al. 1990 The Presence Gonadotropin Receptors in nonpregnant Human Uterus,

Human Placenta, Fetal Membranes, and Decidua J. clin Endocrinol &Metab 70:421-430.

179- Lei Z., Toth P., Rao C., Pridham D.1993 Noval Coexpression of Human Chorionic Gonadotropin (hCG/Human Luteinizing Hormone Receptors and their Ligand hCG in Human Fallopian Tubes. J. Clin Endocrinol & Metab 77:863-872.

180-Rao C. Li X. ,Toth , Lei Z. M., and Cook V.D. 1993 Noval Expression of Functional Hormone Chorionic Gonadotropin /Luteinizing Humane Receptors Gene in Human Umbilical Cords . J Clin Endocrinol & Metab 77:1706-1714.

181-Rao C. 1996 The Beginning of a New Area in Reproductive Biology and Medicine :Expression of Low levels of Functional Luteinizing Hormonal Human chorionic Gonadotropin Receptors in Nongonadal Tissue. J Phsiol Pharmacol 47:41-53

182-Lei Z., Rao C., Kornyei J.L. Licht P., Hiatt E.S. 1993 Novel Expression of Human Chorionic Gonadotropin /Luteinizing Hormone Receptor Gene in Brain Endocirnol 132:2262-2270.

183- Meduri G., Charnaux N., Loosfell H., Jolivet A et.l. 1997 Luteinzing Hormone /Human chorionic Gonadotropin Receptor in Breast Cancer.Cancer Rech. 57:857-864.

184-Lin J. Lei Z. Lojuns S., Rao C.et al 1994 Luteinizing Hormone/Human Chorionic Gonadotropin Receptor Gene in Human Endometrial Carcinomas J.Clin Endocrionol & Metab 79:1483-1491.

185-Huang Z.H., Lei Z., and Rao C. 1995 Immortalized and Erior Pituitary & T3 Gonadotropes Contain Functional Luteinizing Hormone/Humane Chorionic Gonadotropin Receptors . Mol Cell Endocrinol 114:217-222.

186- Pabon J.E., Bird J.S., Li X., Huang Z.H. et.al. 1996 Human Skin Contains Luteinizing Hormone /Chorionic Gonadotropin Receptors J. Clin Endocrinol & Metab .81:2738-2741.

187- Lei Z.M., Reshef E., and Rao C. 1992 The Expression of Human Chorionic Gonadotrpin Luteinizing Hormone Receptors in Human Endometrial and Myometrial Blood Vessels. J. Clin Endocrnol & Metab 75:651-659.

188- Kornyel J., Leiz., and Rao C., 1993 Human Myometrial Smooth Muscle Cells and Human Chorionic Gonadotropin. Biol Reprod. 49:1149-1157.

189- Ziecik A.J..Stanchev P.D., Tilton J.E., 1986 Evidence for the presence of Luteinizing Hormonal/ Human Chorionic Gonadotropin Binding Sites in the Porcine Uterus. Endocrinol .119:1159-1163.

190- Bonnamy P.O. Benhaim A., and Leymaric P. 1989 Human chorionic Gonadotropin Affects Tissue Levels of Progesterone and Cyclic Adenosine 3,5 Monophosphate in the Metestrus Rat Uterus in Vitro . Biol Reprod 40:511-516.

191- Menon K., and Jaffe R.B. 1973 Chorionic Gonadotropin Sensitive Adenylate Cyclase in Human Placenta. J Clin Endocrinol & Metab. 36:1140-9.

192- Demeros L.M. Gabbe S.G., Villee C.A., and Greep Ro. 1973 Human Chorionic Gonadotropin Mediated Glycogenolysis in Human Placental Villi .Biophys Acta 313:202-210.

193- Wolf A., Henrich I., Benz R., and Lauritzen C. 1985 Influence of Human Chorionic Gonadotropin and Prostaglandins on the steroid Metabolism of the perfused placenta Contr Gynecol Obstet 13:162-164.

194- Varangot J. Cedard L., and Yannotti S., 1965 Perfusion of Human Placenta in Vitro Study of the Biosynthesis of Estrogens. Am J Obst Gynecol 92:534-547.

195- Rosenberg S. M., and Bhanager A.S. 1984 Sex Steriod and Human Chorionic Gonadotropin Modulation of in Vitro Prolactin Production by Human Term Decidua. Am J. Obstet Gynecol 148:461-466.

196-Yuen B.H. Moon Y., Shin D. 1986 Inhibition of Human Chorionic Gonadotropin Production by Prolactin From Term Human Trophoblast Am. J. Obstet Gynecol 154:336-340.

197-Lojun S., Bao S., Lei Z.M. , and Rao C. 1997 Presence of Functional Luteinizing Hormone / Chorionic Gonadotropin (hCG) Receptors in Human Breast Cell Lines:Implications Supporting the Premise that hCG Protects Women Against Breast Cancer. Biol Reprod 57:1202-1210.

198-Russo J. and Russo I H. 1995 Hormonally Induced Differentiation :A Novel Approach to Breast Cancer Prevention J. Cell Biochem 22 (suppl 1.):58-64.

199-Alvarado M.V., Russoo J. ,Alvarado N.E., and Russo I., 1994 Human Chorionic Gonadotropon inhibits proliferation and Induces Expression of Inhibin in Human breast Epithelial cells in Vitro .In Vitro Cell &Dev Biol 30A:4-8.

200-lowry O.H.,Resebrough N.J.,Farr A.L. and Randall R.J. 1951 Protein Measurement with the Folin Reagent .J. biol. Chem. 193:265-275.

201- Strickland S., and Richards W.G. 1992 Invasion of the Trophoblast Cells.Cell 71:355-357.

202- Goustin A.C, Bestsholtz, C., Pfeifer- Ohlsson S., Persson H., Rydnert J. Bywater M et al. 1985 Coexpression of the sis and myc Proto-oncogene in Developing human Placenta Suggests Autocrine Control of Trophoblast Growth .Cell 41:301-312.

203- Chassin D. ,.Benifa J.L., Delattre C., fernandez H., Ginist D. , Janneau J., et al. 1994 Identification of Genes Over Expressed in Tumors Through Preferential Expression Screening in Trophoblast. Cancer Res. 54:5217-5223.

204- Bennett W. A., Lagoo Deenadayalan S., Whitworth N.S., stopple J. A., Barber W.H. et. Al. 2000 First -Trimester Human Chorionic Villi Express Both Immuno Regulatory and Inflammatory

Cytokines :a role for interleukin-10 in regulating the cytokine network of pregnanancy. Am. J. Reprod -Immuno 41:70-8.

205-Bonnamy P.J., Benhaim A., Leymarie P. 1990 Estrous Cycle – Related Change of High Affinity Luteinizing Hormone/ Human Chorionic Gonadotropin Binding Sites in the Rat Uterus. Endocrinol 126:1264-1269.

206-Bennett J.P,, and Yamamura H.I Neurotransmitter Hormone, or Drug Receptor Binding Methods .In Yamamura H.I., Enna S.J., Kuhar M.J., (eds) Neurotransmitter Receptor Binding (2ed). Raven Press New York 1985 P61-89..

207-Shao K., Balasubramania S.V., Pope C.M., and Bahl O.P., 1998 Effect of Individual N- Glycosyl chains in the ß-Subunit on the conformation of Human Choriogonadotropin Mol cell Endocrinol.146:39-48.

208-Bellisario R., and Bahl O.P. 1975 Human chorionic Gonadotropin V. Tissue Specificity of Binding and Partial Characterization of Soluble Human Chorionic Gonadotropin – Receptor Complexes J. Biol .chem. 250:3837-3844.

209-Rao C.V., Receptors for Gonadotropins in Human Ovaries In:Muldoon T.G., Mahesh V.B., Peez Ballester B., (eds) Recent Advances in Fertility Research, Part A. New York:Liss; 1982,P123-135.

210-Marsh J.M. 1976 The Role of Cyclic AMP in Gonadal Steroidogensis. Biol Reprod .14:30-53.

211-Huhtaniemi L.T., and Catt K.J. 1981 Differential Binding Affinities of Rat Testis Luteini Hormone (LH) Receptors for Human Chorionic Gonadotropin Human LH, and Ovine LH. Endocrinol. 108:1931-1938.

212-Wittliff J.L., Hormone and Growth Factor Receptor In: Donegan W.L. and Spratte J.S. (ed) Cancer of the Breast, 4ed Philadelphia:w.B. Saunders Co. 1995 pp 346-369.

213-Janson P.O. 1975 Effects of the Luteinizing Hormone on Blood in the Follicular Rabbit Ovary as Measured by Radioactive Microsphers. Acta Endocrinol 79:122-133.

214- Willtbank M.c., Gallagher K.P., Dysko R.C., Keyes P.L., 1089 Regulation of Blood Flow to the Rabbit Corpus Luteum: Effects of Estradiol and Human chorionic Gonadotropin Endocrinol. 124:605-611.

215-Korenman S.G., Krall J.F., 1977 The Role of Cyclic AMP in the Regulation of Smooth Muscle Cell Contraction in the Uterus Biol Reprod. 16:1-17.

216-Joshi S.G., 1983. Progestin Regulated Proteins of the Human endometrium. Sem Reprod Endocrinol 1:21-36.

217-Jacobs S., and Cuatrecasas P. Problems in Studying Hormone Receptor Binding .In Jacob J (ed.) Advances in Pharmacology and Therapeutics Vol.1 Pergamon , New York 1979 PP.215-222.

218-Jacobs S.Chang K.J., and Cuatercases P.1975 Estimation of Hormone Receptor Affinity by Competitive Displacement of Labeled ligand :Effect of Concentration of Receptor and of Labeled ligand Biochem . Biophys Res. Commun .66:687-692.

219-Dufau M.L., Ryan D.W., Baukal A. J., and Catt K.J. 1975 Gonadotropin Receptors: Solubilization and Purification by Affinity Chromatography. J. Biol chem. 250:4822-4824.

220-Keutmann H.T., 1992 Receptor. Binding Regions in Human Glycoprotein Hormones Mol. Cell Endocrinol 86:1-6.

221- Ruddon R.W., Byran A.H., Hanson C.A., Perini F., et. al. 1981 Characterization of the Intracellular and Secreted Forms of the Glycoprotein Hormone Chorionic Gonadotropin Produced by Human Malignant Cells. J. Biol chem. 256:5189-5196.

222-Keutmann H.J., M cllory P.J. Bergert E.R., Ryan R.J., 1983 Chemically Deglycosylated Human Chorionic Gonadotropin

Subunits: characterization and biological properties .Biochemistry 22:3067-3072.

223-Kalyan N.K. and Bahl O.P 1983 Role of Carbohydrate in Human Chorionic Gonadotropin :Effect of Deglycosylation on the Subunit Interaction and on its in Vitro and in Vivo Biological properties J. Biol Chem 258:67-74.

224-Ascoli M. Puett D. 1978 Gonadotropin Binding and Stimulation of Steroidogenesis in Leyding Tumors Cells Proc Natl Acad Sci USA 75:99-102.

225-Dufau M.L., Charreau E.H. and Catt K.J. 1974 Characteristics of a Soluble Gonadotropin Receptor From the Rat Testis J., Biol. Chem 248:6973-76.

226-Haour F., Saxena B.B. 1974 Characterization and Solubilization Gonadotropin Receptor of Bovine Corpus Luteum J. Biol chem. 248: 2195-2199.

227-Vutlai Luu Thi M.T., Misrahi M., Houllier A., Jolivet A, Milgrom E.1992 Variant Form of the Pig Lutropin/ Choriogonadoropin Receptor. Biochemistry 31:8377-8383.

228-Pandian M.R., Bahl O.P.,and Segal S.J. 1975 Labeling of LH/CG Receptor in Rat Ovaries Biochem Biophys. Res. Commun 64:1199-1205.

229-Sexena B.B. In Methods in Receptor Research (Blecher M.,ed) Part I, Marcel Dekker New York 1976, P 251-292.

230-LKB Application Note 306.1977 SDS and Conventional Polyacrymide GEL Electrophoresis with LKB Muitiphor.

231- Segrest J.P. and Jackson R.C. In Method In Enzymology (V.Ginsburg, ed.) Vol XXVIII, Academic press N. Y., 1972, P481.

232-Dubois M.,Gilles K.A., Hamilton J.K., Rebers P.A,and Smith F.1956 Colorimetric method for determination of sugars and related substance.Anal. Chem.28:350-356.

233- Garfin D.E. In Method in Enzymology (Dentschr M. P.,ed) Vol 182, Academoc Press Inc.,1990, P459.

234- Igarashi S., Minegishi T., Nakamura K., Nakamura M., et.al.1994 Functional Expression of Recombinant Human Luteinizing Hormone Human Choriogonadotropin Receptor, Biochem Biohys Res. Commun . 201:248-256.

235-Mc farland K.C. Sprengel R., Phillips H.S., Kohler M. et.al. 1989 Lutropin –Chorio Gonadoropin Receptor:An Unusual Member of the G-Protein Coupled Receptor Family. Science 245:494-499.

236- Rosemblit N., Ascoli M, Segaloff D.L. 1988 Characterization of a Antiserum to the Rat Luteinizing Hormone/Chorionic Gonadotropin Receptor. Endocrinology 123: 2284-2290.

237- Loosfelt H., Misrahi;M. Atger M., Salesse R., et al.1989 Cloning and Sequencing of Porcine LH/CG Receptor CDNA:Variants Lacking Transmembrance Domain .Science 245:525-528.

238- Minegishi T., Nakamura K., Takakura Y., Miyamato K., Hasegawa Y.,Igarashi M. 1990 Cloning and Sequencing of Human LH/CG Receptor CDNA. Biochem Biophys Res Commun.172:1049-1054.

239- Rodrignez M.C., Segaloff D.L. 1990 The Orientation of the Lutropin/Choriogonadotropin Receptor as Revealet by Site-Specific Antibodies Endocrinology 127:674-681.

240- Hipkin R.W, Sanchez. Yague J., Ascoli M. 1992 Identification and Characterization of a Luteinizing Hormone Chorionic Gonadotropin (LH/CG)Receptor Precursor in a Human Kidney Cell Line Stably Transfected with the Rat Luteal LH/CG Receptor Complementary DNA Mol Endocrinol 6:2210-2218.

241- Ascoli M., Segaloff D.L. 1989 On the Structure of the Luteinizing Hormonal/ Chorionic Gonadotropin Receptor. Endocr Rav. 10:27 – 44.

242- Pataja –Repo U.E., Merz W.E. Rajaniemi H.J. 1991 Significance of the Glycan Moiety of the Rat Ovarian Luteinzing Hormone/Chorionic Gonadotropin and Human CG for Receptor-Receptor – Hormone Interaction. Endocrinology 128:1209-1217.

243- Ji I., Slaughter R.G., Ji T.H. 1990 N-Linked Oligosaccharides are not Required for Hormone Binding of the Lutropin Receptor in a Leyding Tumor cell Line and Rat Granulosa Cell. Endocrinology 127:494-406.

244- Zhang R., Tsai Morris C.H., Kitamura M., Buczko E., Dufau M.L. 1991 Changes in Binding Activity of Lutenizing Hormone Receptors by Site Direct Mutagenesis of Potential Glycosylation Sites. Biochem Biophys Res Commun 181:804-808.

245- Liu X., Davis D., Segaloff D.L. 1993 Disruption of Potential Glycosylation Sites for N-Linked Glycosylation Does not Impair Hormone Binding to the Lutropin /Choriogonadotropin (LH/CG) Receptor if Asn 173 is left intact. J Biol chem. 268:1513-1516.

246-Freifrlder D(ed.).Physical Biochemistry, Application to Biochemistry and Molecular Biology 2th(ed.) San Fransico, W.H. Freeman and Company, 1982.

247-Silvestien R.M., Basster G.C., Morril T.C. eds. Spectrophotometric Identification of Organic Compounds New York. Joun Wiley & Sons 1981.

248-Ascoli M.1982 Internalization and Degradation of Receptor-Bound Human Choriogonadotropin in Leyding Tumor Cells Fate of the Hormone Subunits J.Biol. Chem 257:13306 -13311.

249-Tsavaris P.Vonorta K. ,Tsoutsos H.,Kozatsani –Halividi D. et al 1993 Carcinoembryonic Antigen (CEA), α -Feto-Protein, CA 19-9 and CA-125 in Advanced Colorectal Cancer (ACC). In J Biol Markers. 8:88-93.

250-Chevinsky A.H. 1991 CEA in Tumor of Other Than Colorectal Origin .Semin Surg Oncol .7:162-166.

251-DiSaia P. ,Haverback B.J., Dyce B.J. Marrow C.P 1975 Carcinoembryonic Antigen in Patients with Gynecologic Malignancies. Am J Obstet Gynecol .121:159-163.

252-Barrelet V., Mack J.P. 1975 Variations of the Carcinoembryonic Antigen Level in the Plasma of Patients with Gynecologic Cancers During Therapy . Am. J. Obstet Gynecol 121:164-168.

253-Schwartz P.E., Chambers S.K., Chambers J.T., et.al 1987. Circulating Tumor Markers in the Monitoring of Gynecologic Malignancies. Cancer 60:353-361.

254-Tuxen M.K., Soletormos G., Domber-nowsky G.1995 Tumor Marker in the Management of Patients with Ovarian Cancer. Cancer Treat Rev 21:215-245.

255-Stall K.E., Martin E.W. 1981 Plasma Carcinoembryonic Antigen Levels in Ovarian Cancer Patients :A Chart Review and Survey of Published Data. J Reprod Med. 26:75-82.

256-Cohen C., Shulman G., Budgeon L.R. 1982 Endocervical and Endometrial Adenocarcinma :A Immunoperoxidase and Histochemical Study. Am. J. Surg. Pathol .6:151-156.

257-Van Nagell J. R., Donaldson E.S., Gay E.C., et.al 1979 Carcinoembryonic Antigen in Carcinoma of the Uterine Cervix. Tissue Localization and correlation with plasma Antigen Concentration. Cancer 44:944-984.

258-Bychkov V., Rothman M., Bardawill W.A 1983 Immunocytochemical Localization of Carcinoembryonic Antigen(CEA), Alpha –Fetoprotein (AIP), and Human Chorinic Gonadotropin (hCG) in Cervical Neoplasia Am. J. Clin .Pathol. 79:414-420.

259-Berbec H., Paszkawska A., Siwek B., Gradziel K., Cybulski M. 1999 Total Serum Sialic Acid Concentration as a Supporting Marker of Malignancy in Ovarian Neoplasia Eur.J. Gynaec. Oncol 6:389-392.

260-Nathanson L., Fishman W.H 1971 New Obseration on the Regan Isoenzyme of Alkaline Phosphatase in Cancer Patients Cancer 27:1388-1397.

261-Kellen J.A., Bush R.S., Malkin A., 1976 Placenta –Like Alkaline Phosphates in Gynecological Cancer .Cancer Res.36:269-271.

262-Nollwen E.J., Pollet D.E., Schelstraete J.B., et al. 1985 Human Placental Alkalinee Phosphatase in Benign and Malignant Ovarian Neoplasia .Cancer Res.45:892-902.

263-Shetty P.A., Damle S.R. and Shahane A.I 1985 Clinical Significance of heat labile regan isoenzyme variant in Hodgkin's disease. Cancer 55:605-607.

264-Milne D.B., Trace Elements in Tietz Text Book of Clinical Chemistry 3rd ed. W.B. Sannders Company P.1030 -1031, 1999.

265-WHO Trace Element in Human Nutrition and Health 1996.

266-Malk K.L., Bhardway S.D., and Seigh P.P 1989 Diagnostic and Prognostic Value to Serum Zinc, Copper in Different Type of Malignancies .Excerpta Medica.

267-Bates M. Bronchial Carcinoma an Integrated Approach for Diagnosis and Management Springer-Verlag Berlin Heidlberg, New Youk P129,1984.

268-Henderson J.Shuskik C., Kremer R. and et al 1989 Circulating Levels of Parathyroid Hormone –Like Peptide in Patients with Malignancy and Hypercalcemia .Clin Oncol Suppl.8

269-Rieke J.W.Donaldson S.S.,and Horning S.J 1989 Hypercalcemia and Vitamin D Metabolism in Hodgkin's Disease Cancer 63:1700-1707.

270-Waxman J.1990 Hypercalcaemia –New Mechanism for Old Observation Br.J. Cancer 61:647-648.

271-Bondy P.K., and Gitby E.D. 1982 Endocrine Function in Small Cell Undifferentiated Carcinoma of the Lung. Cancer 50:2147-2153.

272-Kaplan L.A., and Pesce A.H. eds.Clinical Chemistry -Theory , Analysis and Correlation 2nd (ed.) Mosby Company ST Louis 1989 P924-930.

273-Valery's Practical Clinical Biochemistry Gowenlock A.H., McMurray J.R., and Mclauchlan D.M (eds) 6th(ed.) Heinmann Medical book,London 1988.

274-Katopodis N., Hirashaut Y., Nancy L.G. and Stock C.C 1982 Lipid Associated Sialic Acid Test for the Detection of Human Cancer.Cancer Res 42:5270-5275.

275-Saif Allah P.H. Biochemical Studies on Prolactin and some Tumor Markers in Breast Cancer Ph. D Thesis .Colleg of Science, Baghdad Univ .2000.

276-Landenson J.H., ,and Bower G.N 1973 Free Calcium in Serum Clin .Chem. 19:575-582.

277-Gordan G.S., Fitzpatrick M.E., and Lubich W.P. 1967 .Identification of Osteolytic Sterols in Human Breast Cancer .Trans .Ass. Amer. Physocians 80:183-186.

278-Greaves M., Ibbotson K.J., Atkins D. and Martin T.J. 1980 Prostaglandins as Mediators of Bone Resorption in Renal and Breast Tumors. Clin Sci 58:201-204.

279- Tashjian A.H., Voeikel E.F., Lazzaro M., Bosma T., and Levine L. 1987 Tumor Necrosis Factors stimulates Bone Resorption in Mouse Calvaria Prostoglandin –Mediated Mechanism. Endocrinology 120:270-275.

280-Lorenzo J.A., Quinton J., Sousa S.,and Raisz L.G 1986 Effect of DNA and Prostaglandin Synthesis Inhibitor on the Stimulation of Bone Resorption by Epidermal Growth Factor in Fetal Rat Long-bone Cultures J. Clin Invest 77:1897-19001.

281-Sato K.,Fujll Y., Kasond K., and et al., 1989 Parathyroid Hormone –Related Protein and Interleukin I a Synergism Stimulate Bone Resorptio in Vitro and Increase the Serum Calcium Concentration in Mice in Vivo. Endocrinol 124:2172-2176.

Part (C)

Human Chorionic Gonadotropin in Breast Tumors

Sami A. AL-Mudhaffar
Zainab M .AL-Azzawi

List of contents
Summary 179
list of appreviations 180

Chapter One: Introduction 182

1. Human Chorionic Gonadotropin 182

   1.1. Structure of hCG 182

   1.2. Human chorionic gonadotropin Forms 183

     a. Blood and urinary forms of hCG 184

- hCG βcf 184
- Intact hCG 185
- **FreeβhCG** 185
- **freeα hCG** 185

     b. Pituitary forms of hCG 186

   1. 3. Production and Action of hCG 186

     a. Normal Production and Action of hCG 186

- **Pregnancy** 187
- **Fetus** 187

- Placenta 187
- Central Nervous System (CNS) 187
- Other organs 188

b. Abnormal Production and Action of hCG 188
- Trophoblastic Tumors 188
- Non trophoblastic Tumors 189
- *Gynecologic Tumors 189*
- *Non gynecologic Tumors 190*

1.4. Potential Therapeutic Uses of hCG 191
- Pregnancy 191
- Central Nervous System (CNS) 192
- Non trophoblastic Tumors 192

2. Human Chorionic Gonadotropin and Breast Cancer 193

2.1. Human Breast Development 193

2.2. Pathogenesis and Site of Origin of Breast Cancer 194

2.3. Effecting Factors on Breast Cancer 194

2.4. Role of hCG in Breast Cancer 196

- **Inhibition** 196
- **Progression** 197

2.5. Mechanism of hCG Action in Breast Cancer 197

- Effect on inhibin expression 198
- **Effect on Programmed Cell Death (PCD) GenesExpression** 198
- **Effect on Insulin-like Growth Factor (IGF) Expression** 198
- **Effect on new genes expression** 198

3. Human Chorionic Gonadotropin Receptor 199

3.1. Structure of hCG Receptor 200

3.2. Distribution and Function of hCG Receptor 201

3.3. HCG Receptor in Breast Tissue 204

Aim of the work 204

Chapter Two: Studies on the Binding of ^{125}I –anti hCG Antibody to the HCG in Human Breast Tumor Homogenate 206

Experimental and methods 207

2.1. Materials 207

- Chemicals 207
- Instruments 208
- Patients 208
- *Blood Sampling 209*
- Collection of Breast Tissue Specimens 209
- Preparation of Breast Tumor Tissue Homogenate 209

2.2. Methods 210

- Determination of hCG Levels in Sera of Benign and Malignant Breast Tumor Patients and Controls 210
- Determination of Total Protein Content in Breast Tumor Homogenate 212

2.3. Binding Studies of hCG in Breast Tumor Homogenate 212

with ^{125}I- anti hCG Antibody

2.3.1. Preliminary Tests of hCG Binding in Breast Tumor Homogenate with ^{125}I-anti hCG Antibody 212

2.3.2. Most Appropriate Conditions of the Binding of hCG in Breast Tumor Homogenate with ^{125}I-anti hCG Antibody 213

- **The Effect of Different Protein Amount of Breast Tumor Homogenate on the Binding of hCG with ^{125}I-anti hCG Antibody 213**

- **The Effect of Different Concentrations of ^{125}I-anti hCG Antibody on the Binding with hCG in Breast Tumor Homogenate   214**

- **The Effect of pH on the Binding of hCG in Breast TumHomogenate with ^{125}I- Anti hCG Antibody 215**

- **Time Course of the Binding of hCG in Breast Tumor Homogenate with ^{125}I-anti hCG Antibody 216**

- *The Effect of halides on the Binding of hCG in Breast Tumor 216 Homogenate with ^{125}I-anti hCG Antibody*

2.4. Determination of Affinity Constant (K_a) and the Maximal Binding Capacity (B_{max}) of hCG in Malignant Premenopausal Breast Tumor Patients Associated with ^{125}I-anti hCG Antibody

217

Results and discussion 219

Determination of hCG Levels in Sera of Breast Tumors Patients 219

Binding Studies of hCG in Breast Cancer Tumor Homogenate with ^{125}I- anti hCG Antibody 221

Preliminary Tests of hCG Binding in Breast Tumor Homogenate
With ^{125}I-anti hCG Antibody 222

Most Appropriate Conditions of the Binding of hCG in breast Tumor Homogenate with ^{125}I-anti hCG Antibody 222

The Effect of Different Protein Concentration of Breast Tumor Homogenate on the Binding of hCG with ^{125}I-anti hCG Antibody 222

The Effect of Different Concentrations of ^{125}I-anti hCG Antibody On the Binding with hCG in Breast Tumor Homogenate 224

The Effect of Different pH on the Binding of hCG in Breast 226 Tumor Homogenate with ^{125}I-anti hCG Antibody

Time Course of the Binding of hCG in Breast Tumo 228 Homogenate wit ^{125}I-anti hCG Antibody

The Effect of Halides on the Binding of hCG in Breast

Tumor Homogenate With ^{125}I-anti hCG Antibody

Determination of Affinity Constant (K_a) and the aximum Binding

Capacity (B_{max}) of hCG in Pre-menopausal Malignant Breast Tumor Homogenate Associated with ^{125}I-Anti hCG Antibody   232

Chapter Three: Studies on the Binding of ^{125}I–hCG to the hCG in Human Breast Tumor Homogenate 236

Abstract 236

Introduction 237

Materials and Methods 238

3.1. Materials 238

- Chemicals 238
- Instruments 238
- Patients 238
- Preparation of Breast Tumor Tissue Homogenate 238

3.2. Methods 238

- Determination of Total Protein Content in Breast Tumor Homogenate 238

3.3. Binding Studies of hCG Receptor in Breast Cancer

238 Tumor Homogenate with ^{125}I-hCG

3.3.1. Preliminary Tests of hCG Receptor Binding in Breast Tumor Homogenate with ^{125}I-hCG   238

3.3.2. Most Appropriate Conditions of the Binding of hCG Receptors in Breast Tumor Homogenate with ^{125}I-hCG 240

- *The Effect of Different Protein Amount of Breast Tumor Homogenate on the Binding of hCG Receptors with ^{125}I-hCG   240*

- *The Effect of Different Concentrations of ^{125}I-hCG on the Binding with hCG Receptors in Breast Tumor Homogenate 240*

- *The Effect of Different pH on the Binding of hCG Receptors in Breast Tumor Homogenate with ^{125}I-hCG 241*

- *Time Course of the Binding of hCG Receptors in Breast Tumor Homogenate with ^{125}I-hCG 242*

- ***The Effect of Different Halides on the Binding of hCG Receptors in Breast Tumor Homogenate with ^{125}I-hCG 243***

3.4. Determination of the Concentration of hCG-Receptor and the Affinity concentration of ^{125}I-hCG Association with its Receptors in Benign and Malignant Breast Tumor   243

Results and discussion   245

Membrane preparation 245

Binding Studies of hCG Receptor in Breast Cancer Tumor Homogenate with ^{125}I-hCG 245

Preliminary Tests of hCG Receptor Binding in Breast Tumor Homogenate with ^{125}I-hCG 245

Most Appropriate Conditions of the Binding of hCG Receptors in Breast Tumor Homogenate with ^{125}I-hCG 246

The Effect of Different Protein Amount of Breast Tumor Homogenate on the Binding of hCG Receptors with ^{125}I-hCG 246

The Effect of Different Concentrations of ^{125}I-hCG on the Binding With hCG Receptors in Breast Tumor Homogenate 247

The Effect of Different pH on the Binding of hCG Receptors in Breast Tumor Homogenate with ^{125}I-hCG 249

Time Course of the Binding of hCG Receptors in Breast Tumor Homogenate With ^{125}I-hCG 252

The Effect of Different Halides on the Binding of hCG Receptors in Breast Tumor Homogenate with ^{125}I-hCG

256

Determination of the Concentration of hCG-Receptor and the 257

Affinity concentration of ^{125}I-hCG Association with its

Receptors in Benign and Malignant Breast Tumor

Chapter four: Kinetic and Thermodynamic Parameters of the Binding of Partially Purified hCG from Malignant Breast Tumor Homogenate

Abstract 259

Introduction 260

Materials and Methods 261

4.1.Materials 261

- Chemicals 261
- Instruments 261
- **Patients**

4.2. Methods 261

- Partially purification of hCG by Gel Filtration Technique 261
- Void Volume Determination 262
- *Sample Addition* 262
- Standard hCG Addition 263

4.3. The Choice of the Appropriate Conditions for the Binding of the Partially purified hCG from Malignant Premenopausal Breast Tissue to ^{125}I-Anti HCG Antibody 264

- **The Effect of Different Protein Concentration** 264
- **The Effect of Different Concentrations of ^{125}I-Anti hCG Antibody** 265
- **The Effect of Different pH on the Binding** 266
- Time Course of the Binding 268

4.4. The Kinetic and The Thermodynamic Studies

- Determination of Affinity Constant (Ka) and the Maximal Binding Capacity (B_{max}) of hCG in Partially purified hCG from Malignant premenopausal Breast Tissue with ^{125}I-Anti hCG Antibody 267
- **Thermodynamic Studies of the Binding of hCG in Premenopausal Patients with Breast Tumor to ^{125}I-Anti hCG Antibody** 269

Results and Discussion 271

Partially purification of hCG by Gel Filtration Technique 271

The Choice of the Appropriate Conditions for the Binding of the Partially Purified hCG from Malignant Premenopausal Breast Tissue to ^{125}I-Anti hCG Antibody 272

The Effect of Different Protein Concentration 272

The Effect of Different Concentrations of ^{125}I-Anti hCG Antibody

The Effect of Different pH on the Binding 274

Time Course of the Binding 275

Determination of Kinetic Parameters of ^{125}I-anti hCG Antibody

Binding with partially Purified hCG From Malignant Breast Tumor 276

The Thermodynamic Studies of ^{125}I-anti hCG Antibody to the

Partially Purified hCG in Malignant Breast Tumor 283

Thermodynamic Parameters of Transition State 283

Chapter Five: Spectral Studies on Stander hCG, PartiallyPurified hCG and (^{125}I- anti hCG Antibody/ hCG) Complexes 286

Abstract 286

Introduction 287

Materials and Methods 288

- Instruments 288

- Patients 288

5.2. Methods

- Gel Filtration Technique for Separation of Free and Bound 288

    ^{125}I -anti hCG Antibody

    5.3. Separation Procedure of (^{125}I-Anti hCG Antibody/hCG)

    Complex 288

- **HCG from Benign (Fibrocyst) Breast Tumor Homogenate** 288

    and its ^{125}I -Anti hCG Antibody

- **HCG from Premenopausal Malignant (IDC) Breast Tumors** 289

    Homogenate and Its ^{125}I-anti hCG Antibody

- **Standard hCG and Its ^{125}I-anti hCG Antibody Reagents** 290

- ^{125}I-anti hCG **Antibody** Reagent 291

    5.4. Partially purification of hCG by Gel Filtration Technique 291

- Preparation of the Column, Gel and Filtration Technique 291

- Eluent Buffer Preparation 291

- Samples Addition 292

    5.5. The UV Spectrum 292

- The UV Spectrum of (^{125}I-Anti hCG Antibody / hCG) Complex of Benign Breast Tumor 292

- The UV Spectrum of (^{125}I-Anti hCG Antibody / hCG) Complex of Malignant Breast Tumor 292

- The UV Spectrum of (^{125}I-Anti hCG Antibody / hCG) Standard Complex 292

- The UV Spectrum of ^{125}I-Anti hCG Antibody 293

- *The UV Spectrum of Partially Purifiesd hCG from malignant Premenopausal Breast tumor tissue* 293

- The UV Spectrum of the Standard HCG 293

5.6. Factors Affecting the Absorption Properties of Benign, Malignant, Standard (^{125}I-Anti hCG Antibody/hCG) Complex, Stander and Partially Purified hCG. 293

- *pH Effect* 293

- *Effect of Solvents Polarity* 294

- *Effect of Urea, KCl and (Urea, KCl) Mixture* 294

- *Spectrophotometric pH Titration of the Complex* 295

Results and Desiccation 0 296

Gel Filtration Technique for Separation of Free and Bound 296

^{125}I-Anti hCG Antibody

The UV Spectra of (^{125}I-Anti hCG ntibody/hCG)Complexes, 298

Standard, Partially Purifed hCG and ^{125}I-Anti HCG Antibody

Factors Affecting the Absorption Properties of Standard, Benign

,Malignant, (^{125}I-Anti hCG Antibody/hCG) Complex, standard

and partially Purified hCG 299

pH Effect 300

Effect of Solvents Polarity 301

Effect of Urea, KCl and (Urea, KCl) Mixture 303

Spectrophotometric pH Titration of the Complex 306

Conclusion **308**

Futurework 309

References 310

Summary

1. The level of hCG in sera of patients with breast tumor (benign, pre- and post-menopausal malignant) with age matched healthy individuals as a control was determined by immunoradiometric Assay (IRMA). The sera hCG level in malignant groups was significantly elevated when compared to those of benign and control groups ($p<0.05$).

2. The presence and characterization of hCG in cytosolic homogenate of breast tumor tissues were investigated by a modified immunoradiometric Assay (IRMA). This method was found to be a suitable for the assessment of hCG in benign and malignant breast tumor.

3. Different factors affecting the binding of hCG and ^{125}I-ant hCG antibody, affinity constant and hCG concentration (B_{max}) were studied.

4. The presence and characterization of hCG in cytosolic homogenate of breast tumor tissues were investigated by using a modified radioreceptor assay (RRA). This method was found to be a suitable for the assessment of hCG-receptor in benign and malignant breast tumor.

5. Different factors affecting the binding of hCG receptor and ^{125}I-hCG, affinity constant and hCG concentration (B_{max}) were studied.

6. hCG was partially purified from cytosolic fraction of human premenopauasl malignant breast tumor homogenate by gel filtration technique using Sephadex G-100. The binding of (partially purified hCG/^{125}I-anti hCG antibody) complex was investigated. The results indicated that binding reaction is time and temperature dependent process.

7. Spectroscopic studies in the U.V. region was carried out on standard, partially purified hCG, standard complex, benign and malignant complexes. The effect of different factors affecting the absorption band, such as pH, solvent perturbation and pH titration were studied.

Abbreviations

Ab	Antibody
Ab*	Labeled antibody
Ag	Antigen
(Ab-Ag)	^{125}I-anti hCG antibody /hCG) Complex
B	Bound
B$_{max.}$	Maximal binding capacity
BSA	Bovine Serum Albumin
(B/T) %	Percentage of bound over total
CAMP	3`,5` Cyclic adenosine mono phosphate
Cpm	Counts per minute
CNS	Central nervous system
CIS	Carcinomas in situ
COX	Cyclooxygenase
DMBA	7,12-dimethylbenz[α]anthracene
DHEAS	Dehydroepiandrostrone sulfate
EDTA	Ethylenediammine tetraaceticacid
EG	Ethylene Glycol
ER	Estrogen receptor
F	Free
FSH	Follicle stimulating hormone
GnRH	Gonadotrophin releasing hormone
HCG	Human chorionic gonadotrophin
HCG βcf	Human chorionic gonadotrophin beta core fragment
HLH	Human Luteinizing hormone

HI-1	Hormone-induced 1
IDC	Infiltrating ductal carcinoma
IRMA	Immunoradiometric Asaay
IDPs	Intraductalproliferations
IGF	Insulin like growth factor
IGFBPs	Insulin like growth factor binding proteins
Ka	Affinity constant
Kd	Equilibrium dissociation constant
KD	Kilodalton
KJ	Kilo joule
LH	Luteinizing hormone
MCF-7	Malignant human breast epithelial cell line
NSB	Nonspecific binding
PEG	Polyethylene glycol
P53	Tumor suppressor gen
PR	Progesterone receptor
PCD	Program Cell death
RT-PCR	Reverse transcription-polyemrase chain reaction
RIA	Radioimmuno assay
SB	Specific binding
T	Total count
TSH	Thyroid stimulating hormone
WAP	Whey acidic protein

Chapter one   Introduction

1. Human Chorionic Gonadotropin[1-3]

Human chorionic gonadotropin (hCG) is the signature hormone of the placenta. It's a hetrodimeric glycoprotein hormone that also belongs to the cystine–knot growth factor family. hCG is a member of closely related pituitary glycoprotein hormones (TSH, FSH, and LH), present in mammals, which are important to the correct functioning of the reproductive system. In mammals the glycoprotein hormones consist of a common alpha subunit which in human is encoded by a single gene. Each hormone has different β subunit and this subunit is responsible for the different target specificity of each hormone.

1.1. Structure of hCG

The alpha subunit of hCG, which is identical in sequence to the alpha subunit of the pituitary glycoprotein hormones, is composed of 92 amino acids; while the target-receptor-specific β subunit of hCG has significant sequence homology, with 80% identity to β-LH subunit. The β-hCG subunit contains 145 amino acids with 30 additional amino acids at the carboxyterminus[4]. This modification is an essential facet of hCG biology allowing sera concentration in the pregnant mother to reach peak level of 50-100000mIU/ml (1-10mg/l). hCG also varies from LH by virtue of increased glycosylation, internal disulfide bond and sialic acid content[5].

Carbohydrates constitute approximately 30% by weight of each subunit[6]. Their moieties play a role in the secretion, stability, folding, subunit assembly of hormone and also it seems to be important to maintain the proper conformation of the hormone[7-9]. Each carbohydrate moiety terminates in sialic acid, which accounts for 10% of the weight, of the molecule and confers a high degree of resistance to degradation and consequently a long plasma half- life[6,10,11]. The solid crystal structure of hCG showed that each of the gonadotropin subunits is rich in disulfide bonds (the α subunit have 10 half-cystines residues that form five intrachain disulfide linkages and 12 half-cystines that form six conserved disulfide bridges. It also has structural homology to the disulfide-knot growth factor proteins[2,12]. The three dimensional model of hCG, proposed a structure which predominantly composed of three helical segments ,two in the α and one in the β subunit forming antiparallel strands in each subunit. They are joined by three hairpin loops giving the hormone only a small hydrophobic core with a large interfacing area[2,13-16], as in figure (1-1).

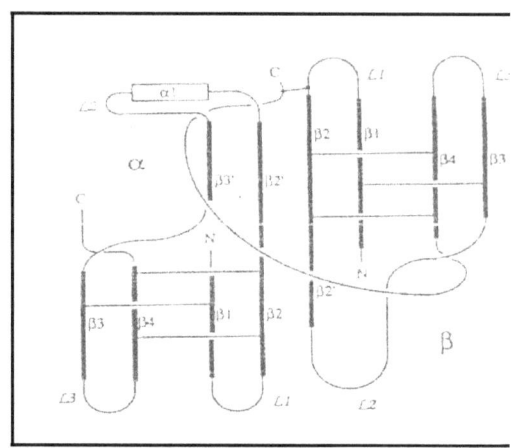

Figure (1-1): A schematic drawing of the hCG dimer Topology[17].

The β subunit, like the γ subunit of hCG, is composed of three loops. The β subunit contains six disulfide bridges which hold the molecule together when peptide bond cleavages take place in loop 2.

1.2. Human Chorionic Gonadotropin Forms

The combination of multiple subunits, multiple N-linked and O-linked oligosaccharide side chains causes significant heterogeneity in hCG structure[17].

hCG free subunits, degraded molecules, molecules with irregular N-and O-linked oligosaccharide side chains and fragments, which well informed, are present in sera, urine and other body fluids[18,19].

a. Blood and Urinary Forms of hCG

Although most measurements of hCG are made in serum, the source of hormone used for clinical treatment of infertility and other medical problems is usually a purified urinary fraction containing the biologically active forms of the hormone. All standard hCG reference preparations are all from the urine of normal pregnant women, that hCG serum measurements were based on[10,18,20].

Although a number of isoforms of hCG circulate in blood, most of these forms vary only by PI differences, due to variable content of the sialic acid residues. Only a small quantity of free subunit and various quantities of nicked hCG circulate in blood; in contrast, a much greater variety of molecular forms of hCG is present in urine due to proteolytic processing of hCG passing through the kidney. Many efforts to separate the several forms of hCG from different variety raw urine sources were evaluated[21-23].

- **Nicked hCG:**

The urinary metabolites of hCG and which may also exist to a lesser extent in blood are nicked form of hCG as well as nicked β subunit. Nicked hCG is simply a heterodimer with M.wt.(~36500D), it has peptide bond cleavages in loop 2 of the β subunit (between beta residues 44-49). Loop 2 is known to be exposed to solvent and is easily eliminated by protease . Dissociation of nicked hCG will result in free nicked β subunit, which is also found in a small extent in urine[17,24-26]. The existence of nicked hCG was confirmed by isolation and gel electrophoresis[20,22]. Reliable immunoassays for nicked forms of hCG were not available, due to diminished immunopotency after cleavages in β loop 2,until the development of a fairly specific nicked hCG immunometricassay[27] and immunoassay systems[28]. Anumber of studies suggested that nicked forms of hCG may have clinical significance as markers of certain cancers[24,27,29].

- **hCG βcf :**

Is the main urinary metabolite that produced by proteolysis of hCG or its β-subunit in the kidney since very little hCG βcf is detected in the blood[30]. Core molecules are generally present in urine in much greater molar concentration (2-10 times) than dimeric gonadotrophins or free subunits[31]. Other studies have shown that the molecule may be produced directly by placental cells in tissue culture and pituitary tissue[32,33]. The structure of this fragment was reported since 1983[34] and was the first to be studied and discovered upon solution of the crystal structure of hCG[35]. It is derived from hCG β subunit and is composed of residues 6-40 disulfide bridged to residues 55-92 and contaning tirmned carbohydrate groups with no sialic acid and the polypeptide chains are head together by disulfide bridges[35]. Highly specific assays were developed to the hCG βcf and applyed in verious clinical situations[19,18,21]. Extensive studies of its compartment distribution and concentration throughout pregnancy were conducted by de Medeirosetal[36], finding that it is increased in parallel to hCG throughout pregnancy, making its measurement a useful marker of ecto pregnancy. Furthermore, it has diagnostic applications in cancer tests including ovarian, lung, bladder and various gynecological cancers[37-40].

- **Intact hCG:**

It is the major hCG-related molecule present in sera that was separated as purified standard from different raw urine sources. It is composed of a heterodimer with intact polypeptide backbone. Composing of α subunit with 92 amino acids and β subunit with 145 amino acids residue polypeptide, mono and biantennary oligosaccharides and mostly O- linked trisaccharides [17]. Detectable level of intact hCG in sera and urine has been reported, observing that its level in sera of nonpregnant women increase with age, and higher than in men[30,40]. It rises during the first tirmester (8-10 weeks) to maintain the steroid environment necessary for the pregnancy;

following by a rapid decrease until 15 weeks[6]. Intact hCG levels have been also measured in non trophoblastic tumor, observing a slight elevation in its levels as it is compared to a pregnant woman[42].

- **freeβ hCG:**

It is defined as mono subunit drived from dissociation of hCG. In this form, with M.Wt about (~22000D), only β subunit is present and no α-subunit; having biantennary N-linked oligosaccharides and mostly trisaccharides O-linked oligosaccharide[17,43]. Using different techniques, the presence of βhCG in sera, urine, tissues, cyst fluid and cell line in different normal, benign and malignant tumor have been demonstrated. Free βhCG was found to be the most secreted form in the non trophoblastic tumors; while its level was found to be low or in normal range in the normal individuals[44-46].

- **free α hCG:**

The α-subunit of hCG was found to be synthesized and released by normal placental tissue as a small precursor form that convertes to a larger form prior to secrete; with finding that the free α form do not bind to purified βhCG subunit[47]. It is derived from the dissociation of hCG having identical structure of the combined α subunit excepte its carbohydrate composition which prevents recombination. Many studies were undertaken for studying the presence of the free α subunit in different organs, its immunological and biological property and its anti α monoclonal antibodes[48-50]. It was found to be in a small extent in urine and sera of healthy individuals. It has also been detected in sera, tissues and cytosol of the pregnant individuals, several endocrine and nonendocrine, and cell line tumors[51-53].

b. Pituitary Forms of hCG

Although hCG is not considered to be a pituitary hormone, there have been many reports of the presence of it in the urine and blood of healthy nonpregnant individuals[54-56]. While the immunological studies indicated the likely presence of hCG in pituitary tissues[57], it was never isolated until 1996 by Birken.S,et.al[58]. They have been able to isolate sufficient quantities of pituitary hCG and examine its primary structure, carbohydrate composition and in vitro bioactivity. The structure of pituitary hCG appears to be of an intermediate hormone between the structure of the placental hCG and that of the pituitary hLH, non essential hormone. Other forms of hCG from the pituitary were been studied. Western blotting techniques shown that the pituitary forms of hCG did not exhibit nicking[590]. While immunological evidence reported that the hCGβcf is present in pituitaries at level about 1% that of hLH βcf ;such low levels make it difficult to be isolateed[59].

1.3. Production and Action of hCG
a. Normal Production and Action of hCG

Although trophoblast is the major source of hCG, a wide variety of normal tissues, including anterior pituitary, can produce hCG[60-72]. Non trophobastic hCG is not glycosylated and its levels vary in the circulation due to rapid clearance. Some non trophoblasti tissues may not even release hCG, so that it will serve as a local ligand for hCG receptors[73,74].

- **Pregnancy:**

The appearance of hCG both in sera and urine soon after conception and its subsequent rise in concentration during early gestational growth make it an excellent marker for the early detection of pregnancy[75]. Sera level of hCG rise rapidly, increasing with conspectus number, after implantation of the trophoblast and reach a peak at 8-10 weeks of gestation then decline and remain at a low level until term[75,76]. hCG has multiple tropic effect during pregnancy; it has long been known to act on the ovaries to cause synthesis of estradiol and progesterone, which regulate reproductive tract function[(69,77,78]. New evidence has revealed that hCG can also act directly on reproductive tract organs themselves and regulate their function[78-83].

- **Fetus:**

hCG levels are very low in the fetal circulation, less than 5% of these in mother (maximum 50 mIU/ml), suggesting that hCG secretion is directed into the maternal circulation and is prevented from entering the fetus[74]. It may control fetal adrenal androgen synthesis, gonadal steroid production, brain growth and differentiation, protect the fetus from certain viral infection and may relax the umbilical vessels, keeping it from becoming too rigid[74,84-86].

- **Placenta:**

hCG was previously originated from the placenta. So, the production and presence of hCG and its subunits on placenta cell surface was detected by using various methods. These methods demonstrate hCG as a continuous layer on the surface of syncytiotrophoblast of the early and term placenta[87-89]. hCG is responsible of giving the signals of invasion from syncytiotrophoblasts to the trophoblast cell layer of the placenta to invade the uterus to establish a vacuolar connection with the maternal circulation during pregnancy[74].

- **Central Nervous System (CNS):**

The Central Nervous System (CNS) is also one of the specific target tissues for hCG, by which it is able to elicit multiple effects in the (CNS) through binding to their receptors, which have been identified[86,90]. hCG is involved in a multiple effect on the gonadotropin-releasing hormone (GnRH), constitute a negative short-loop feedback regulation of synthesis and secretion. Many observations administrated that gonadotropins can induce behavioral changes such as decreased feeding, exploratory activity and electrical activity of the brain[91].

- **Other organs:**

Studies on unicellular animals, some microorganisms and normal human tissue (ovary, testes, pituitary, lung, liver, kidney, colon and stomach) observed the presence of molecules immunologically similar to hCG[30,41,92,93].

b. Abnormal Production and Action of hCG

hCG belongs to the family of embryonically related marker proteins that include carcinoembryonic antigen and α-fetoprotein. Thus, a wid variety of trophoblast and nontrophoblast cancers and cancer cell contain intact hCG and/ or one of its subunits[94-96]. The presence of them is probably due to synthesis rather than sequestration. Their expression increases in advanced cancer, suggesting that they might be involved in the progression of the disease[97]. The regulatory mechanisms involved in the expression of hCG subuints genes in cancer cell are not known. Studies suggest that hCG may have dual role in cancers[81,98-100]. It promotes some whereas it inhibits other, due to whether they produce intact hCG or just its subuint which may have a stimulatory effect, probably due to the formation of homodimers[101].

♦ Trophoblastic Tumors

Detectable levels of hCG have been reported in conditions other than normal pregnancy, as originally described by Vaitukaitis and co-workers[102], included ectobic pregnancy, threatened abortion and trophoblastic disease. Its level is used for diagnosis of Down syndrome and genetically abnormal pregnancy[103,104]. In ectobic pregnancy and threatened abortion hCG level are progressively decrease and it is not known whether it is a cause or consequence[5,74]. Abnormally high levels of hCG are a risk factor for trisomy 21 and for the later onset of pre-eclampsia. Why hCG elevated is unclear, but such elevation could reflect either general trophoblast immaturity or increased syncytiotrophoblast turnover, both of which have been described in this disease[5]. hCG and/or its subunits (in blood and urine) also play an important role in the diagnosis of gestational trophoblastic disease, as well as, for monitoring the success or failure of chemotherapy in these disease, making it a most effective tumor marker, including partial or complete moles and choriocarcinomas[5,80,88,105]. Studies have demonstrated that choriocarcinomas produce excessive amounts of hCG, reach almost 100% sensitivity and specificity, due to the loss of self-regulation of biosynthesis and these high amounts promote tumor growth and metastasis in the host body. These findings suggest a potential treatment for this disease by selectivity inhibiting tumor hCG synthesis[80].

♦ Non trophoblastic Tumors

A great variety of non torphoblastic (gynecologic and nongynecologic) malignant tumors express hCG with a range of 19-30% of all tumors studied; moderately increased and showed variable correlation to tumor stage and histological grading[106-108]. In sera, studies with highly specific sandwich procedures, indicated that βhCG is mainly elevated while intact hCG and αhCG are slightly elevated or within the normal range; in contrast with that observed in pregnant women who secrete intact hCG in larg excess[109,110].

The expression of hCG and/or one of its subunits increases in advanced cancer suggesting that they might be involved in the disease[111]. Thus hCG measurement and other several circulating substances were used to determine whether patients have a high probability of metastasis, with up to 97% positive for hCG. The combined use of hCG and CEA together with cytology for better discrimination of benign from malignant effusions have been recommended. There is usually a parallel relation between these sera levels and the clinical evaluation of the disease under chemotherapy[112].

Employing two solid phase capture antibodies technique for urinary βhCG and hCGβcf ; demonstrated the presence of these forms in different malignancies[113,114]. Using flowcytometry and immunohistochemical techniques ,showed a higher expression of βhCG than intact hCG and there was no relationship between the βhCG postivity and the histological type of tumor[115,116].

All above studies suggest that βhCG ,may be the major form, that serves as a valuable tool in diagnosis of malignant nontrophoblastic tumors in general.

- **Gynecologic Tumors:**

The early production of hCG by epithelial ovarian carcinoma in tissue culture has arise the possibility that this hormone may be a useful marker in gynecologic tumors[117]. Most studies have focused on sera determination of hCG in this tumor. Low percentage, within the range of 18-25%, of total sera hCG level was found in early studies on patients with gynecologic cancer, using cut off level 5mIU/ml[7,108,118,119]. While other recent studies have demonstrated that the percentage of intact and βhCG elevation was (48%)[42] in comparable to that reported by Grossmann (37%) for gynecologic cancer; based on immunoradiometric design[108].

The elevation of βhCG was investigated with other tumor markers in women with primary epithelial ovarian cancer. This investigation has found a correlation between markers levels and cancer stage. It has also reported the importance of βhCG in following the ovarian cancer[7,112,120]. While, the examination of selected epithelial ovarian cancer found no relation between the βhCG positivity and the histological type, or even the histological grade of the malignant tumors[121].

The usefulness of βhCG and hCGβcf has been demonstrated too in assessing the prognosis of primary cervical cancer[114,119]. A detectable elevation of βhCG in 30% of cervical cancer has been abserved while low elevation of intact hCG in the same patient (1.3%) has been seen. Non of the cervical adenocarcinoma has had elevated hCG level, but βhCG positive has been found in adenocarcinoma of the uterine cervix in which the cell of tumor have some histological resemblance to trophoblastic cell[7,122,123]. Increasing of βhCG synthesis during progression of vulva cancer and its elevation has been found in 50% of patients with that cancer. This Increasing indicating that the patient with elevated sera βhCG had a worse progressive tumor compared with the group of normal βhCG[124].

For endometrial cancer, no base was found in the distribution of hCG level in patients with poorly and well differentiated endometrial cancer; suggesting that it might be involved in the progression, as promoter, of this disease.

- ***Non gynecologic Tumors:***

Like other hormones, hCG acts via binding to its receptors. Until about 18 years ago, hCG receptors had not been shown to be present in nongonadal tissues. The observation of their presence, including some cancers arising from nongonadal tissues, forced a change in the widely –held belief that hCG is only a gonadal regulating hormone[74,80].

In non gynecologic tumors, hCG has two different effects on tumor growth. It promotes some cancers, having a promoter property, such as cancer of the lung, whereas it inhibits prostate cancer. In fact, contraceptive hCG vaccine is now being tested, especially against cancer of colon and pancreas[126-129].

By employing radioimune assay (RIA), a great variety of malignant tumors have been found to be expresed hCG. Its levels were only moderately increased and showed variable correlation to tumor stage, histological grading and clinical course. The production of hCG or its subunits was found to be associated with tumors of high prognosis and greatest incidence such as tumors of (lung, liver, gastrointestinal tract tumor, meloma, testis, bladder, stomach, colon, pancreas, head and neck)[106,113,107].

Different studies estimated a different percentage relation between βhCG level and variable tumor origin. Nevertheless, in all of them low elevation of intact hCG was demonstrate. It was found in 55% of patients with (pancreas, colon, stomach, bladder, and hepatocellular tumor)[107,130]. Other studies have observed the level of βhCG to be in the range of (30-47%) for bladder cancer, (32-72%) for pancreatic cancer and less than 15% in lung cancer[113,130,131]. Studies on Urinary βhCG and hCGβcf revealed an over expression of them in malignant and benign tumors (bladder, pancreatic and Urothelial cancer)[113,114,132].

The βhCG has been found to be a promising tumor marker in early diagnosis and following the cancer of (testis, bladder, liver, stomach, colorectal, pancreas and lung), indicating that patients with pancreatic cancer having highly elevated hCG sera levels than patients with normal hCG sera level are having wors progressive tumor[131,133,134].

Using immunohistochemical techniques various investigators have detected positive staining for hCG in many origin, such as, (colorectal cancer and Urothelial carcinoma)[135,136]. Other investigations have reported that extracts of cancer tissues from different origin tumors contained hCG-like material higher than normal tissue[115,137]. More than 80 different cancer cell lines examined for the production of hCG and/or its subunits; demonstrated its expression in these different histological types and origins of cancer cell lines[115,138,139]. The stimulation of growth of several of those cell lines by hCG has been reported[140]. Thus cancer cells are able to regulate independently their own growth.

These findings with other results which confirmed that hCG has also chemical and physiological properties of growth factor[15]; have provided scientific basis for studies of prevention and control of cancer by active or passive immunization against hCG and its subunits.

1.4. Potential Therapeutic Uses of hCG

Although still greater understanding is needed, gain made in the last decade has demonstrated that nongonadal actions of hCG are physiologically important and may have relevance to better understanding several diseases and their treatment.

- **Pregnancy:**

The statement on improving pregnancy rates in assisted reproductive technologies was based on the finding that hCG treatment of coculture of cow oviduct epithelial cells with two-cell cow embryos resulted in greater embryonic development. Coculture has been shown to improve pregnancy rate in women who failed to become pregnant in more than three previous cycles[80,141].

hCG treatment may save pregnancy if the threatened loss is not anatomic defects, infection or fetal anomalies. hCG treatment may work by increasing the placental endocrine activity, by preventing immunologic mechanisms that promote fetal rejection, by increasing uterine blood flow, by decreasing uterine activity, and so forth[74]. Clinical trials are being conducted to assess objectively the therapeutic value of hCG treatment of women with threatened and habitual abortion[80].

The ability of hCG to maintain myometrial quiescence suggests it may be used in the treatment of preterm labor and delivery, unless it is caused by infection and premature rupture of membranes. In fact, administration of hCG has a tocolytic effect in a mouse preterm-labor model[74]. This effect seems to be mediated by down-regulation myometrial gap junction[80].

- **Central Nervous System (CNS):**

HCG treatment has been shown to improve recovery of spinal cord-injured rats. Motor neurons, among other cells in spinal cord, contain hCG receptors. Although how hCG working through these receptors in healing spinal cord injuries is not known yet, it is noteworthy that hCG belongs to the same family as nerve growth factor, hCG neurotropic and neurotransmitter properties, and may suppress the immune responses perhaps through its action on Tcell, monocytes, and macrophages[91,80].

- **Non trophoblastic Tumors:**

Potential hCG isused in prostate cancer treatment is based on the finding that tumors contain hCG receptors and that hormone has antiproliferative and anti-invasive actions in prostate cancer cell. These finding suggest a potential use for hCG in the treatment of castrated prostate cancer patients[74].

hCG also has multiple anti-breast cancer action, which we will detail, that may explain the decreased breast cancer incidence in some women whom complete full-term pregnancy at a young age.

2. Human Chorionic Gonadotropin and Breast Tumor

The view on the relation between hCG and breast cancer partially turned from disease and tumor marker to disease and hormonal therapy relation in the last decade.

The main focus of many laboratories investigation is the prevention of breast cancer through the understanding of the endocrinological and molecular aspects of inhibition of cancer initiation and progression. So they focused on the inhibitory effect of pregnancy, which is mediated by the placental hormone (hCG). These studies encompass from the endocrinological influences that modulate the normal development of the mammary gland to the role that these influences play in determining the susceptibility or resistance of this organ to undergo malignant transformation when exposed to exogenous carcinogens[142-144]. Breast cancer, like other cancers, is not one disease and has multiple etiologies and for these reasons, expectation that hCG treatment will help every women will be lowered.

2.1. Human Breast Development

Two important concepts, in breast development, which has been and continued being a major biological puzzle are that, this organ is one of few that is not completely developed at birth and it reaches its full differentiation only through the hormonal stimuli induced by pregnancy and lactation[145].

The development of the breast, which is vigorously controlled by the ovary, placenta and pituitary hormones, can be defined by several parameters, such as its external appearance, total area, volume, degree of branching, number of structures present in the mammary gland, and degree of differentiation of individual structures, i.e., lobules and veoli[147]. The study of breast development reveals that the breast is composed of lobular structures reflecting different stages of development. It composes of lobules type 1, which are the most undifferentiated ones and lobules type 2 evolve from the previous ones and have a more complex morphology. During pregnancy, lobules type 1 and type 2 progress to lobules type 3. Lobules type 4, which are present only during the lactational period of the mammary gland, regress to type 3 after weaning[145]

There is a significant difference in the content and relative percentage of lobular structures present in the breast according to the parity status of a woman. In nulliparous women, lobules type 1 and 2 are almost constantly present throughout their lifespan; the lobules type 2 decrease in number after menopause. In the parous woman's breast the lobules type 3 are the most frequent structures present. Only after the fourth decade of life there is an increase in the number of type 1 due to the regression of the more differentiated lobules type 3 at the end of the fifth decade of life, the breast of both nulliparous and parous women contains predominately lobules type 1[145,146].

2.2. Pathogenesis and Site of Origin of Breast Cancer

It is not known when in the lifetime of a woman breast tumor initiates. The term tumor is applied indistinctly to either benign or malignant lesions, with notifying that the breast is the most frequently diagnosed malignancy in the female population[147,148]. Studies of chemically induced carcinogenesis in an experimental animal model and primary cultures of human breast epithelial cells have shown that the initiation of the neoplastic process is inversely related to the degree of differentiation and *in vivo* cell proliferation of the mammary gland[149-151]. An important concept that emerged from these studies is that the lobules type 1 have been identified as the site of origin of preneoplastic lesions such as atypical ductal hyperplasia, which evolve to ductal carcinoma *in situ*, progressing to invasive carcinoma. Lobulars type 2 give rise to the origin of atypical lobular hyperplasia and lobular carcinoma in situ, whereas lobules type 3 and 4 originate more in benign breast lesions[145,147].

2.3. Effecting Factors on Breast Cancer

Despite the numerous uncertainties surrounding the origin of cancer and no clear understanding of the cause of the worldwide breast cancer incidence increase, several studies indicated several factors that cause or prevent breast cancer[152,153].

Several experimental studies on some of these factors postulate that they seem to affect the/or effect by architectural pattern of the breast[145,147,149].

Although there is no explanation as yet for higher breast cancer risk exhibited by nulliparous and late parous women, experimentally fact induced rat mammary carcinomas model develop only when the carcinogen interacts with the undifferentiated and highly proliferating mammary apithelium of young nulliparous rats[146,149]. These observation also support the hypothesis that the presence of lob1 explains the higher breast cancer risk of nulliparous women, as they represent the population with highest concentration of undifferentiated structures in the breast. Suggesting that these lobules are biologically different form those of early parous women.

The direct association of breast cancer risk with nulliparity, as well as the protection afforded by early first full-term pregnancy, has been in great part explained in many studies[153,154], figure(1-2). It has been observed in the rodent experimental model and in human, that cells derived from the differentiated lob 3 are resistant to growth in vitro and do not express transformation phenotype upon carcinogen treatment, as do cells from lob1. These observations support the hypothesis, which postulated that the induction of differentiation of the breast by the reproductive process is responsible for the inhibition of carcinogenic initiation[145-147,149,154], figure (1-3).

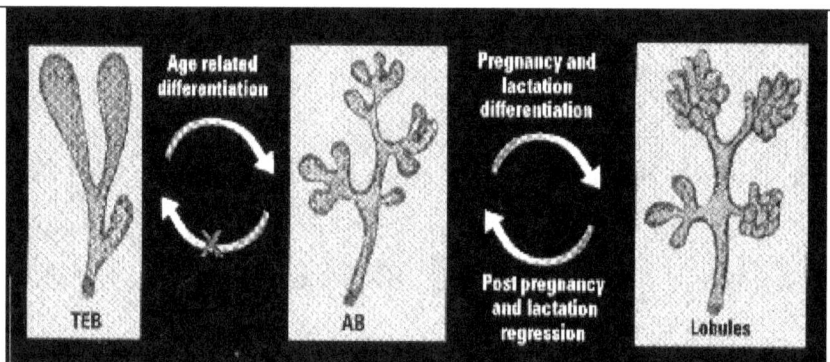

Figure (1-2):Terminal end buds (TEB) differentiation to alveolar buds(AB) in the nulliparous female under the regular hormonal stimuli of the menstrual cycle, pregnancy and lactation [145].

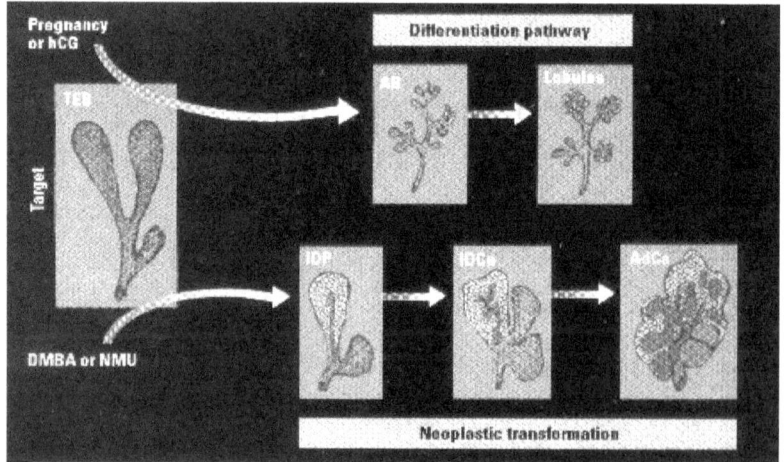

Figure (1-3): Terminal end buds (TEB) evolve to alveolar duds (AB) or lobules if pregnancy or hCG stimulate them towards the differentiation pathway. If a carcinogen reaches the target (TEB) during the susceptibility period, it diverts this evolution to yhe neoplastic transformation pathway, developing instead into intraductal proliferation (IDP), intraductal carcinoma (IDCa), and invasive adenocarcinoma (AdCa) [145].

2.4. Role of hCG in Breast Cancer

Several studies have established many evidence that there is very little probability for doubt that breast is another nongonadal target of hCG action and its action may be relevant to physiological changes in breast on preparation for lactation and at the same time confers protection against cancer[154-156].

- **Inhibition:**

Since pregnancy is a state of chorionic elevation of many hormones, the protective effect of pregnancy can be due to any of these hormones. Russo et.al., initiated a number of studies using a rodent breast cancer experimental model since women cannot be used to study. In this model, the administration of 7,12-dimethylbenz[α]anthracene (DMBA) resulted in a consistent induction of mammary tumors and was more effective in nulliparous than parous animals. It has been demonstrated that the inhibitory effect of pregnancy on mammary cancer initiation is mediated by hCG, depending on the observation of nulliparous animals which were greatly reduced prior to carcinogen administration when they ware pretreated or simultaneously treated with hCG[157-160,155].

This hormone may act on breast to promote nonreversible differentiation of proliferation-competent epithelial cell into secretory cells in terminal end buds. Coincidentally, this differentiation, which is a physiologic phenomenon to prepare the breast for lactation, also makes the cells less susceptible to carcinogenic transformation[146,147,149].

The effect of hCG on the development of mammary gland, not under the influence of pregnancy, has been evaluated to assess the effect of this hormone on two important endocrine organs: the ovary and pituitary gland. hCG showed no effect on ovarian-size, sera level of its hormones or sera level of pituitary hormones after cessation of the treatment; indicated that the hormonal milieu induced by hCG sufficed for differentiating the mammary epithelium[161,162]. These findings predict that early hCG treatment of women plan to delay their first pregnancy may reduce their breast cancer risk.

- **Progression:**

The studies of the productive effect of hCG-induced differentiation on experimental mammary carcinogensis led to postulate the possibility that hCG may be useful in preventing the development of breast cancer in women[155,159,163]. The Russo et.al., tested the effect of hCG on tumor progression, using rat experimental animal. They have found that treatment with hCG inhibited mammary carcinogenesis. This inhibition occurred by stopping the progression of early lesions (i.e, IDPs, CIS). Finding that indicated hCG has a significant potential as a chemoprventive agent not only before the cells were initiated by carcinogen, but after the carcinogenic process was vigorously progressing[155]. Also, emphasizing the importance of hCG receptor in mediating anticancer actions of hCG in breast cancer cell[153,164]. It has been shown that hCG can inhibit the tumor growth and invasion, increase the cyclic AMP and decrease ER levels in breast cancer cell lines which are positive ER, and can decrease the growth of ER negative breast cancer cell lines[153,164].

2.5. Mechanism of hCG Action in Breast Cancer

The differentiation effect, by hCG on the mammary gland, alone may not explain the protective role of this hormone. hCG may directly inhibit cell growth, invasion and proliferation[143,165]. The protective role phenomenon of hCG was found to be in a great part mediated by several changes that have been observed in the mammary gland of rat treated with hCG, either alone or after DMBA treatment, but they were absent in the animals treated with DMBA alone, figure (1-4), such as:

- **Effect on Inhibin Expression:**

The hCG-induced differentiation of the mammary gland was found to be associated with synthesis of inhibin. The expression of both inhibin α and β was increased and there synthesis was accompanied by a significant activation of *c-myc* and *c-jun* gene, which remained activated even after the cessation of hCG treatment[155,166]. The finding that *c-myc* and *c-jun* gene were also elevated indicates that early responsegenes could be involved in the pathway of hCG-inhibin-induced synthesis[155].

- **Effect on Programmed Cell Death (PCD) Genes Expression:**

It has been found that inhibition of progression of mammary carcinomas by hCG has been associated with the activation of genes known to be responsible of programmed cell death, cell growth arrest and apoptosis. There were remarkable inductions of several apoptotic genes such as P53, ICE [155,164,167,168]. The effect of hCG on the activation of PCD gene was specific for mammary gland, since the hormonal treatment did not modify their expression in the ovary, even though it is the target organ of hCG action[167]. In mammary and other epithelial cells in culture, both P53 dependent and P53 independent pathways have been identified. In hCG treatment the P53 dependent pathway was involved in the PCD process[167].

- **Effect on Insulin-like Growth Factor (IGF) Expression:**

hCG treatment decreases IGF and increases some of IGF binding proteins (IGFBPs). Since IGFs are potent mitogenes and antiapoptotic agents for breast cancer; thus decreasing of these factors, by hCG treatment, reduce the availability of IGF and its action[169].

- **Effect on New Genes Expression:**

The induction of the differentiated genes (β casein and whey acidic protein (WAP)), two of the major milk proteins in most specie, was observed in the treatment of mammary gland by hCG. Also a third gene called hormone-induced 1 (HI-1) was express in this process[155]. hCG treatment was found to be enhancing the cellular DNA repair mechanisms of the mammary epithelium and increasing the expression of the above genes that might prevent and/or decrease the expression of other that might promot the carcinogenic transformation of breast epithelial cell[164,167].

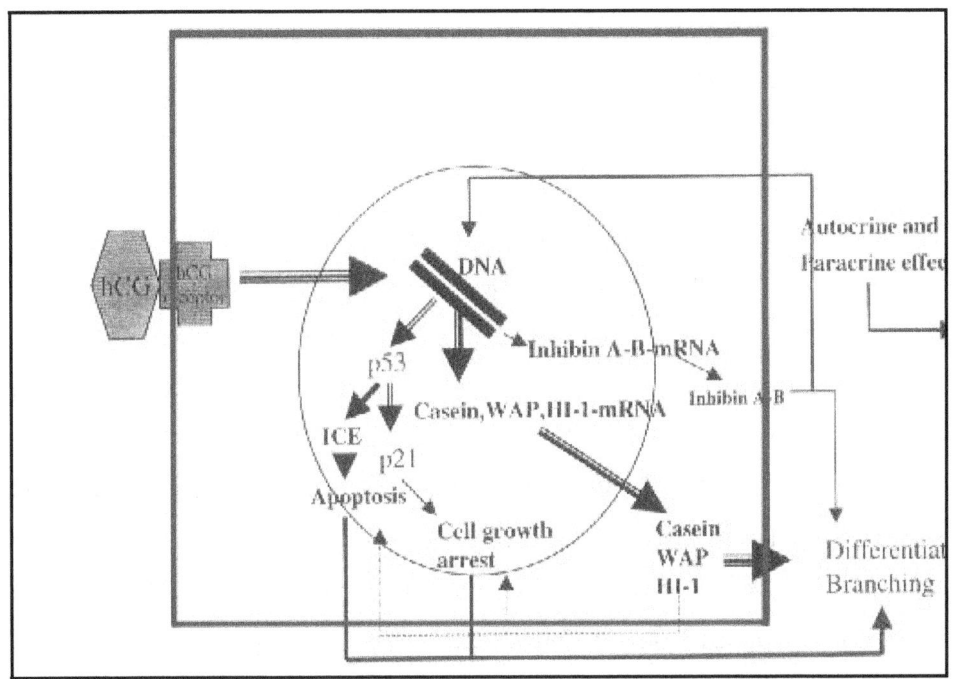

Figure (1-4): The postulated mechanism of action of hCG. The hormone binds to a specific membrane receptor, activating genes identified to be specific for pregnancy-or hCG-induced differentiation, and that have been found to be correlated with the lobular development of the mammary tissue. Thus, a pathway of activation of p53 and ICE may lead to apoptosis, or through p21 to cell growth arrest. Activation of inhibin α and β, and od the milk proteins, casein, whey acidic protein(WAP), and HI-1, may lead to the differentiation.

3. Human Chorionic Gonadotropin Receptor

There are two receptors for gonadotropins, one for FSH and another for hCG and LH. Although hCG and LH share a common receptor, not all the receptors binding characteristics of these two hormones are similar. This may be due to subtle structural difference between the two hormone molecules. Only intact hCG and LH but not any other hormones including alpha and beta subunits of hCG and LH, can bind to the receptors[170].

3.1. Structure of hCG Receptor

hCG receptor is a single chain transmembrane glycoprotein, which is a member of group II of the G protein-coupled receptor family[170]. In human, the hCG receptor gene is composed of 11 exons and 10 introns and its coding region is over 60 kb long. The receptor protein contains 696 amino acids.

It composes of highly conserved seven transmamberane domains (I through VII) contain six potential sites for N-linked glycosylation and three sites for protein kinase C phosphorylation, as in figure (1-5). The mature receptor contains an extra cellular domain of 333 amino acids, a transmembrane domain of 266 amino acids and intra cellular domain of 70 amino acids. The extracellular domain is rather large in keeping with the large size of the ligand[170,171].

hCG receptors are present not only in cell surface membranes but also to a variable degree in lysosomes, rough and smooth endoplasmic reticulum, cis and trans Golgi elements, nuclear membranes and interface of dispersed and condensed chromatin. Some of the properties of intracellular organelle receptors are similar while others are different from those in plasma membranes. These differences may be partly due to differences in membrane environment of the organelles[172,173].

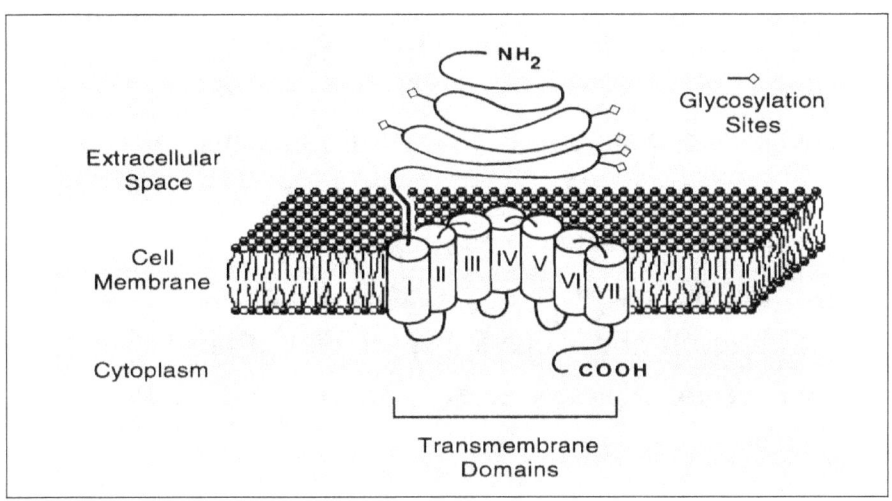

Figure (1-5): G-protein-coupled hormone receptor. HCG receptor is composed of seven highly conserved transmembrane domains (I to VII), a large extracellular domain with six potential glycosylation sites, and a relatively short cytoplasmic domain.,

3.2. Distribution and Function of hCG receptor

It was assumed for along time that only gonadal tissues could respond to hCG from human placenta and LH from the pituitary, because of the belief that the receptors which are required for their action were only present in the gonadal tissues.

This belief was challenged after the presnce of hCG receptor in the porcine uterus was first demonstrated[174].

Subsequently, several laboratories have employed many techniques for detecting hCG receptor by using Northern blotting, western blotting, ligand blotting, covalent receptor cross-linking, ligand binding, insitu hybridization, imunocytochemistry, gel mobility shift, gene transfecation, and laser scanning contocal microscopy. Not all techniques were used on all tissues; however more than one was used in most cases[80,172,175-177].

Tissues also contain lower levels of functional hCG receptors forcing a new look at old concepts that have been around for more than 50 years in reproductive biology and medicine[178].

Nongonadal hCG receptors have been described in a number of species; with identical affinity and specificity of binding to the hCG as gonadal receptors. These species include human, monkey, pig, cow, sheep, rat, rabbit, mice and turkey[80,172].

Even though more tissues than anyone ever imagined are now known to contain the receptors, other tissues like liver, kidney, lung, skeletal smooth muscle, heart and spleen are receptor negative[80,179].

Recent data have shown, while these in adult are receptor negative, some of these tissues in human fetus are receptor positive, suggesting that hCG may have been a developmental role[80].

Malignant transformation of nongonadal endocrine tissue can result in an inappropriate appearance of functional hCG receptor[180].

However not all kinds of tissues from all the species have yet been examined. The human nongonadal tissues are by far the most extensively studied among the species as lists in table (1-1), which included all tissues to date and some of their hCG receptor function.

Table (1-1): Nongonadal tissue distribution of hCG receptors and their functions.

Tissue	Function*	Reference
Oviduct	Release sperm bound to epithelial cell and enhance growth of early embryos in cocultures with epithilial cells	181
Uterus	Differentiation of stromal cells	182
Cervix	Increase COX expression	183
Placenta	Regulation of hCG biosynthesis	81
Fetal mambranes	Weaking of membranes through increase in COX-1 expression	83
Umbilical cord	Relax umbilical vessels	85
T-cells, monocytes, and macrophages	Increase monocyte chemoattractant protein-1 expression	184
Urinary bladder	Maintain normal itsfunctions	185
Skin	Regulate androgen metabolism and action	90

Bone	Turnover	186
Adrenal cortex-zona reticularis	Increase DHEAS secretion	187
Brain	Regulation of LH synthesis	90
Neural retina	Visual processing of information	188
Breast	Promote nonreversible differentiation of proliferative to secretory-type epithelial cells	154
Blood vessels in target tissues	Vasodilation through increasing prostacyclin	189
Prostate	Androgen metabolism and action	99
Male reproductive tract	Secretion and sperm maturation	190

*In many cases, functions are based on in vitro or in vivo data. In few cases, they are drevied from logical deduction

3.3. hCG Receptor in Breast Tissue

Several laboratories have now demonstrated that rat and sow breast tissues, normal and malignant human breast tissues and human breast cancer cell lines contain hCG receptor transcript and receptor protein which can bind ^{125}I-hCG[177,191-193].

The receptor levels are the highest in breast epthelium, followed by blood vessels, stromal cell, smooth muscles and higher in the luteal phase than in proliferative phase, suggesting that ovarian steroid hormones may regulate breast hCG receptor levels[191,192]. Sow breast receptor levels increase from the proliferative to the secretory phase, suggesting that progesterone may upregulate and/or estrogen may downregulate the receptors.

MCF-7 cells contain higher hCG receptor levels than other human breast cancer cell lines tested. This was due to an increased transcription of the gene of hCG receptor in the MCF-7cells[177].

In the breast cancer cell, hCG anticancer actions were found to be mediated by hCG receptor. So the present study was undertaken as a complementary study to characterize the binding condition between hCG and its receptor.

Aim of the work

1. Investigation the presence of hCG level is sera of patients with benign and malignant breast tumors.

2. Development of IRMA assay for the determination of hCG from cytosolic tissues homogenate of benign and malignant breast tumors.

3. Development a quantitative radioreceptor assay for detection and analysis of membrane-associated hCG-receptor of human breast cancer tissues.

4. Partially purification of hCG from pre-menopausal malignant breast tumor tissue homogenate and characterization its binding and study the effect of various factors on this binding (hCG concentration, ^{125}I-anti hCG antibody, pH, temperature, time and different halides).

5. Determination of kinetic and thermodynamic parameters of the binding of partially purified hCG with ^{125}I-anti hCG antibody.

6. Spectroscopic studies on standard and partially purified hCG, standard, benign and malignant complexes of hCG.

Chapter Two
Studies on the binding of I-anti hCG antibody to the hCG in human breast tumor homogenates

Experimental
Materials and Methods

2.1. Materials
☐ **Chemicals**

All laboratory chemicals and reagents were of analytical grade and were used without further purification and tabulated in the following table.

Table (2-1): Chemicals used and Companies.

Chemicals	Company
1. Immunoradiometric assay kit for hCG levels	Immunotech(France)
2. Radiometric assay kit for hCG levels	Immunotech(France)
3. Tris (hydroxy methyl) aminomethane, $CuSO_4.5H_2O$, NH_4Cl, $LiCl$, $ZnCl_2$, $MnCl_2$, $MgCl_2$, Na-K tartrate, Urea, $CaCl_2$, KCl, CsCl, ethylendiamine-tetraaceticdisodium salt (EDTA).	Fluka: (Switzerland)

4. NaCl, NaF, NaBr, NaI, glycerol, Na$_2$CO$_3$, Sucrose, NaOH, hydrochloric acid, Folin-ciocolteau, Ethanol and Bovine serum albumin (BSA)	BDH, limited, Poole (UK)

☐ **Instruments**

The instruments used in this work were tabulated in the following table

Table (2-2): Instruments used and Companies.

Instruments	Company
1. Gamma counter type 1270-rack gamma II 2. Spectrophotometer ultraspace type 4050	LKB
3. Double beem cintra 5	GBC Scintific

spectrophotometer	Equipment(Australia)
4.pH-meter	Pye-Unicam
5.Cooling centrifuge; with a maximum speed 5000 r.p.m.	Hettich
6.Memmert water bath, memmert incubator	West Germany
7. SM-shaker	England
8. Combicold rack	LKB

Patients

Three groups of breast tumor patients were included in this study.

Group I : Consisted of 25 patients with benign breast tumors

Group II : Consisted of 21 premenopausal patients with breast cancer.

Group III : Consisted of 14 postmenopausal patients with breast cancer.

Group IV : Consisted of 12 controls.

The patients were newly diagnosed and were not undergone any type of therapy. Patients suffered from any disease that may interfere with our study were excluded.

All patients were admitted for diagnosis and surgery to University Hospital (Al-Nahrain College of Medicine), Nursing Home Private Hospital, Al-Yarmuk Teaching Hospital, Al-Jaderia private hospital and Al-Arabi Private Hospital.

The host information of all patients and normal healthy subjects are summarized in table (2-3).

Table (2-3): The host information of breast tumor patients and healthy subjects studied.

Group	Patients	No.	Age	Type of tumor
I	Benign breast tumor	25	14-41	Fibrocystic chandes (adenosis)
II	Premenopausal malignant breast tumor	21	30-47	Infiltrattive ductal carcinoma
III	Postmenopausal malignant breast tumor	14	54-72	Infiltrattive ductal carcinoma
IV	Control	12	14-37	-

Blood Sampling

Blood samples (7ml) were obtained from the patients just before surgery by veinpuncture. Age matched sera were obtained from (12) healthy premenopausal women.

Blood samples were centrifuged at 1500xg for 10 minutes after allowing the blood to clot at room temperature. The sera were aliquoted and frozen at –20 °C until assaying.

Collection of Breast Tissue Specimens

The tumor tissues were surgically removed from breast tumor patients by either mastectomy (cancer patients) or lumpectomy (benign tumor patients). The specimens were cut off and immediately rinsed with ice-cold isotonic saline solution. They were collected individually in plastic receptacles and stored at −20 ºC until homogenization.

☐ Preparation of Breast Tumor Tissue Homogenate

The frozen tissue were thawed, weighed, pulverized finely with a scalpel in petri dish standing on ice bath, and then homogenized at 4ºC in a buffer solution with a ratio of 1:3 (weight-volume) by using a manual homogenizer[208].

The buffer used was Tris/HCl (0.05M, pH 7.2) containing (0.25M sucrose, 5mM EDTA and glycerol 20%). The homogenate was filtered through several layers of nylon gauze to eliminate fibers of connective tissue, then centrifuged at 1000xg for 15 minutes at 4ºC. The supernatants were used throughout our study[209,210].

Solutions

Tris/HCl 0.05M, pH 7.2 containing (0.25M sucrose, 5mM EDTA and glycerol 20%) was prepared by dissolving (0.606 gm) of Tris (hydroxymethyl) amino methane and (8.345 gm) of sucrose, (0.08 gm) EDTA and 20 ml glycerol in (95 ml) distilled water. Then the pH was adjusted to pH 7.2 using HCl (0.2 M). The volume was completed to (100ml) with distilled water.

2.2. Methods

☐ Determination of hCG Levels in Sera of Benign and Malignant Breast Tumor Patients and Controls

Sera levels of hCG were measured in samples collected from pre- and post-menopausal patients and healthy premenopausal women by immunoradiometric assay (IRMA).

The assay details were carried out according to the following procedure:

1. A set of coated tubes containing 50μl of hCG standards or sera of patients and controls, 200 μl of ^{125}I-anti hCG antibody was added to each tube, then mixed well by hand.
2. Two additional non-coated tubes containing only 200 μl of ^{125}I-anti hCG antibody, for total activity computation, were set aside until counting.
3. All tubes were incubated for 60 minutes at 25°C on a horizontal shaker.
4. Aspirate thoroughly the contents of all the tubes except of those for total activity (T) measurement.
5. The tubes were wash twice with 2ml of washing solution and aspirate the solution immediately.
6. The radioactivities of all tubes were measured by gamma counter.
7. The assay protocol of sera hCG was described in Table (2-4).

Table (2-4) IRMA assay protocol of sera hCG (IU/L)

Step 1 Pipetting	Step 2 Incubation	Step 3 Counting
Add to coated	Incubate the	Aspirate the

tubes Successively: 50 µl of calibrator or sample and 200 µl of tracer then Mix.	mixture 1 hour at 18-25°C with shaking.	content of the tubes carefully, except 2 tubes for total activity (T). Rinse the tubes with 2ml of wash solution and aspirate twice, then count activity (cpm).

Reagents

The reagents used in the assay were provided with the kit, and described as follows: -

1. Tracer: one vial contains 22ml antibody to hCG (mouse monoclonal IgG) labeled with ^{125}I, buffer, protein, preservatives and inert red dye. Radioactivity is 650 KBq.

2. hCG standards: The vials contain increasing amount of human hCG standard, serum and preservatives. The standards concentrations are: 0.0-750 IU/L.

3. Coated tubes: The inner surface of each tube is coated with IgG to hCG (mouse monoclonal) directed against an hCG epitope.

4. Wash solution: contains saline solution and detergents.

5. Control sera: the vial contains hCG, human sera and preservatives.

Calculations

1. The mean of c.p.m was determined for each pair of duplicate tubes of standard and unknown sample.

2. A standard curve was drawn by plotting the c.p.m for each standard on the Y-axis against the corresponding concentration on the X-axis, as shown in figure (2-1).

3. The unknown concentration of the sample was calculated from the standard curve.

☐ Determination of Total Protein Content in Breast Tumor Homogenate

The total protein content of breast tissue homogenate was determined by the method of Lowry et.al, using bovine serum albumin (BSA) as the standard protein and the absorbence of the developing color was read at 600 nm against

the appropriate blank[211].

2.3. Binding Studies of hCG in Breast Tumor Homogenate with ^{125}I- anti hCG Antibody

2.3.1. Preliminary Tests of hCG Binding in Breast Tumor Homogenate with ^{125}I-Anti HCG Antibody

1. One hundred microliters of breast tumor homogenate containing 200µg protein for benign, pre- and post-menopausal malignant breast tumor respectively was added to 5 µl (for benign breast tumor) and 10 µl (for pre- and post-menopausal breast tumor) of ^{125}I-anti hCG antibody (1470µg.ml^{-1}), the volume of the mixtures were made up to 250 µl with Tris/HCl buffer (0.05M, pH7.4) containing 0.1% BSA.

2. Two additional tubes containing only 5,10 µl of ^{125}I-anti hCG antibody, were set aside until counting for total activity.

3. The tubes were incubated at 25°C for 60 min.

4. After incubation, 500 µl of PEG 6000(10%) were added to the tubes and incubated again for 150min at 4°C.

5. After incubation, the tubes were centrifuged at 1500xg for 30min at 4°C.

6. The supernatant was discarded by decanting the assay tubes, then the tubes were inverted on a filter paper for 10 min.

7. The rims of the tubes were swabbed with a cotton piece and the amount of bound radioactivity (c.p.m) was counted using gamma counter.

Solutions

1. Tris/HCl buffer (0.05M, pH 7.4) containing 0.1% BSA was prepared by dissolving (0.606gm) of Tris (hydroxymethyl) amino methane and (0.1gm) of BSA in (95ml) distilled water. Then the pH was adjusted to pH 7.4 using HCl (0.2M). The volume was completed to (100ml) with distilled water.

2. PEG 6000(10%) was prepared by dissolving (10gm) of PEG 6000 in 100ml of Tris/HCl buffer (0.05M, pH 7.4)

Calculations

1. The bound fraction (B) represents the counted radioactivity in each tube, expressed in c.p.m i.e (^{125}I-anti hCG antibody/hCG) complex.

2. Total activity (T) represents the counted radioactivity in the tubes containing only ^{125}I-anti hCG antibody.

3. The (B/T)% ratio for each tubes were calculated as follows:

$$(B/T)\% = \frac{\text{Sample mean counts (B)}}{\text{Total activity mean counts (T)}} * 100$$

2.3.2. Most Appropriate Conditions of the Binding of hCG in breast Tumor Homogenate with ^{125}I-anti hCG Antibody.

☐ ***The Effect of Different Protein Amount of Breast Tumor Homogenate on the Binding of hCG with ^{125}I-anti hCG Antibody***

1. A volume of 5 μl (for benign breast tumor) and 10 μl (for malignant pre- and post- menopausal breast tumor) of ^{125}I-anti hCG antibody were added to 100μl containing increasing amounts (100, 200, 300, 400, 500, 600, 700 and 800 μg protein) for benign, malignant pre- and post- menopausal homogenate of breast tumor respectively in a final volume of 250 μl which made up with Tris/HCl buffer (0.05M, pH 7.4) containing 0.1% BSA.

2. Two additional tubes containing only 5 and 10 μl of ^{125}I-anti HCG antibody were set aside until counting for total activity.

3. The tubes were incubated at 25°C for 60 minutes.

4. After incubation, the (^{125}I-anti hCG antibody/hCG) complex was estimated by following the steps 4, 5,6 and 7 in section (2.3.1).

Solutions

Tris/HCl buffer (0.05M, pH 7.4) containing 0.1% BSA and PEG 6000 (10%) were prepared as described in section (2.3.1).

Calculations

1. The (B/T) percent values were determined according to section (2.3.1).

2. The percent of binding values (B/T)% were plotted against the concentration of ^{125}I-anti hCG antibody.

☐ ***The Effect of Different Concentrations of ^{125}I-anti hCG Antibody on the Binding with hCG in Breast Tumor Homogenate***

1. One hundred microliters of the optimum amounts of breast tumor homogenate (700, 600 and 600μg proteins of benign, malignant pre-

and post-menopausal, breast tumors homogenate respectively), were added to increasing volumes 2, 3, 4, 5, 6, 8 and 10μl (0.012-0.059 μg.ml^{-1}) for benign and malignant postmenopausal breast tumors and 10, 15, 20, 25, 30, 35, 40 and 45μl (0.059-0.2646 μg.ml^{-1}) for malignant premenopausal breast tumors of ^{125}I-anti hCG antibody (1470μg.ml^{-1}) then the volume were made up to 250 μl with Tris/HCl buffer (0.05M, pH 7.4) containing 0.1% BSA.

2. A set of tubes containing only the same increasing volumes of ^{125}I-anti HCG antibody (2, 3, 4, 6, 8, 10, 15, 20, 25, 30, 35, 40 and 45 μl) were set a side until counting for total activity computation.

3. The tubes were incubated at 25°C for 60 minute.

4. After incubation, the (^{125}I-anti hCG antibody/hCG) complex was estimated by following the steps 4, 5,6 and 7 in section (2.3.1).

Solutions

Tris/HCl buffer (0.05M, pH7.4) containing 0.1% BSA and PEG 6000 (10%) were prepared as described in section (2.3.1).

Calculations

3. The (B/T) percent values were determined according to section (2.3.1).

4. The percent of binding values (B/T)% were plotted against the concentration of ^{125}I-anti hCG antibody.

☐ *The Effect of pH on the Binding of hCG in Breast Tumor Homogenate with ^{125}I-anti hCG Antibody*

1. One hundred microliters of the optimum amounts of breast tumor homogenate (700, 600,600 μg Proteins of benign, malignant pre- and

post-menopausal, breast tumors homogenate respectively), were added to (3 μl (0.01764μg.ml^{-1}); 25 μl (0.147μg.ml^{-1}) and 4 μl (0.02353μg.ml^{-1}) for benign and malignant pre-, post-menopausal breast tumors homogenate respectively) of ^{125}I-anti hCG antibody (1470μg.ml^{-1}) the mixtures volumes were made up to 250 μl with Tris/HCl buffer (0.05M) containing 0.1% BSA of different pH (6.8, 7.0, 7.2, 7.4, 7.6, 7.8 and 8.0).

2. Three additional tubes containing only (3, 4 and 25) μl of the ^{125}I-anti hCG antibody were set aside until counting for total activity computation.

3. The tubes were incubated at 25°C for 60 minute.

4. After incubation, the (^{125}I-anti hCG antibody/hCG) complex was estimated by following the steps 4, 5,6 and 7 in section (2.3.1).

Solutions

Tris/HCl buffer (0.05M) containing 0.1% BSA and PEG 6000 (10%) in different pH were prepared as shown in section (2.3.1).

Calculations

1. The values of (B/T) % were determined according to section (2.3.1).

2. The percent of binding values (B/T)% were plotted against their corresponding pH values.

□ *Time Course of the Binding of hCG in Breast Tumor Homogenate with ^{125}I-anti hCG Antibody*

1. One hundred microliters of the optimum amounts (700, 600,600 μg for benign, malignant pre- and post-menopausal, breast tumors homogenate respectively), were added to 3μl (0.01764μg.ml^{-1}), 25μl (0.147μg.ml^{-1}) and 4μl (0.02353μg.ml^{-1}) for benign and malignant pre-, post-menopausal breast tumors homogenate respectively of ^{125}I-

anti hCG antibody (1470µg.ml⁻¹). The volumes were completed to 250 µl with Tris/HCl buffer (0.05M, pH 7.2) containing 0.1% BSA.

2. All tubes were incubated at 25°C for different time interval (30, 60, 90, 120, 150 and 180) min.

3. Three additional tubes containing only (3, 4 and 25) µl of the ^{125}I-anti HCG antibody were set aside until counting for total activity computation.

4. After incubation, the (^{125}I-anti hCG antibody/hCG) complex was estimated by following the steps 4, 5, 6 and 7 in section (2.3.1).

5. To determine the time course of hCG binding to ^{125}I–anti hCG antibody at different temperatures. Steps 1, 2 and 3 in the same experiment were repeated at different temperatures (5, 37, 45 °C).

Solutions

Tris/HCl buffer (0.05M, pH 7.2) containing 0.1% BSA and PEG 6000 (10%) in pH 7.2 were prepared as described in section (2.3.1).

Calculations

1. The (B/T) percent values were determined according to section (2.3.1).

2. The values of (B/T)% were plotted against the corresponding incubation time.

☐ **The Effect of Halides on the Binding of hCG in Breast Tumor Homogenate with ^{125}I-anti hCG Antibody**

1. One hundred microliters of the optimum amount of (700, 600, 600 µg proteins of benign, malignant pre- and post-menopausal, breast tumors homogenate respectively), were added to 3 µl (0.01764 µg.ml⁻¹), 25µl (0.147µg.ml⁻¹) and 4µl (0.02353µg.ml⁻¹) for benign and malignant pre-, post-menopausal breast tumors

homogenate respectively) of ^{125}I-anti hCG antibody in a final volume of 250 μl (completed with Tris/HCl buffer (0.05M, pH 7.2, 0.1% BSA) containing 0.1M of each of the following halides: NaF, NaCl, NaBr and NaI. A sample without the addition of any halides was used as a control.

2. Three additional tubes containing (3, 4 and 25 μl) of ^{125}I-anti hCG antibody only, for total activity computation, were set aside until counting.

3. The tubes were incubated for 150 min at 4°C (benign breast tumor homogenate), 120 and 150 min at 45°C (pre-and post-menopausal malignant breast tumor homogenate).

4. After incubation, the (^{125}I-anti hCG antibody/hCG) complex was estimated by following the steps 4, 5,6 and 7 in section (2.3.1).

Solutions

1. Tris/HCl buffer (0.05M, pH 7.2) containing 0.1% BSA and PEG 6000 (10%) in pH 7.2 were prepared as described in section (2.3.1).

2. The halides stock solutions (0.1M) were prepared by dissolving each of the following amounts of salts in 250ml Tris/HCl buffer (0.05M, pH 7.2, 0.1% BSA): 1.049g of NaF, 1.46g of NaCl, 2.57g of NaBr and 3.74g of NaI.

Calculations

1. The values of (B/T)% were determined according to section (2.3.1).

2. The values of (B/T)% were plotted against halides concentrations.

2.4. Determination of Affinity Constant (K_a) and the Maximal Binding Capacity (B_{max}) of hCG in malignant Premenopausal Breast Tumor Patients Associated with ^{125}I-anti hCG Antibody

1. One hundred microliters of the optimum amount of (600μg protein) of malignant premenopausal homogenate was incubated with increasing volumes (5, 10, 15, 20 and 25 μl) of ^{125}I-anti hCG antibody (0.0294-0.147μg.ml^{-1} protein). The final volumes were made up to 250 μl with Tris/HCl buffer (0.05M, pH 7.2, 0.1% BSA).

2. A set of additional tubes containing only increasing volumes (5,10,15,20 and 25μl) of ^{125}I-anti hCG antibody were set aside until counting for total activity computation.

3. The tubes were incubated for 120 min at 45°C.

4. After incubation, the steps 4,5,6 and 7 of the experiment (2.3.1) were repeated.

5. The previous steps were performed at different temperature (4, 25 and 37°C), the time of incubation needed to get the equilibrium state were (120min at 5 °C, 90 min at 25°C and 150min at 37°C).

Solutions

Tris/HCl buffer (0.05M, pH 7.2, 0.1% BSA) prepared as mentioned in section (2.1.7).

Calculations

1. The values B/F ratio were determined:-

   B: The bound radioactivity (mean of counts c.p.m), which represents the (^{125}I-anti hCG antibody/hCG) complex.

   F: The free radioactivity (mean of the counts c.p.m), which represented the non-bound ^{125}I-anti hCG antibody.

   T: The total radioactivity mean of the counts.

   F = Total counts (T) – Bound radioactivity (B)

2. The concentration of the (^{125}I-anti hCG antibody/hCG) complex in (mg/ml) that formed after time (t) was calculated from the

$$B(ug/ml) = \frac{B(c.p.m)}{T(c.p.m)}$$

following equation:-

3. The affinity constant and the maximal binding capacity were determined according to Scatchard equation: -

$$\frac{B}{F} = \frac{1}{Kd}(B_{max} - B)$$

$$Ka = \frac{1}{Kd}$$

Where:
   Ka = Affinity constant
   Kd = Dissociation constant
   B_{max} = Maximal binding capacity

4. The values of the ratio B/F were plotted against the values of B in (μg /ml), gives a linear relationship. The values of the affinity constant of the binding (Ka) at each temperature can be calculated from the slop of the straight line, while the value of the total concentration of hCG (B_{max}) in breast tumor tissue was calculated from the intercept with the x-axis.

Results and Discussion

Three groups of breast tumor patients were included in this study. These groups were classified according to the type of the tumor, as confirmed by histopathological examination. Tissue homogenization was carried out in 0.25M sucrose, 5mM EDTA, 20% glycerol and in cold medium in order to avoid protein denaturation and to decrease the proteolytic enzymes activity.[208,209]

Sucrose is a hypotonic solution that enhances the rupture of plasma cell membranes and preserves other cell organelles. The proteolysis of the proteins, by endogenous protease, has been inhibited by the inclusion of EDTA as an inhibitor in extraction buffer, while supplementation of the buffer with glycerol will improve the stability of the crude membrane receptor[210,211].

Determination of hCG Levels in Sera of Breast Tumors Patients

With a variety of techniques, hCG has been stated to be an appropriate product of human breast carcinoma. It has been proposed as a marker for breast cancer on the basis that raised levels were found in the peripheral blood of some patients. The first known reference of an association of hCG with breast cancer was made by McArthur[212]. But, at that time no biological test specific for hCG was available to substantiate the claim. With the generation of antisera to the β-subunit of hCG, a radioimmunoassay (RIA) was set up which has been stated not to cross-reaction with hLH and enable small amounts of hCG to be assayed in blood[194]. A number of studies using peripheral blood of breast cancer patients with different clinical tumor status have since been published in which a wide variety in the prevalence of raised levels was noted[195-198]. In this study, the hCG levels in sera of patients with benign and pre and post-menopausal malignant breast tumor were measured by IRMA method with matching with a control group subjects. Table (2-5) summarized the groups and the mean concentration of hCG in the four groups. For the mean serum hCG level showed a slight elevation (5.8 ± 0.5) as compared to the control mean serum. For malignant pre-menopausal, hCG level showed detectable elevation than control group but lower than postmenopausal malignant group, which showed highest elevation than the other groups. These results are similar, in general, to previous studies in which the presence, distribution and levels of hCG and/or its subunits and its genes have been previously studied in sera[195,213-214], tissues[198,199,200,203], cyst fluid[215,216], cytosol[51] and cell lines[203,217] of different types of breast tumors. In these studies, high levels of hCG or its subunits have been found in post-menopausal in comparison with pre-menopausal women. Other study demonstrated that all βhCG producing tumors were ductal carcinomas of the breast[196]. But in general these studies have found no positive correlation neither between the levels of hCG and/or its subunits and clinical stage nor the mass of the tumor. For sera investigation, different percentage of hCG levels from these studies have been estimated. Some studies gave a percentage with the range of (13-17%)[195,213,197], while other give it with the range (30-50%)[209,214,218]. The explanation for the synthesis of the hCG hormone by non endocrine tumors is uncertain. Many hypothesis appeared to explain this phenomena. One, that ectopic production of the hormone represent gene depression associated with malignancy[195,198]. Other hypothesis that in a few cases the tumor might produce hormone of random peptide synthesis. Immunohistochemical localization of hCG and its γ and β subunits in breast cancer cell suggests that most of the βhCG measured in the protein extracts was tumorigenic in origin[87,219]. Because of continuing need to derive markers demonstrable in breast tumor or in the sera of breast cancer patients, recent RT-PCR assay for multiple markers has been used to detect circulating breast cancer cells using hCG[202,220]. This method has shown that positive RT-PCR signal in blood samples of affected patients correlated with stage, in particular for βhCG. Also recent multimarker revers trancription-PCR assay has shown that a combination of βhCG and other marker correlated significantly with tumor size

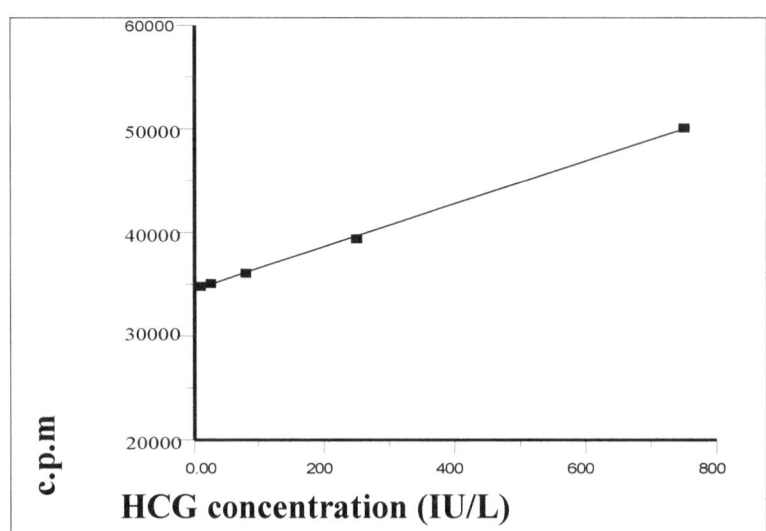

Figure (2-1): Standard curve of hCG determination in human sera by IRMA method. (all other details are explained in the text).

Table (2-5): Sera hCG levels (IU/L) in patients with benign and malignant breast tumors. (All details are explained in the text).

Group	Patients	No. of cases	Age range (Year)	Sera HCG IU/L (mean ± SD)	P values
I	Benign breast tumors	25	14-41	5.8 ± 0.5	$P>0.05$
II	Premenopausal malignant breast tumors	21	30-47	9.3 ± 1.9	$P<0.05$
III	Postmenopausal malignant	14	54-72	18.7 ± 2.1	$P<0.05$

	breast tumors				
IV	Control	12	14-37	5.2±0.8	--

Binding Studies of hCG in Breast Cancer Tumor Homogenate with ^{125}I-anti hCG Antibody

Preliminary Tests of hCG Binding in Breast Tumor Homogenate with ^{125}I-anti hCG Antibody

Binding study of radioliabled anti hCG antibody and hCG in breast tumor tissues was tested by preliminary binding. This test was made by incubation of ^{125}I-anti hCG Antibody with homogenate for 60 min at 25°C.

These conditions, as a beginning, were used for incubation according to the conditions that mention in the kit. For the separation (precipitation) of the complex from binding mixture, at first we used the centrifugation at 1500xg for 30 min only without using any precipitating reagent.

The percent of binding values (B/T%) were very low, about (1.5-2.0 %) for the three groups. So, the try to use precipitating reagent was made. PEG 6000 was used to precipitate the complex with testing its precipitation effect on ^{125}I-anti hCG Antibody alone.

It has been found that percent of binding values (B/T%) were rise to (3.8%) for benign, (22.8%) for malignant premenopausal and (22.5%) for malignant postmenopausal breast tumors, without precipitation of the free tracer.

In these experiments, the obtained data revealed that precipitation of complex by centrifugation alone is not efficient, which lead to use another way to obtained better yield. It also reveals that malignant breast tumors included higher incidence of hCG than those of benign breast tumors.

Most Appropriate Conditions of the Binding of hCG in breast Tumor Homogenate with ^{125}I-anti hCG Antibody.

The Effect of Different Protein Concentration of Breast Tumor Homogenate on the Binding of hCG with ^{125}I-anti hCG Antibody

The binding of antigen(hCG) to antibody (^{125}I-anti hCG Antibody) is not static, it is instead an equilibrium reaction that proceeds in three phases as in the following equation:

$$Ag_n + Ab \underset{k_{-1}}{\overset{k_{+1}}{\rightleftharpoons}} Ag_n Ab \underset{k_{-2}}{\overset{k_2}{\rightleftharpoons}} Ag_a Ab_b \quad \cdots\cdots\cdots\cdots (1)$$

Depending on the relative concentration of Ag and Ab, the complex may cross-link in the third phase to form larger complexes, which then precipitate out of the solution, which namely more precipitation complex, higher binding. So, to determine the effect of different Ag(hCG) concentration from breast tumor homogenate on the binding, increasing amount of homogenate was incubated with ^{125}I-anti hCG Antibody. The binding of hCG in breast tumor with ^{125}I-Anti hCG Antibody, as shown in figure (2-2), was detected only in highly concentration of the homogenate (300 µg) for benign tumor and increases with homogenate increasing until it reaches the saturability at (700 µg) with little increase of the next concentrations. In malignant tumors, bindings were higher from the primary concentration, with the same curve behavior. Previous studies have shown that besides circulatory hormone, mammary glands may synthesize hCG or similar peptides which may act in an autocrine and paracrine manner. They found that human breast cancer tissues immunostain for hCG and breast cyst fluid contains very high levels of biological active hCG as compared to blood[221,222].

Six hundred microgram of malignant pre- and postmenopausal breast tumors and (700 µg) for benign breast tumors were used in the next experiments since they gave maximum value of binding (B/T%).

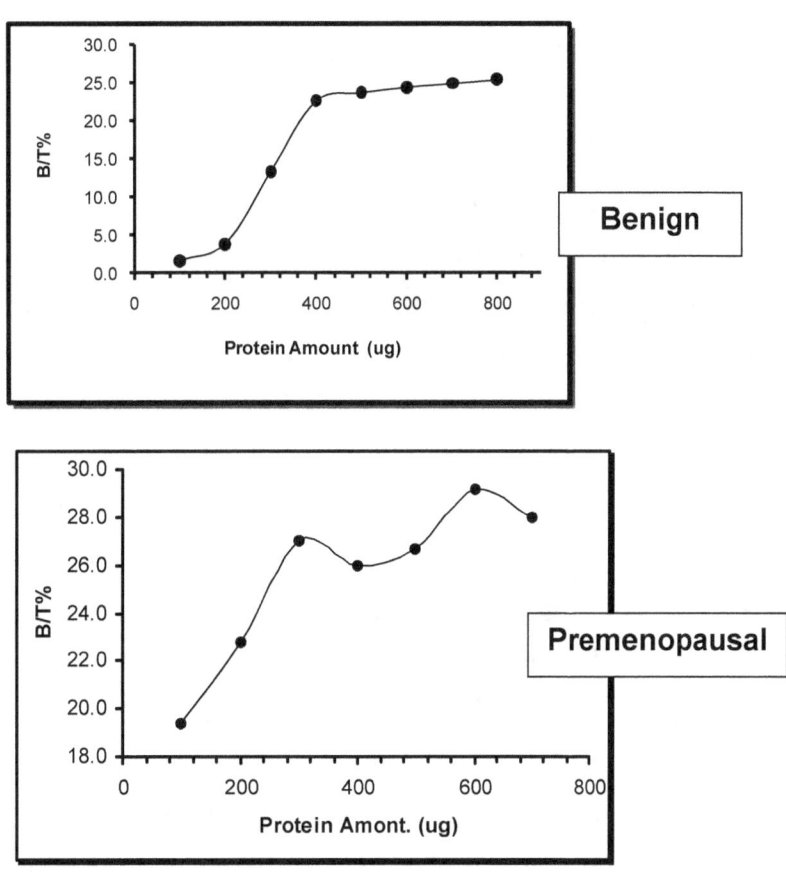

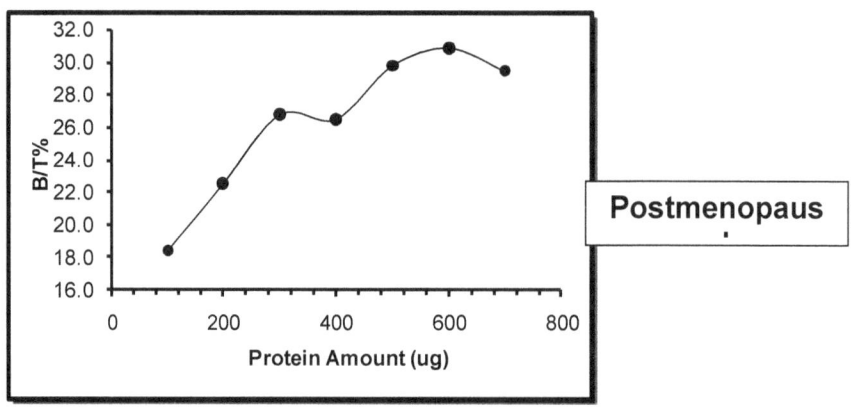

Figure (2-2): Influence of increasing protein amount on the binding with ^{125}I-anti hCG Antibody. (All other details are explained in the text).

The Effect of Different Concentrations of ^{125}I-anti hCG Antibody on the Binding with hCG in Breast Tumor Homogenate

As in the previous equation (1), the binding reaction may also be antibody (^{125}I-anti hCG Antibody) dependent process. To investigate that, increasing concentration of antibody, presents by ^{125}I-anti hCG antibody, was incubated with fixed amount of homogenate protein. Binding curve behavior for the three groups is similar, as shown in figure (2-3). For benign tumor, binding reached its maximum level from the primary ^{125}I-anti hCG antibody concentration (3μl, 0.0176μg.ml^{-1}); clarifying that the presence of hCG in benign breast tumor is very little and its detection need high concentration level of homogenate with low concentration of ^{125}I-anti hCG antibody to reaches equilibrium, which presents by maximum binding. This finding may explain some previous works, which have shown the presence of hCG in urine and normal tissue only be concentration of extracts to achieve hCG levels sufficient for measurement[196]. Also unable to detect measurable level of hCG in many breast tumors by using the standard condition of the hCG kits which use low amount of sera and much higher excess amount of the ^{125}I-anti hCG antibody (as in IRMA kit). Premenopausal group has shown higher ^{125}I-anti hCG antibody concentration need (25μl, 0.147μg.ml^{-1}) with less maximum binding (32.4%) than postmenopausal group (33.2%, 4μl(0.0235μg.ml^{-1})). In malignant breast tumor, it was not arrived in the previous studies to a conclusion if breast cancer females at postmenopausal age have a larger level or number of hCG secreting tumor than of younger age group[195,198]. Beside that, although the relation between the size and stage of tumor with the secretion of hCG is not significant, some studies declared that some breast tumors with large volumes secreted large amount of hCG[198]. These finding may explain the need of premenopausal tumor group, which having a large and bigger tumor volume than postmenopausal group, to more amount of ^{125}I-anti hCG antibody concentration.

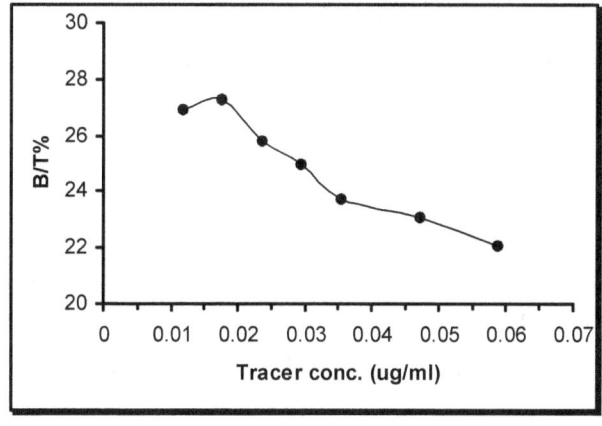

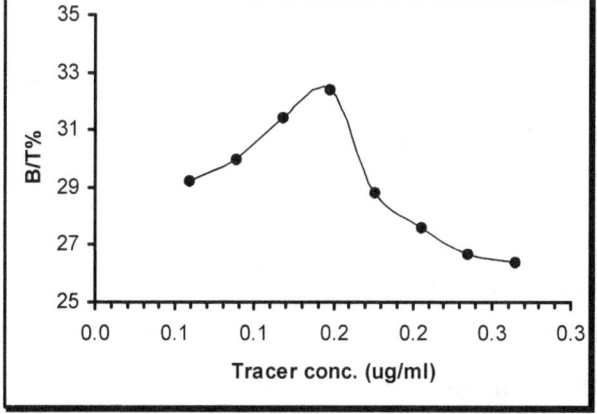

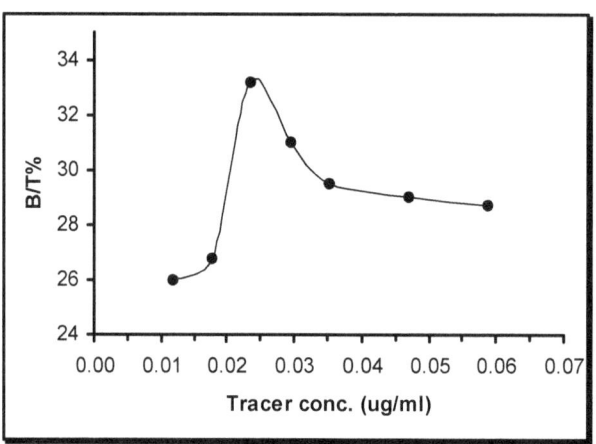

Figure (2-3): Influence of increasing ^{125}I-Anti hCG Antibody concentration on the binding with hCG in the homogenate. (All other details are explained in the text.

The Effect of Different pH on the Binding of hCG in Breast Tumor Homogenate with ^{125}I-anti hCG Antibody

One of the factors that affect complex precipitation is pH. To determine the optimal pH for hCG activity, the binding of fixed concentration of ^{125}I-anti hCG antibody was performed with fixed amount of tumors homogenate at different pH.

The binding value is stated in figure (2-4), showing that the three groups reached maximum binding at pH 7.2 with decrease in binding percent at pH higher or lower than optimum pH.

The attraction between antigen and antibody molecules involves electrostatic attraction, hydrogen bonding, vander waals forces and hydrophobic interactions.

In previous studies, a short exposure of hCG solution to pH less than pH 5.0 caused a slight reduction in the biologic activity, indicating that neuraminic acid was not released[223].

In solution, optimum pH is the most important contributors to nonequivalent attraction between antigen and antibody making it one of the factors that influence reaction speeds[224].

The optimum pH in this experiment may charged polar groups on the amino acids residues of hCG and/or ^{125}I-anti hCG Antibody, making them strongly attracted to each other.

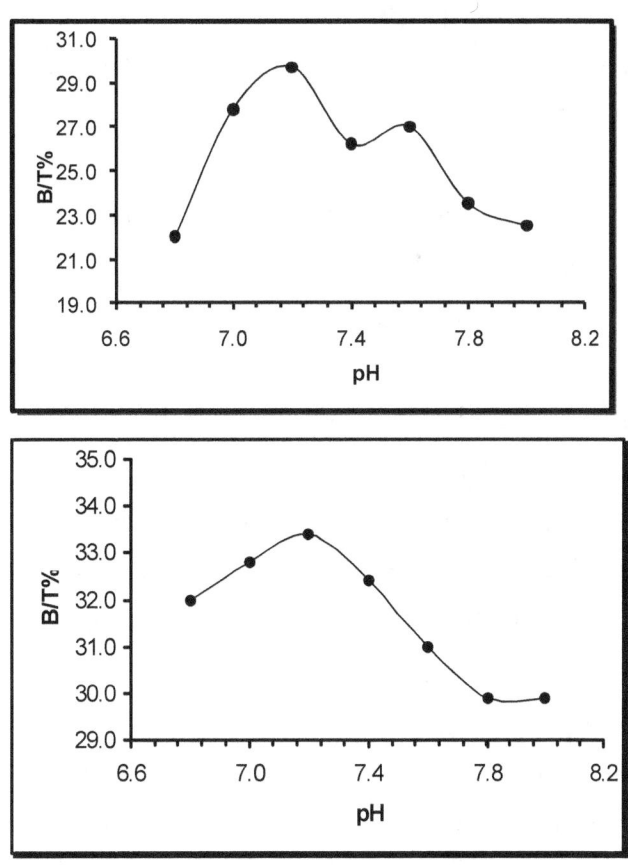

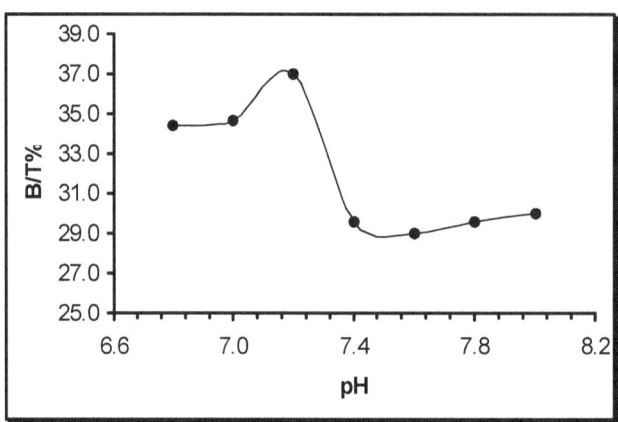

Figure (2-4): Influence of pH on the binding of ^{125}I-anti hCG Antibody with hCG in the homogenate. (All other details are explained in the text).

Time Course of the Binding of hCG in Breast Tumor Homogenate with ^{125}I-anti hCG Antibody

To determine whether the rate of hCG binding to ^{125}I-anti hCG antibody is temperature and time dependence, the bindings were cared out at different incubation time in four temperatures (5,25,37,45) °C; as shown in figure (2-5).

For benign tumors group, the binding shows inverse relation between elevation of temperature and percent of maximum binding, which was greatest (35.9%) at 5°C in 120 min. Malignant tumor groups exhibited variant behavior, the relation between elevation of temperature and present of binding was in conformance. The maximum binding for premenopausal was (46.2 %) at 45 ° C in 120min and was (41.7%) for postmenopausal at 45°C in 160min of incubation.

The behavior of hCG binding in benign tumor may be explain that it needs less energy than malignant tumor to over come energy barrier and give the maximum binding or may be because of the degradation of hCG form benign tumor with increasing temperature[225]. Previous study has shown an agreement with this hypothesis by declare that elevation of temperature markedly reduced biologic and immunologic activity of hCG[223].

For the three groups the binding at 25°C do not follow the order mentioned before for these groups.

Incubation of binding mixture for time periods longer than that required for maximum binding resulted in decrease binding; this may be due to reversible dissociation of the complex after reaching the equilibrium state.

From these results, the binding studies of the subsequent experiments were carried out at 5 °C for 150min for benign tumor and at 45°C for 120 and 150 min of incubation for pre- and postmenopausal respectively.

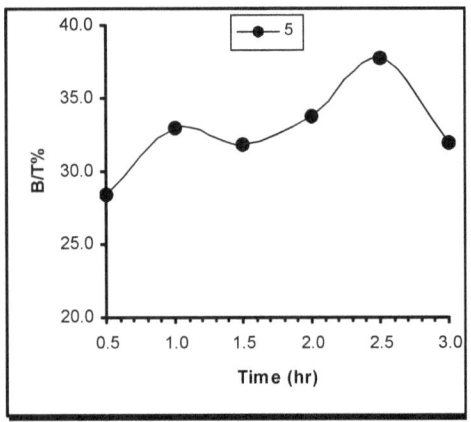
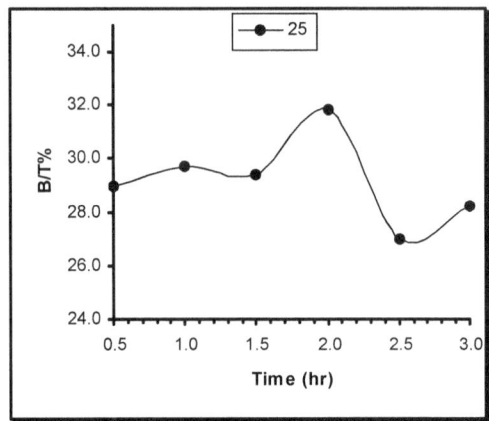
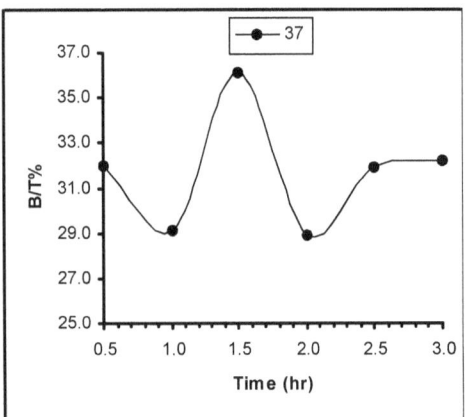
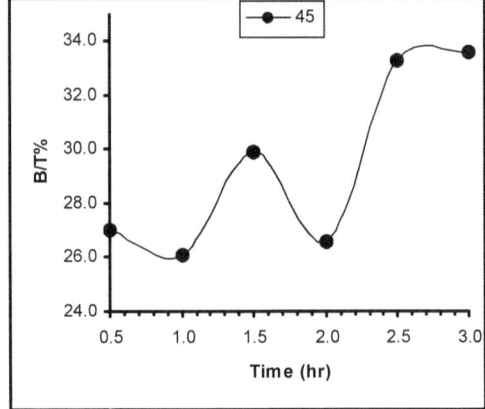

Figure (2-5): Time course of the binding of ^{125}I-anti hCG Antibody with hCG in benign breast tumors. (All other details are explained in the text).

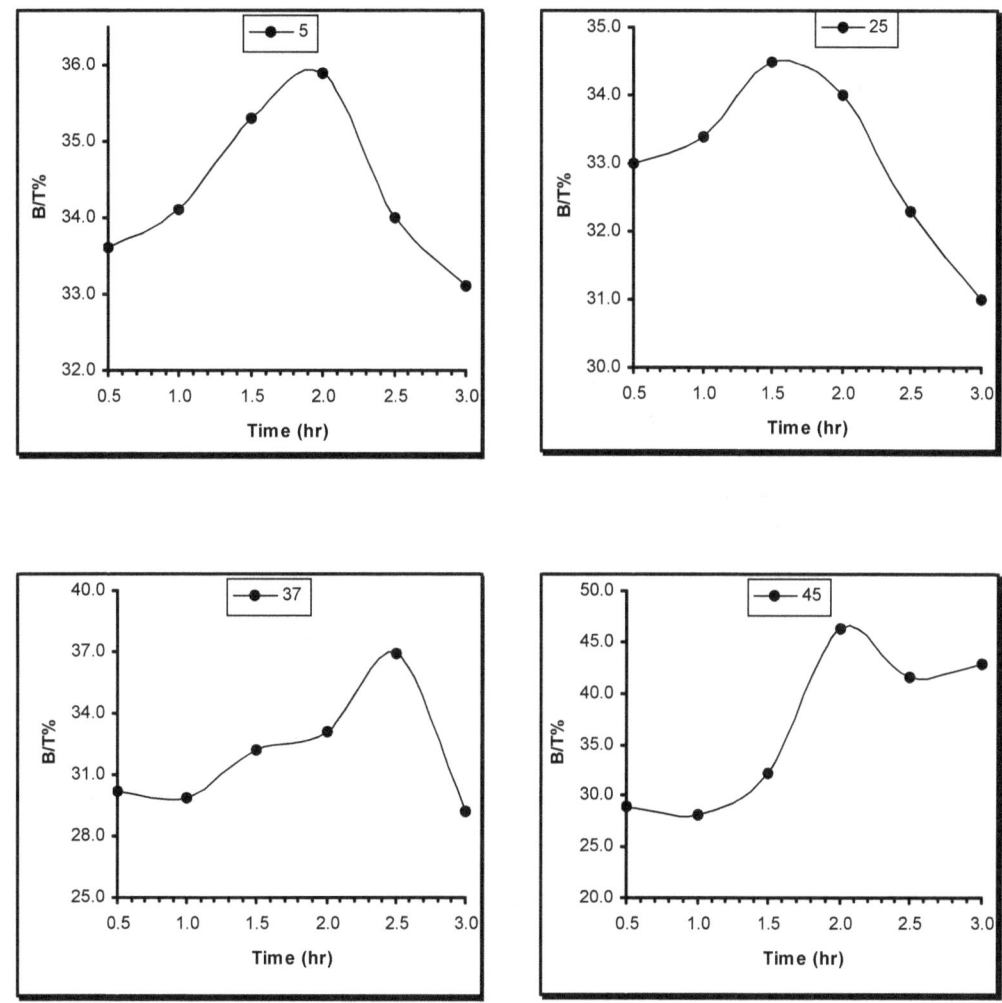

Figure (2-6): Time course of the binding of ^{125}I-anti hCG Antibody with hCG in malignant premenopausal breast tumors. (All other details are explained in the text).

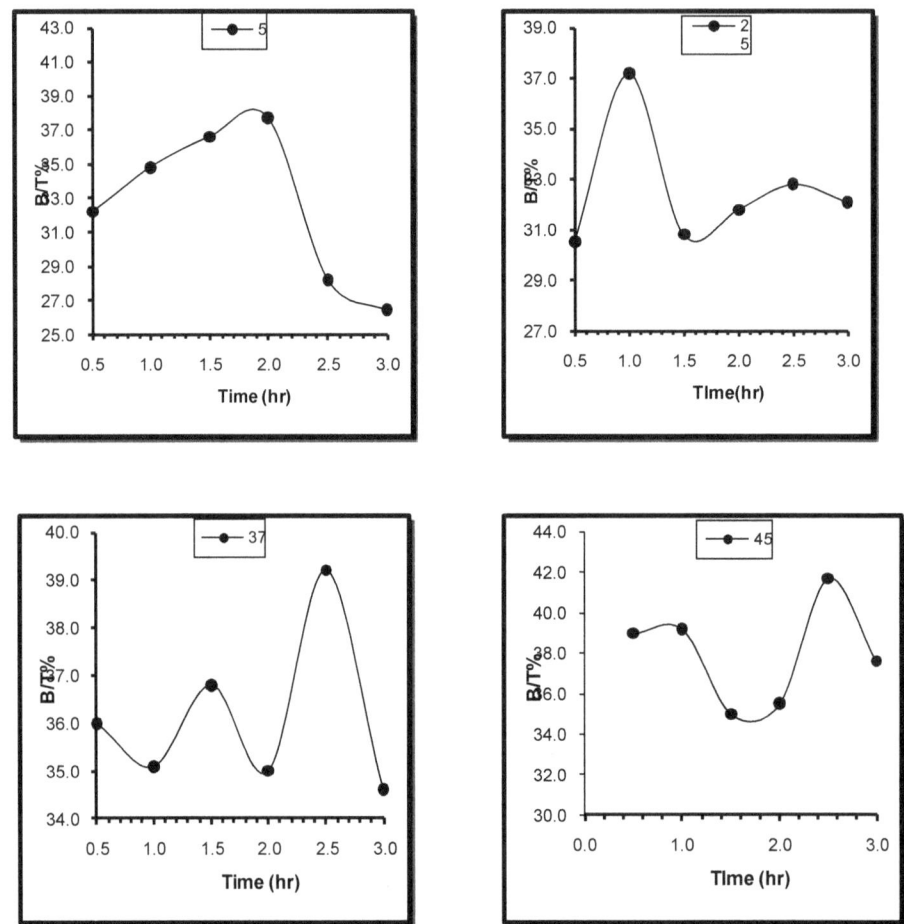

Figure (2-7): Time course of the binding of ^{125}I-anti hCG Antibody with hCG in malignant postmenopausal breast tumors. (All other details are explained in the text).

The Effect of Different Halides on the Binding of hCG in Breast Tumor Homogenate with ^{125}I-anti hCG Antibody

Ionic species and ionic strength affect the binding of the antigen(^{125}I-anti hCG antibody) and antibody(hCG). To investigate this effect on the binding of hCG and ^{125}I-anti hCG antibody, different sodium halides at 0.1M concentration were added to the binding mixture. It seems that sodium halides decrease the maximum binding of (hCG/^{125}I-anti hCG antibody) complex in the three groups, as shown in figure (2-8), according to the following order:

NaI > NaBr > NaCl > NaF

These results are in greement with pervious study made by Lanja.E.O. on LH hormone, which has identical α subunit and 80% identical β subunit of hCG[226]. The decreasing of maximum binding could be as a result of decrease of ionic radius and increasing radius of hydration for anionic salts, leading to greater interaction of the salt, having lower degree of hydration, with an ionic group located in the antibody or antigen combining site[227]. From these results it seems that decreasing of maximum binding could be due to the large size of iodine ion, which could inhibit the interaction between hCG and ^{125}I-anti hCG antibody.

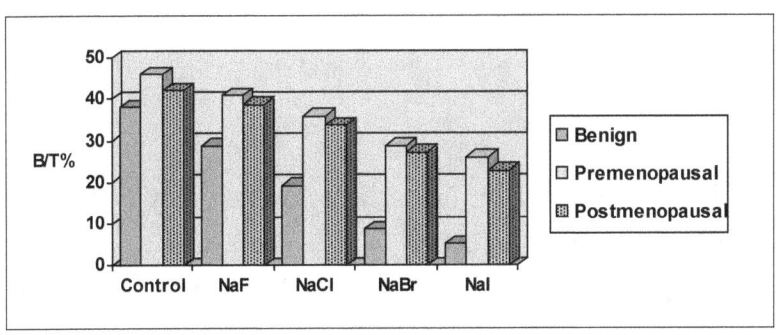

Figure (2-8): The Effect of Different Halides on the Binding of hCG in Breast Tumor Homogenate with ^{125}I-anti hCG Antibody.

(All other details are explained in the text).

Determination of Affinity Constant (K_a) and the Maximum Binding Capacity (B_{max}) of hCG in Pre-menopausal Malignant Breast Tumor Homogenate Associated with ^{125}I-anti hCG Antibody

The simplest proposed model representing this interaction is:

$$^{125}\text{I-anti hCG Antibody} + \text{hCG} \underset{K_{-1}}{\overset{K_{+1}}{\rightleftharpoons}} [^{125}\text{I-anti hCG Antibody/hCG}] \quad \ldots\ldots\ldots(1)$$

Where:

K_{+1}: is the association rate of ^{125}I-anti hCG to hCG.

K_{-1}: is the dissociation rate of (^{125}I-anti hCG/hCG) complex formed.

At equilibrium:

$$K_a = \frac{[^{125}\text{I-anti hCG Antibody/hCG}]}{[^{125}\text{I-anti hCG Antibody}][\text{hCG}]} \quad \ldots\ldots\ldots(2)$$

$$K_d = \frac{[^{125}\text{I-anti hCG Antibody}][\text{hCG}]}{[^{125}\text{I-anti hCG Antibody/hCG}]} \quad \ldots\ldots\ldots(3)$$

Thus

$$K_a = \frac{1}{K_d} = \frac{K_{+1}}{K_{-1}} \quad \ldots\ldots\ldots(4)$$

Where:

K_a: is the equilibrium constant of the association (affinity constant).

K_d: is the equilibrium constant of the dissociation ^{125}I-anti hCG antibody/hCG) complex.

In this experiment, scatchard plot analysis was used to measured the concentration of hCG in pre-menopausal malignant breast tumor homogenate (B_{max}) and the affinity constant (K_a) of the binding with

^{125}I-anti hCG antibody, as in figure (2-9), by using the optimal conditions, which were obtained in previous experiments.

Results in table (2-6) show that affinity constant (Ka) is temperature depended. It increased with increased temperature from (8.8809 µg^{-1}.ml) at 5°C to (13.417µg^{-1}.ml) at 45 °C.

Whereas the values of dissociation constant (K_a), which calculated by using equation (4), shows the lowest K_d value at 45 °C with the following order 25<5<37<45 °C.

The straight line which obtained from scatchard plot analysis, as shown in figure (2-9), indicate the presence of only one species of hCG site, or more but with the same affinity and number of binding site.

Table (2-6): the kinetic parameter of ^{125}I-anti hCG antibody binding to hCG in pre-menopausal malignant breast tumor homogenate. (All other details are explained in the text).

Temp °C	Binding Capacity B_{max}(µg.ml^{-1})	K_a (µg^{-1}.ml)	K_d (µg.ml^{-1})
5	0.1405	8.8809	0.1125
25	0.1474	7.6865	0.1300
37	0.1481	10.613	0.0942
45	0.1905	13.417	0.0745

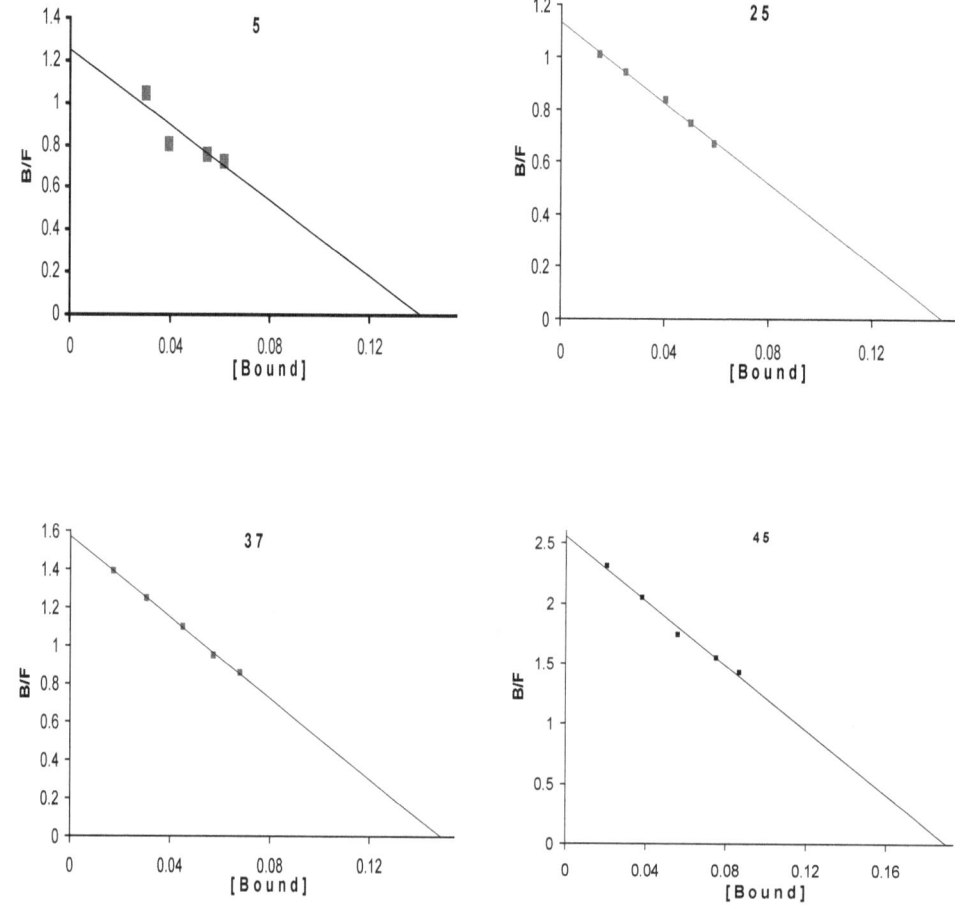

Figure (2-9): Scatchard plot of ^{125}I-anti hCG antibody binding to the partially isolated hCG in pre-menopausal malignant Breast Tumor Homogenate. (All other details are explained in the text).

Chapter Three
Studies on the binding of I-hCG to the hCG in human breast tumor homogenates

A quantitative radioreceptor assay for detection and analysis of membrane-associated hCG receptor of human breast tissues was developed by binding highly specific and biological active radiolabed hormone (^{125}I-hCG) to particulate membrane preparations. The optimum conditions for ^{125}I-hCG-receptor binding were obtained as follows: the optimum protein amount was (2.5, 10 and 10 µg) for benign, pre- and post-menopausal malignant breast tumor homogenate respectively. ^{125}I-hCG concentration was (18×10^{-4}ng.ml^{-1}) for benign, while it was (16×10^{-4}ng.ml^{-1}) for both pre- and post-menopausal malignant breast tumor homogenate. The optimum pH was 7.0 for benign group while it was 7.6 for the two malignant groups. It was found that the binding was time and temperature dependence that the optimum binding temperature and time for benign group were at 45°C in 60 minutes. For pre-menopausal malignant group it was at 37 °C for 30 min while it was at 5 °C and 60 min for post-menopausal malignant group. The use of different halides result in the decrease of maximum binding of benign and malignant groups. The concentration of hCG-receptor (B_{max}) and affinity constant (K_a) in benign and pre-menopausal malignant groups were measured by Scatchard analysis indicating that they are higher in benign group than malignant group.

Introduction

hCG and LH are structurally and functionally similar hormones[1]. They bind to a common receptor. The receptor is a single chain transmembrane glycoprotein that belongs to the family of G-protein-coupled receptors. It is consists of a large extracellular ligand-binding domain, seven transmembrane-spanning regions and a short cytoplasmic tail[228].

For many years, the knowledge about the localization and function of hCG receptors were constrained to only the gonadal tissues[229]. Neither their localization nor function in nongonadal tissues of human body was suspected until about 18 years ago[174]. The first reports on nongonadal hCG receptors initiated further studies to characterize them in detail[80,230-233].

The nongonadal receptors were investigated by using a wide range of traditional ligand binding as well as ligand blotting and covalent receptor cross-linking, that detect the receptor from gene transcription to signaling pathways and ultimately the biological response[178,179,234-237].

The hCG receptor mRNA and protein were found and characterized in the human myometrium[234], placenta[81], fallopian tube[73], breast[153], brain[90], as well as other tissues. Nongonadal distribution of hCG receptors are not species specific, as they have been found in human, monkey, rat, rabbit, mice, pig, cow, sheep, and even turkey[7,230].

The presence of these nongonadal receptors raised the possibility of their functions and many subsequent studies were directed to address this possibility[73,77,79]. These studies revealed the presence of these receptors not only in tumor tissue but also found association between receptor density and oncologic process. The association between structural change in the receptor and the defective regulation was also found.

In the breast cancer cell, hCG anticancer actions were found to be mediated by hCG receptor[90]. So the present study was undertaken as a complementary study to develop a radioreceptor assay for hCG receptor determination in benign and malignant breast tumors. These determinations were undertaken to characterize the optimum binding condition between hCG and its receptor.

Materials and Methods
3.1. Materials

- **Chemicals**

All chemicals and reagents mentioned in chapter two were used in the experiments of this chapter.

- **Instruments**

All instruments mentioned in chapter two were also used in the experiments of this chapter.

- **Patients**

The same patients tissues mentioned in section chapter two were used in the following experiments.

- **Preparation of Breast Tumor Tissue Homogenate**

The same Preparation of breasts tumor tissue Homogenate mentioned in chapter two were used in the following experiments.

3.2. Methods

- **Determination of Total Protein Content in Breast Tumor Homogenate**

The same procedure mentioned in chapter two was used in the experiments of this chapter.

3.3. Binding Studies of hCG Receptor in Breast Cancer Tumor Homogenate with ^{125}I- hCG

3.3.1. Preliminary Tests of hCG Receptor Binding in Breast Tumor Homogenate with ^{125}I-hCG

1. One hundred microliters containing (30 µg of protein for benign, malignant pre-and post-menopausal breast tumors respectively) was added to 40 µl of ^{125}I- hCG (10ng.ml^{-1}), the volume of the mixture was made up to 250 µl with Tris/HCl buffer (0.05M, pH 7.2) containing 0.1% BSA.
2. One additional tube containing only 40 µl of ^{125}I-hCG were set aside until counting for total activity.
3. The tubes were incubated at 25°C for 60 min.
4. After incubation, the tubes were centrifuged at 1500xg for 30 min at 4°C.
5. The supernatant was discarded by decanting the assay tubes, then the tubes were inverted on a filter paper for 10 min.
6. The rims of the tubes were swabbed with a cotton piece and the amounts of bound radioactivity (c.p.m) were counted in a gamma counter; this (c.p.m) is refer to the total binding.

7. Non specific binding was accounted for by preparing the same incubation with addition of 250 fold excess of unlabeled hCG and following the steps 2-6.

Solutions

Tris/HCl buffer (0.05M, pH 7.2) containing 0.1% BSA was prepared by dissolving (0.606gm) of Tris (hydroxymethyl) amino methane and (0.1gm) of BSA in (95 ml) distilled water. Then the pH was adjusted to pH 7.2 using HCl (0.2 M). The volume was made up to (100 ml) with distilled water.

Calculations

1. The counted of radioactivity in each tube (expressed in c.p.m) represents the total binding fraction (TB), i.e (^{125}I-hCG /hCG receptor) complex.

2. The counted of radioactivity in each tube containing ^{125}I-hCG and excess of unlabeled hCG represents the nonspecific binding (NSB).

3. The specific binding (SB), expressed in c.p.m, was calculated by subtracting the radioactivity, expressed in c.p.m, obtained in the presence of unlabeled hCG from that produced in the absence of unlabeled hCG.

$$SB(c.p.m) = TB(c.p.m) - NSB(c.p.m)$$

4. The precent of specific binding (SB%) can be calculated from the following formula:

$$SB\% = \frac{SB(c.p.m)}{T(c.p.m)} \times 100$$

Where: T is the total count of the ^{125}I- hCG, expressed in c.p.m,

3.3.2. Most Appropriate Conditions of the Binding of hCG Receptors in Breast Tumor Homogenate with ^{125}I- hCG.

- *The Effect of Different Protein Amount of Breast Tumor Homogenate on the Binding of hCG Receptors with ^{125}I-hCG*

1. Forty microliters of ^{125}I-hCG (10ng.ml^{-1}) to 100 µl containing increasing protein amounts (1.5, 2, 2.5, 5, 10, 15, 20, 25, 30 µg of benign and 2.5, 5, 10, 15, 20, 25, 30, 35 µg.ml^{-1} of pre- and post-menopausal malignant breast tumors respectively). The volume of the mixture made up to 250 µl with Tris/HCl buffer (0.05M, pH 7.2) containing 0.1% BSA.

2. One additional tube containing only 40 µl of ^{125}I-hCG was set aside until counting for total activity computation.

3. The tubes were incubated at 25°C for 60 min.

4. After incubation, the (^{125}I-hCG /hCG receptor) complex was estimated by following the steps 4, 5, 6 and 7 in section (3.3.1).

Solutions

Tris/HCl buffer (0.05M, pH7.2) containing 0.1% BSA was prepared as described in section (3.3.1).

Calculations

1. The percent of specific binding (SB%) was calculated according to the formula mentioned in section (3.3.1).

2. The percent of specific binding value (SB%) were plotted against the increasing protein amounts.

- *The Effect of Different Concentrations of ^{125}I-hCG on the Binding with hCG Receptors in Breast Tumor Homogenate*

1. One hundred microliters containing (2.5, 10, 10 µg proteins of benign, malignant pre- and post-menopausal breast tumors homogenate respectively) were added to increasing volumes (10, 20, 25, 30, 35, 40, 45, 50 and 55 µl) containing (4×10^{-4} - 22×10^{-4}

ng.ml^{-1}) of ^{125}I-hCG (10ng.ml^{-1}) then the volume were made up to 250 µl with Tris/HCl buffer (0.05M, pH 7.2) containing 0.1% BSA.

2. A set of tubes containing only the same increasing volumes of ^{125}I- hCG (10, 20, 25, 30, 35, 40, 45, 50 and 55 µl), were set a side until counting for total activity computation.

3. The tubes were incubated at 25°C for 60 min.

4. After incubation, the (^{125}I-hCG /hCG receptor) complex was estimated by following the steps 4, 5, 6 and 7 in section (3.3.1).

Solutions

Tris/HCl buffer (0.05M, pH 7.2) containing 0.1% BSA was prepared as described in section (3.3.1).

Calculations

1. The percent of specific binding (SB%) was calculated according to the formula mentioned in section (3.3.1).

2. The percent of specific binding values (SB%) were plotted against the concentration of ^{125}I-hCG.

- *The Effect of Different pH on the Binding of hCG Receptors in Breast Tumor Homogenate with ^{125}I-hCG*

1. One hundred microliters containing (2.5, 10, 10 µg proteins of benign, malignant pre- and post-menopausal breast tumors homogenate respectively) were added to 45µl(18x10^{-1} ng.ml^{-1}), 40µl(16x10^{-1} ng.ml^{-1}) and 40µl(16x10^{-1} ng.ml^{-1}) for benign and malignant pre-, post-menopausal breast tumors homogenate respectively) of ^{125}I-hCG (10ng.ml^{-1}), the mixtures volumes were made up to 250 µl with Tris/HCl buffer (0.05M) containing 0.1% BSA of different pH (6.8, 7.0, 7.2, 7.4, 7.6, 7.8, 8.0).

2. Two additional tubes containing only (40 and 45) µl of the ^{125}I-hCG were set aside until counting for total activity computation.

3. The tubes were incubated at 25°C for 60 min.

4. After incubation, the (^{125}I-hCG /hCG receptor) complex was estimated by following the steps 4, 5, 6 and 7 in section (3.3.1).

Solutions

Tris/HCl buffer (0.05M) containing 0.1% BSA was prepared as shown in section (3.3.1).

Calculations

1. The percent of specific binding (SB%) was calculated according to the formula mentioned in section (3.3.1).

2. The percent of specific binding values (SB%) were plotted against their corresponding pH values.

- *Time Course of the Binding of hCG Receptors in Breast Tumor Homogenate with ^{125}I-hCG*

1. One hundred microliters containing (2.5, 10, 10 µg proteins of benign, malignant pre- and post-menopausal breast tumors homogenate respectively) were added to 45µl(18x10^{-1} ng.ml^{-1}), 40µl(16x10^{-1} ng.ml^{-1}) and 40µl(16x10^{-1} ng.ml^{-1}) for benign and malignant pre-, post-menopausal breast tumors homogenate respectively) of ^{125}I-hCG (10ng.ml^{-1}), the mixtures volumes were made up to 250 µl with Tris/HCl buffer (0.05M, pH 7.0 and 7.6) containing 0.1% BSA for benign and malignant tumor respectively.

2. All tubes were incubated at 25°C at different time interval (30, 60, 90, 120, 150 and 180) min.

3. Two additional tubes containing only (40 and 45) µl of the ^{125}I-hCG were set aside until counting for total activity computation.

4. After incubation, the (^{125}I-hCG /hCG receptor) complex was estimated by following the steps 4, 5, 6 and 7 in section (3.3.1).

5. To determine the time course of HCG receptors binding to ^{125}I–hCG at different temperatures. Steps 1, 2 and 3 in the same experiment were repeated at different temperatures (5, 37, 45 °C).

Solutions

Tris/HCl buffer (0.05M, pH 7.2) containing 0.1% BSA was prepared as described in section (3.3.1).

Calculations

1. The percent of specific binding (SB%) was calculated according to the formula mentioned in section (3.3.1).

2. The percent of specific binding values (SB%) were plotted against the incubation time.

- *The Effect of Different Halides on the Binding of hCG Receptors in Breast Tumor Homogenate with ^{125}I-hCG*

1. One hundred microliters containing (2.5, 10, 10 µg proteins of benign, malignant pre- and post-menopausal breast tumors homogenate respectively) were added to 45µl(18×10^{-1} ng.ml^{-1}), 40µl(16×10^{-1} ng.ml^{-1}) and 40µl(16×10^{-1} ng.ml^{-1}) for benign and malignant pre-, post-menopausal breast tumors homogenate respectively) of ^{125}I-hCG (10ng.ml^{-1}), the mixtures volumes were made up to 250 µl with Tris/HCl buffer (0.05M, pH 7.2) containing 0.1% BSA containing 0.1M of each of the following halides: NaF, NaCl, NaBr and NaI). A sample without the addition of any halides was used as a control.

2. Two additional tubes containing only (40 and 45) µl of ^{125}I-hCG were set aside until counting for total activity computation.

3. The tubes were incubated for 60min at 45°C (benign breast tumor homogenate), 30min at 37°C and 60min at 5°C (pre-and post-menopausal malignant breast tumor homogenate).

4. After incubation, the (^{125}I-hCG /hCG receptor) complex was estimated by following the steps 4, 5,6 and 7 in section (3.3.1).

Solutions

1. The halides stock solutions (0.1M) were prepared by dissolving each of the following amounts of salts in 250ml Tris/HCl buffer (0.05M, pH 7.2, 0.1% BSA): 1.049g of NaF, 1.46g of NaCl, 2.57g of NaBr and 3.74g of NaI.

2. Tris/HCl buffer (0.05M, pH 7.2, 0.1% BSA) was prepared as described in section (3.3.1).

Calculations

1. The percent of specific binding (SB%) was calculated according to the formula mentioned in section (3.3.1).

2. The percent of specific binding values (SB%) were plotted against halides concentrations.

3.4. Determination of the Concentration of hCG-Receptor and the Affinity concentration of ^{125}I-hCG Association with its Receptors in Benign and Malignant Breast Tumor.

1. One hundred microliters containing (2.5, 10, 10 µg proteins of benign, malignant pre- and post-menopausal breast tumors homogenate respectively) were incubated with increasing volume 25, 30, 35, 40, 45µl (1-1.8ng) and 20, 25, 30, 35, 40µl (0.8-1.6ng) of ^{125}I-hCG (10ng.ml^{-1}) for benign and pre-malignant breast tumors

homogenate respectively, the mixtures volumes were made up to 250 μl with Tris/HCl buffer (0.05M, pH 7.0 and 7.6) for benign and malignant tumor respectively.

2. A set of tubes containing only (20, 25, 30, 35, 40, 45μl) of ^{125}I-hCG were set aside until counting for total activity computation.

3. The tubes were incubated for 60min at 45°C for benign breast tumor homogenate and 30min at 37°C for pre- menopausal malignant breast tumor homogenate.

Solutions

Tris/HCl buffer (0.05M, pH 7.2, 0.1% BSA) was prepared as described in section (3.3.1).

Calculations

The values of ^{125}I-hCG which is bound specifically were calculated by using the following formula:

$$B = \frac{\text{Total binding - Non specific binding}}{\text{Total count}} \times \text{Concentration of } ^{125}\text{I-HCG in each assay tube}$$

4. The concentration of receptors and affinity constant were determined according to Scatchard equation: -

$$\frac{B}{F} = \frac{1}{Kd}(B_{max} - B)$$

$$Ka = \frac{1}{Kd}$$

Where:

Ka = Affinity constant

Kd = Dissociation constant

B_{max} = Maximal binding capacity

4. The values of the ratio B/F were plotted against the values of B in (ng /ml), gives a linear relationship. The values of the affinity constant of the binding (Ka) can be calculated from the slop of the straight line, while the value of the total concentration

of hCG-receptors (B_{max}) was calculated from the intercept with the x-axis.

Results and Discussion

Membrane preparation

As described the experiment of Preparation of breast tumor tissue homogenate in chapter two, the method used for preparation the crude homogenate for hCG binding studies was the same for membrane preparation for obtaining hCG receptor. In previous studies the extracted hCG receptor that used for binding studies could be obtained as either particulate membrane by homogenate centrifugation at different speeds or soluble receptors extracted from the membrane by using a nonionic detergent[238-240].

Binding Studies of hCG Receptor in Breast Cancer Tumor Homogenate with ^{125}I-hCG

Preliminary Tests of hCG Receptor Binding in Breast Tumor Homogenate with ^{125}I-hCG

Benign and malignant breast tumor were investigated by using breast tumor homogenate as a source of hCG receptors. These receptors were detected through the incubation of crude homogenate with ^{125}I-hCG with and without added of non-radioactive hCG in order to demonstrate whether the specific binding was proportional to the concentration of hCG receptor. The incubation was carried out at 25°C for 60 minute, as a beginning, according to the condition that mention in the RIA kit; then separate the formed complex at 1500xg for 30 minute. Although many previous studies used PEG 6000 for precipitate (^{125}I-hCG /hCG receptor) complex, but in this experiment the use of this precipitating reagent was neglected due to its precipitation of the free ^{125}I-hCG. The specific binding was found to be (3%) in benign tumor, (2.2%) and (3.3%) for malignant pre- and post-menopausal breast tumor respectively. These data revealed the presence of hCG receptor in human breast tumors. The low binding value probably due to the relatively large size of the complex that restricted the separation procedure[238].

Most Appropriate Conditions of the Binding of hCG Receptors in Breast Tumor Homogenate with ^{125}I-hCG.

The Effect of Different Protein Amount of Breast Tumor Homogenate on the Binding of hCG Receptors with ^{125}I-hCG

Increasing amount of breast tumor homogenate were incubated with fixed amount of ^{125}I-hCG with and without added of non-radioactive hCG. The specific binding percent was increased with the increasing concentration of the homogenate in the incubation mixture, as shown in figure (3-1). This nonlinearly may be due to the heterogeneously between the ligand and receptor preparation. In this experiment benign tumor required lower amount (2.5µg) than other groups for reaching the maximum binding (4.6%), while malignant pre- and post- menopausal groups required (10 µg) to reach there maximum binding (4.1%, 3.8%) respectively. Although many previous studies preformed that malignant breast cell lines contacting significantly higher level and higher relative transcription rate of hCG receptors genes than normal breast cell line[166,177]; but from the initial researches until now, there was no sign to the presenting values of hCG receptor binding with ^{25}I-hCG in each kind. In other hand, other previous studies estimated that hCG level are higher in malignant breast tumor (infiltrated ductal carcinoma) than benign breast tumor. So, the obtunded results could be explained that even the malignant breast tumors have higher hCG receptors, it also have higher hCG level. This may lead to a conclusion that the number of hCG receptors that occupied by hCG in malignant tumor is higher than benign tumor in which higher free receptors are available for binding with ^{25}I-hCG. According to these results the optimum proteins amount in this experiment were used in all subsequent experiments.

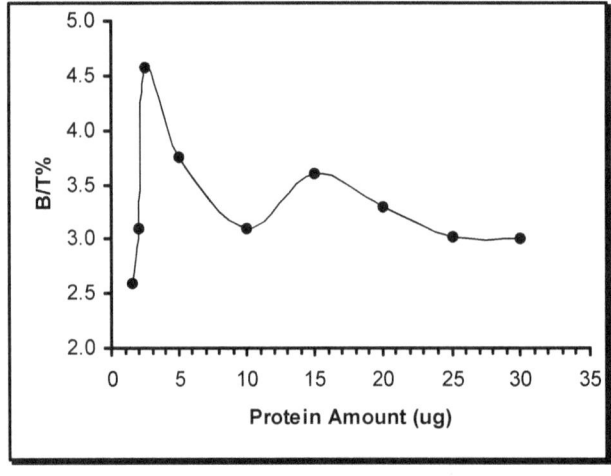

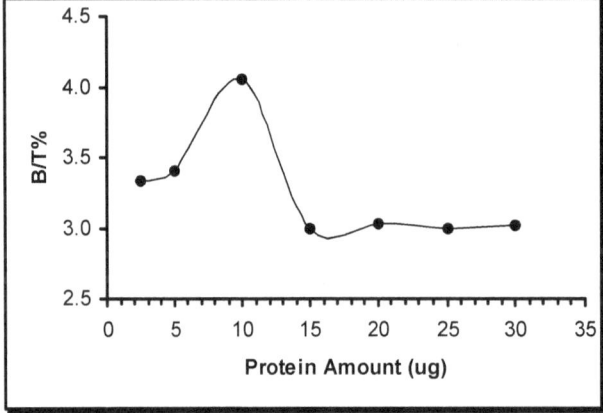

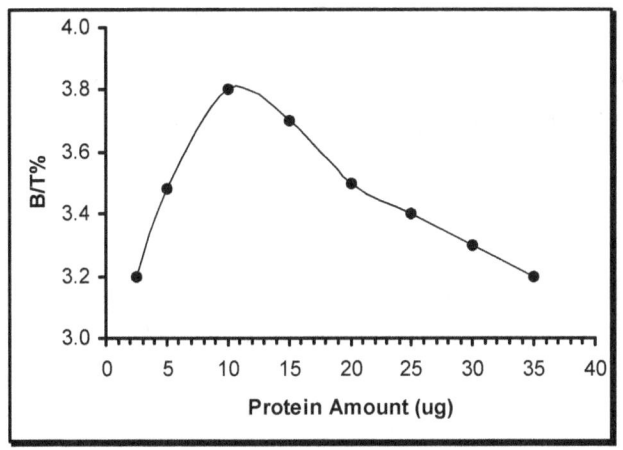

Figure (3-1): Influence of protein amount on the binding with ^{125}I-hCG. (All other details are explained in the text).

The Effect of Different Concentrations of ^{125}I-hCG on the Binding with hCG Receptors in Breast Tumor Homogenate

To estimate another factor affects the (^{125}I-hCG /hCG receptor) complex formation, fixed concentration of benign (2.5 µg), malignant pre- (10 µg) and post-menopausal (10 µg) breast tumor homogenate were incubated with increasing concentration of ^{125}I-hCG for 60min at 25°C, as shown in figure (3-2).

In this figure, binding increased with the added amount of the ^{125}I-hCG in the three groups. This is probably due to the cross-linking of hCG and ^{125}I-hCG is more likely with the increasing ^{125}I-hCG in the incubation mixture to perform large complex until reached its maximum binding.

For benign tumor maximum binding was (5.5%) at (1.8 ng.ml^{-1}) ^{125}I-hCG antibody concentration, (4%) and (3.7%) for malignant pre-and post-menopausal breast tumors receptively at (1.6 ng.ml^{-1}).

After the maximum binding the increasing of ^{125}I-hCG concentration lead to decrease in binding percent, probably because all the hCG receptors sites are covered with ^{125}I-hCG and inhibiting the formation of the complex[241]. This indicates the dependence of the binding on ^{125}I-hCG in binding mixture[242].

According to these results the above concentration of ^{125}I-hCG was used in the subsequent experiments.

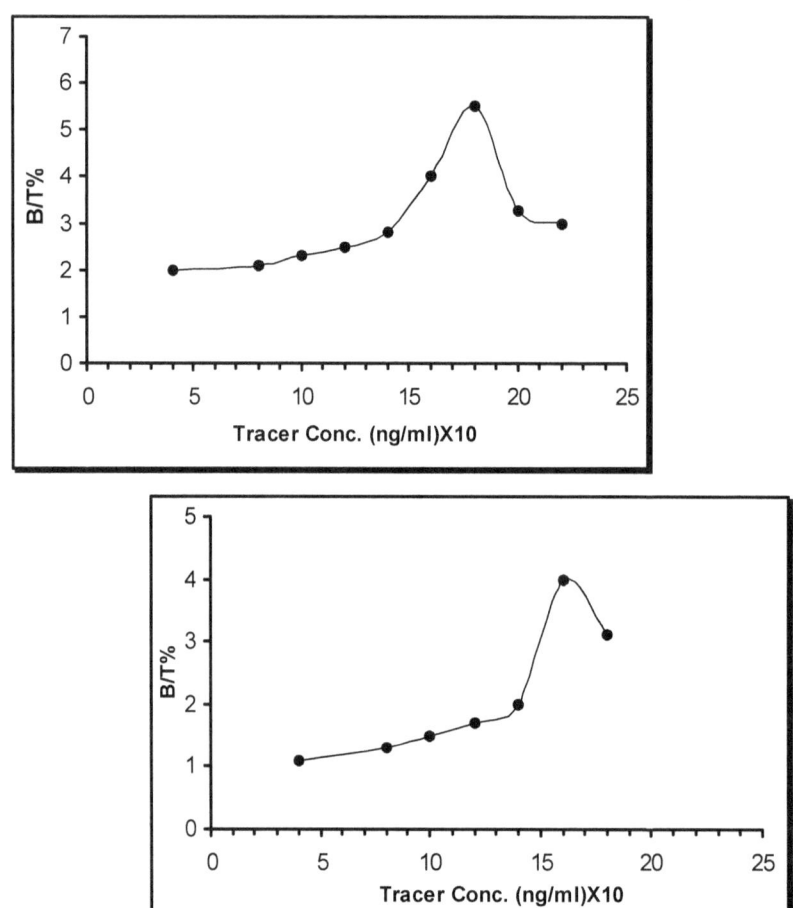

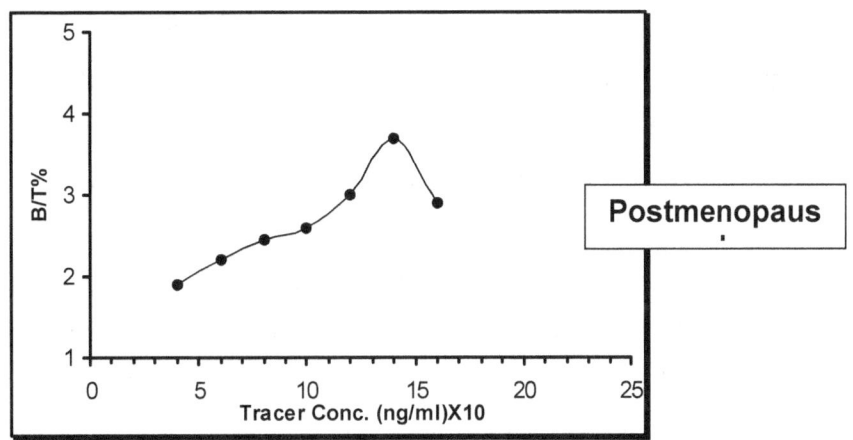

Figure (3-2): Influence of increasing ^{125}I-hCG concentration on the binding with hCG receptors.

The Effect of Different pH on the Binding of hCG Receptors with ^{125}I-hCG in Breast Tumor Homogenate.

To determine the optimal pH for receptor activity, the binding was preformed at different pH values, as shown in figure (3-3). In this figure, benign tumor showed two sharp peaks obtained at pH 7.0 (6.7%) and pH 7.4 (6.0%); while the malignant breast tumors showed two sharp peaks at pH 7.2 (pre- 4.1%, post- 3.7%) and pH 7.6 (pre- 5.0%, post- 4.6%). From these results, the bindings were found to be pH dependent and the shift in the pH of the environment may affect the properties of the macromolecules involved in the binding. It was not determined whether the decreased binding at suboptimum pH values resulted from a decrease in binding affinity, binding rate or inactivation of hCG receptor[239,243]. The presence of two peaks for each group may be refers to the presence of more than one kind of hCG receptors which have different property. Previous studies support this hypothesis by revealing that human breast cell lines contain multiple hCG receptors transcripts and three proteins of different molecular sizes which can bind ^{125}I- hCG [165]. Whether the multiple transcripts were the products of the different transcription initiation sites or from alternate splicing of a single transcript or due to the differences in polyadenylation is not known. Also unknowns is whether the receptor proteins were translational products of different transcripts or are the result of proteolytic degradation despite careful handling and the presence of protease inhibitors in the buffers. The presence of multiple receptor transcripts and more than one receptor protein have previously been found in other hCG receptor-positive tissues[73,79,83,85]. Other previous study revealed disparate results of pH effect on hCG receptors from different sources. In bovine corpus luteal plasma membrane preparation, the maximum binding was at pH 7.2-7.4 and declined both at high and low pH values[239]. In rat testis the binding exhibited a broad pH optimum between pH 6.8 and 7.3[243], while in another study the equilibrium binding of rate testis homogenate with ^{125}I-hCG showed a relatively sharp pH optimum at 7.4[238]. Rat ovarian hCG receptor binding have been carried out almost exclusively at 7.4[240]. In uterus hCG receptor, the effect of pH on the binding exhibited optimum pH between 7.0-7.4[42]. According to these results, the above optimum pH were used in the subsequent experiments.

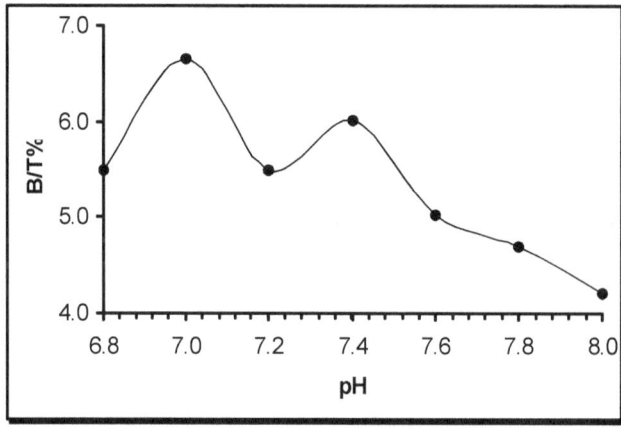

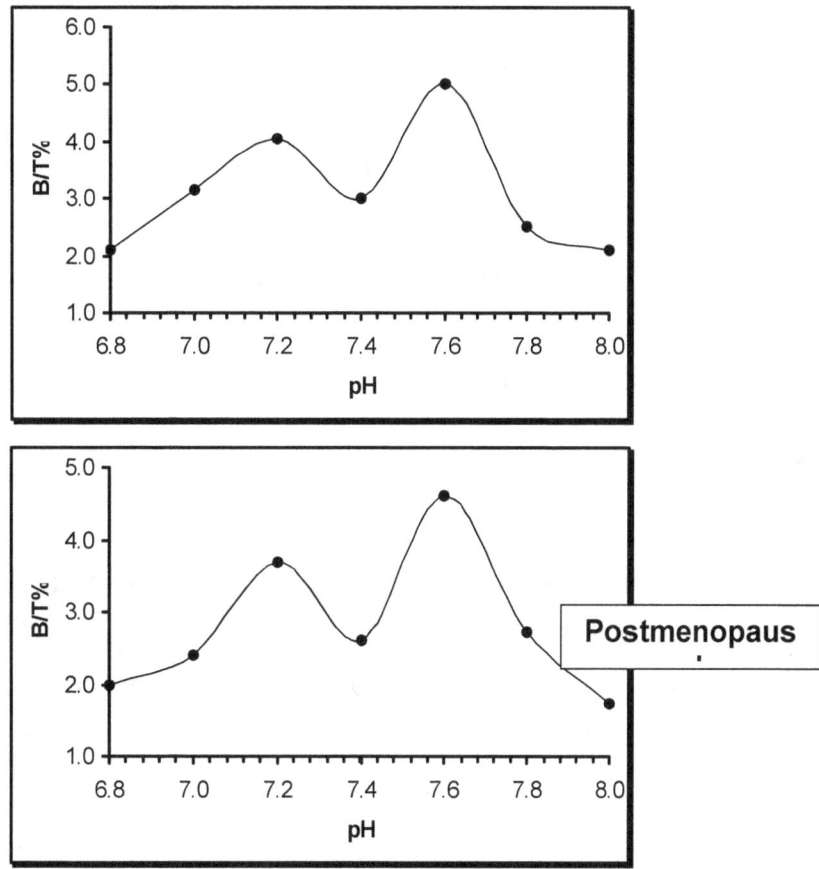

Figure (3-3): Influence of pH on the binding of ^{125}I-hCG with hCG receptors. (All other details are explained in the text).

Time course of The Binding

Time course pattern for ^{125}I-hCG binding with hCG receptor in breast tumors homogenate at different temperatures (5, 25, 37, 45) was found to be markedly time and temperature dependent, as stated at figures (3-4), (3-5), (3-6). The maximum binding was occurred in less temperature when binding move from benign to postmenopausal group. The maximum binding (8.1%) in benign tumor was obtained at 45°C after incubation for 60 minute. For malignant premenopausal tumor it was (6.2%) at 30 min., whereas it was (5.8%) at 60 minute for postmenopausal breast tumor. This could be due to the presence of different active forms of the receptor that each form is active in different kind of tumors. In the three groups, incubation of ^{125}I- hCG with hCG receptor in breast tumor homogenate for time periods longer than required for maximum binding result in decreasing the bonding. This could be due to partial inactivation of hCG receptors or partial degradation of ^{125}I-hCG occurring during the binding assay[243]. For benign group the maximum binding was obtained at 45°C and this could be that hCG receptor need more energy than malignant tumor to over come energy barrier and give the maximum binding. For malignant premenopausal group, the loss of binding was more rapid at 45 °C than 37°C and this could partially explain the lower maximum values obtained for binding at 45°C compared with 37°C[238,243]. While in malignant postmenopausal group, greater binding occurred at the lower temperature 5°C, this may be due to the more rapid degradation of hCG receptors during incubation at higher temperatures previous studies showed variable results depending on the source of hCG receptors. In studying the presence of hCG receptors in Porcine uterus, maximum binding was at 24°C in 16 hour[174], while the maximum binding of hCG receptors for cervical porcine was at 36°C in 4 hours[244]. For rat testis hCG receptor, the maximum binding in the crude homogenate was at 25°C in 4 hours[243], while the maximum binding for soluble rat testis hCG receptor was found to be at 25°C in 10 hours[238]. In human ovarian and uterine homogenates, hCG receptor reached its maximum binding at 37°C for 1 hour for each tissue[42].

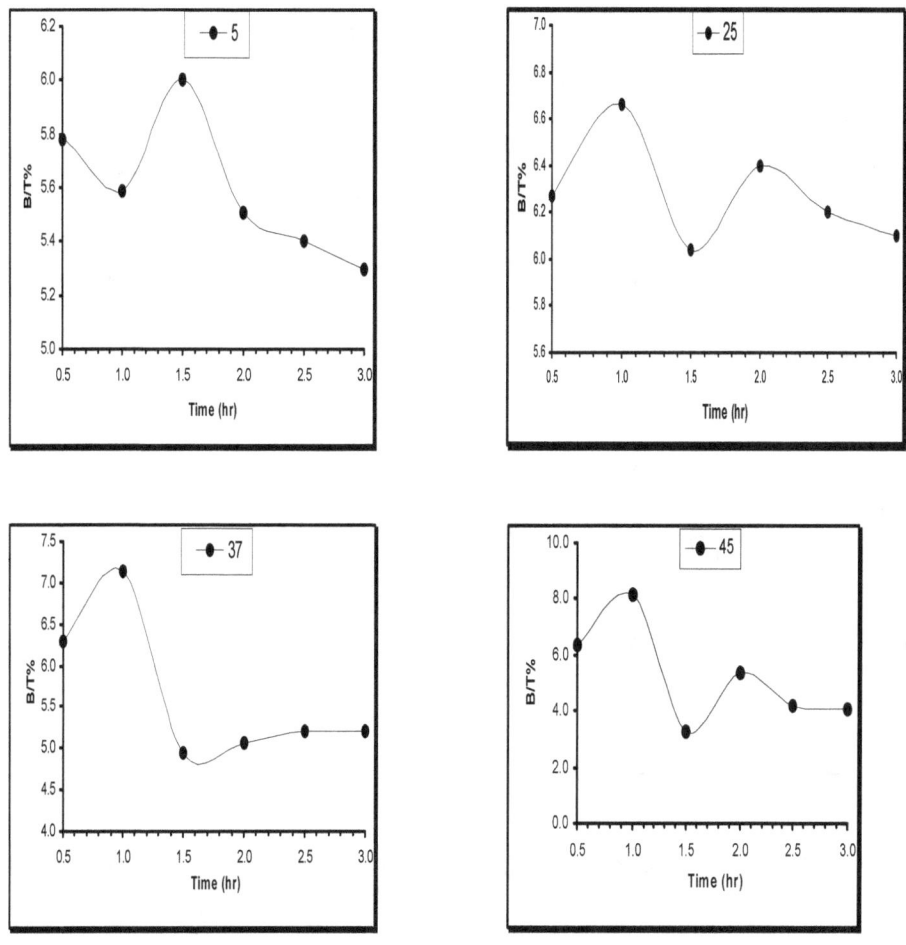

Figure (3-4): Time course of the binding of ^{125}I- hCG with hCG receptor in benign breast tumors. (All other details are explained in the text).

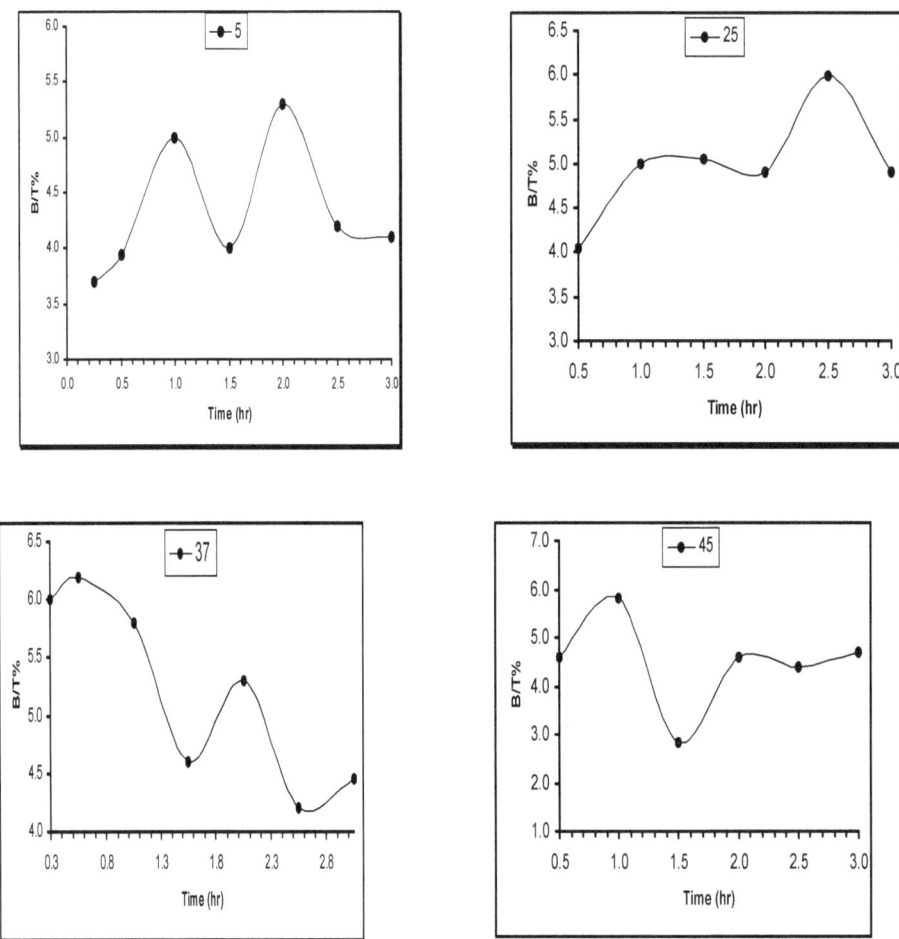

Figure (3-5): Time course of the binding of ^{125}I- hCG with hCG receptor in pre- malignant breast tumors. (All other details are explained in the text).

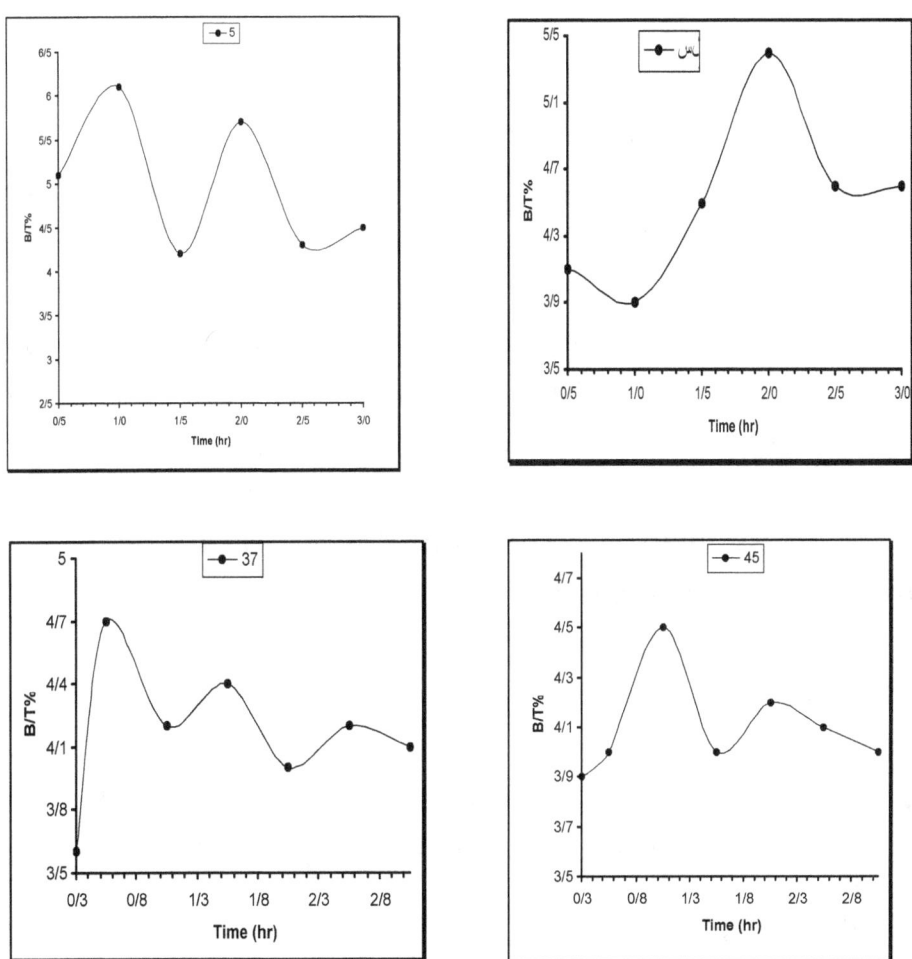

Figure (3-6): Time course of the binding of ^{125}I- hCG with hCG receptor in post- malignant breast tumors. (All other details are explained in the text).

The Effect of Different Halides on the Binding

Different sodium halides at 0.1M concentration were investigated to study their action on the binding of ^{125}I- hCG with hCG receptor in breast tumor homogenate, as shown in figure (3-7).Even It seems that sodium Florid made a slight increase, but in general the sodium halide decrease the binding of (^{125}I- hCG/hCG receptor) complex in the two groups according to the following order:

Benign breast tumor   NaF < NaBr < NaI < NaCl

Malignant premenopausal breast tumor   NaF < NaBr < NaCl < NaI

Sodium iodide causes lower specific binding in the two groups; this could be due to that NaI has less degree of hydration permits. So, greater interaction of Iodine salt with an ionic group located in the hCG receptor or ^{125}I- hCG combining site happen [227].

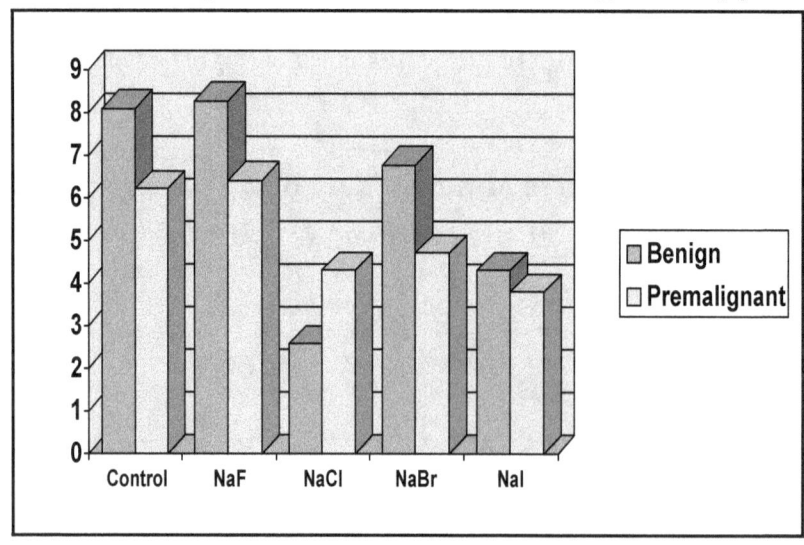

Figure (3-7): Effect of Halides on the binding of ^{125}I-hCG with hCG receptors. (All details are explained in the text).

Determination of the Concentration of hCG-Receptor and the Affinity concentration of ^{125}I-hCG Association with its Receptors in Benign and Malignant Breast Tumor.

The equilibrium association constant (Ka) values were observed for the whole homogenate, plasma membrane fraction and solubilized hCG receptor in many origin. It gives different values depending on the origin kinds and the method[232,245-247].

The concentration of hCG receptors in benign and malignant breast tumor homogenate (B_{max}) and the affinity constant (Ka) of the binding with ^{125}I-hCG has been measured in the optimum condition of the binding, as in figure (3-8, A and B).

As in table (3-1), the affinity constant (Ka) was found to be tumor type dependent (i.e, benign or malignant). It is found to be higher in benign (1ng^{-1}.ml) than malignant tumor (0.7ng^{-1}.ml). The concentration of hCG receptor in benign breast tumor was found to be (0.2ng.ml^{-1}), which is higher than malignant breast tumor(0.1838ng.ml^{-1}).

Table (3-1): The kinetic parameters of ^{125}I-hCG binding to hCG receptor in breast tumor homogenate.(All other details are explained in the text).

Tumor type	Temp °C	Binding Capacity B_{max} (ng.ml^{-1})	K_a (ng^{-1}.ml)	K_d (ng.ml^{-1})
Benign breast tumor	45 °C	0.2339	1.0029	0.9971
Malignant breast tumor	37 °C	0.1838	0.7747	1.2908

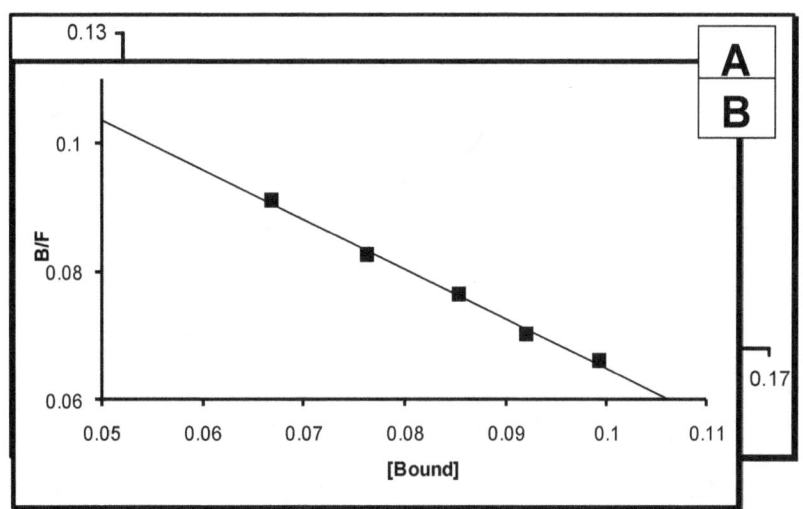

Figure(3-8): Scatchard plot of ^{125}I-hCG binding to hCG receptor in

A: benign breast tumor

B: malignant breast tumor

ChapterFour
kinetics and thermodynamics parameters of the binding of partially purified hCG from malignants breast tumor homogenates

Abstract

Gel filtration chromatography technique was used for partially purification of hCG from premenopausal malignant breast tumor homogenate with using standard hCG as a comparison on Sephadex G-100. The elution volume (V_e) for hCG from Sephadex G-100 were calculated and revealed the presence of one form of hCG. The optimum binding condition between the partially purified hCG and ^{125}I-anti hCG antibody were determined. Kinetic and thermodynamic parameters associated with the binding of ^{125}I-anti hCG antibody to partially purified hCG were investigated. The results show that the reaction follow pseudo-first order reaction kinetics. The maximum binding (B_{max}) of partially purified hCG in pre-menopausal malignant breast tumor was (0.014312µg.ml^{-1}) after 60 minutes incubation at 45ºC. The B_{max} was increase with increasing temperature. The affinity constant (K_a) were found to be depended on the temperature and it was found to be 7.756 µg.ml^{-1}). The association constant k_{+1} was increased with temperatures. The Van`t Hoff plot demonstrated a linear relationship between K_a and 1/T. Arrhenius plot indicate that there was a linear relationship between Ln k_{+1} and 1/T. the transition state thermodynamic *parameters ($E_a, \Delta H^*, \Delta G^*, \Delta S^*$) for the formation of ^{125}I-anti hCG antibody/hCG) complex were determined.*

Introduction

Human chorionic gonadotropine is a member of a family of four closely related glycoprotein hormones. In mammals, hCG consists of two subunits. The alpha subunit encoded by a single alpha gene, in contrast, multiple gene exist for the beta subunit of hCG[248,249]. HCG is produced primarily by the placenta or cells destined to become placental tissue[250]. However, It has been demonstrated that the pituitary also produces a small quantity of hCG[58]. It is found in the urine and blood of pregnant women, in pituitary tissue, blood, the urine of postmenopausal women and in the urine from some cancer patients in a variety of forms whose concentration have clinical importance[17,58,223].

So, many efforts to isolate and purification the several forms of hCG and prepare purified standards forms from it was evaluated. These efforts have been used different techniques to purify hCG and its forms, such as, gel filtration [3,4,31,47,56,223,251,252], ion and anion exchange chromatography [3,227,251,253], affinity and immunoaffinity chromatography [58,252], hydrophobic interaction chromatography[254], HPLC chromatography[3] and PAGE electrophoresis[4,58,253,255].

Many reports have been made to characterize the purified hCG by studying its three dimensional structure, the reproductive physiology, receptor interaction, biosynthesis, metabolism, functional and biochemical properties, the relation between biological, immunological activity and the chemical composition of hCG[2,15,31,35,223,251,256].

These characterization, of hCG composition, have been made by many techniques, such as, proteases analysis, various hCG fragments, enzymatic activities, electrophoresis, SDS-gel electrophoresis, amino-terminal sequence analysis, amino acid analysis, mass spectrometric analysis, immunoassay, bioassay, carbohydrate analysis[3,18,31,58,223,251-254, 257].

In this chapter hCG was partially purified from premenopausal breast cancer patient and characterized by immunoassay.

Materials and Methods
4.1. Materials

- Chemicals

The same chemicals and reagents mentioned in chapter two were used in the experiments of this chapter, beside the Sephadex G-100 and Blue dextran (2000) from Pharmacia fine chemicals (Sweden).

- Instruments

All instruments mentioned in chapter two were used in the experiments of this chapter.

- Patients

The same malignant premenopausal patients tissues mentioned in chapter two were used in the following experiments.

4.2. Methods

- **Partially purification of hCG by Gel Filtration Technique**

Gel filtration chromotography technique was used for partially purification of hCG from malignant breast tumor homogenate.

The dimensions of the column were chosen according to the following equation[258].

$$\text{Diameter} = \sqrt{\frac{m}{10}}$$

Where:

m= amount of protein in mg.

$$L = 30 \times \text{diameter}$$

Where:

L: length of the column

The gel (Sephadex G-100) was allowed to swell in excess of Tris/HCl buffer (0.05M, pH 7.2) containing 5mM EDTA and (0.02%) NaN$_3$ (20 ml of buffer per gram of gel) and left to stand for three days at room temperature without stirring to equilibrate with the buffer. The buffer was decanted and the gel was resuspended in excess volume of eluent buffer three times. The de-gassed slurry was carefully mixed before pouring into the vertical column which contains 5 ml of eluent buffer using a glass rod attached to the inner surface of the column. After the gel has settled, the column outlet was opened. Packing was continued and column was equilibrating with Tris/HCl buffer for 24 hr until the gel reached a stable bed height 30cm.

Eluent Buffer Preparation

Tris/HCl buffer (0.05M, pH 7.2) containing 5mM EDTA and (0.02%) NaN$_3$ was prepared as follows:

Tris 0.05M (0.606 gm), Tris (hydroxymethyl) amino methan, (0.08 gm) EDTA and Sodium azid (0.02%) (weight:volume) as antibacterial agent in 95 ml distilled water then the pH was adjusted to pH 7.2 using HCl (0.2 M). The volume was completed to 100 ml with distilled water.

- Void Volume Determination

The elution volume of blue dextran 2000 is equal to the column void volume (V_o) and it was determined as follows:

A fresh solution of blue dextran (2mg/ml) was prepared in the eluent buffer. 555ml of blue dextran solution was carried out with the same buffer, using a flow rate of 10ml/hr. Fractions of 1ml were collected and their absorbence were measured at 600 nm.

Eluent Buffer Preparation

Tris/HCl buffer (0.05 M, pH 7.2) containing 5mM EDTA and (0.02%) NaN_3 was prepared as in partially purification of hCG by gel filtration technique section.

- Sample Addition

A volume of 0.555ml of the tissue homogenate containing approximately 10 mg protein was prepare and then applied to the column equilibrate with Tris/HCl buffer.

The fractions were eluted with same flow rate (10ml/hr), 1 ml for each fraction, gel filtration was carried out at 10 °C. The fractions contained hCG were identified by the assay method.

The binding of each fraction was calculated and plotted against the elution volume. The fractions that gave maximum binding was poled together and used in the next experiment. The degree of isolation (folds) of hCG was calculated from the following formula.

$$\text{Yeild\%} = \frac{\text{Total protein content of partially purified hCG}}{\text{Total protein content of crude hCG}} \times 100$$

Then yield % was determined as follows:

$$\text{Purification fold of hCG} = \frac{\text{Specific binding of partially purified hCG}}{\text{Specific binding of crude hCG}}$$

Calculations

1. The radioactivity (c.p.m) and the absorbency were plotted against the fraction number.

2. The fractions under each peak were poled and the absorption spectrum was measured in the (280nm) using a 1cm cuvette against Tris/HCl buffer (0.05 M, pH 7.2), in reference beam.

Eluent Buffer Preparation

Tris/HCl buffer (0.05 M, pH 7.2) containing 5mM EDTA and (0.02%) NaN_3 was prepared as in partially purification of hCG by gel filtration technique section.

- **Stander hCG Addition**

A volume of 0.555ml of stander hCG containing approximately 5 mg protein was prepare and then applied to the column equilibrate with Tris/HCl buffer.

The fractions were eluted with same flow rate (10ml/hr), 1 ml for each fraction, gel filtration was carried out at 10°C. The fractions contained hCG were identified by the assay method.

The binding of each fraction was calculated and plotted against the elution volume. The fractions that gave maximum binding was poled together and used in the next experiment.

Calculations

1. The radioactivity (c.p.m) and the absorbency were plotted against the fraction number.

2. The fractions under each peak were pooled and the absorption spectrum was measured in the (280 nm) using a 1cm cuvette against Tris/HCl buffer (0.05 M, pH 7.2), in reference beam.

Eluent Buffer Preparation

Tris/HCl buffer (0.05M, pH7.2) containing 5mM EDTA and (0.02%) NaN_3 was prepared as in partially purification of hCG by gel filtration technique section.

4.3. The Choice of the Appropriate Conditions for the Binding of the Partially purified hCG from malignant premenopausal Breast Tissue to ^{125}I-anti hCG Antibody

- *The Effect of Different Protein Concentration*

1. A volume of 20 µl (0.1176µg.ml^{-1}) of ^{125}I-anti hCG antibody (1470 µg.ml^{-1}) was added to 100ml containing increasing amount (6.8, 13.7, 20.6, 27.4, 34.3, 41.1, 47.9, 54.8, 61.7µg protein) form the poled fractions that containing (685.7 µg.ml^{-1} protein) of breast tumor homogenate in a final volume of 250µl complete with Tris/HCl buffer (0.05M, pH 7.2) containing 0.1% BSA.

1. One additional tube containing only 20 µl of ^{125}I-anti hCG antibody was set aside until counting for total activity computation.

2. The tubes were incubated at 25°C for 60 min.

3. After incubation, 500µl of PEG 6000 (10%) were added to the tubes and incubated again for 15 min at 4°C.

4. After incubation, the tubes were centrifuged at 1500xg for 30 minute at 4°C.

5. The supernatant was discarded by decanting the assay tubes, then the tubes were inverted on a filter paper for 10min.

6. The rims of the tubes were swabbed with a cotton piece and the amounts of bound radioactivity were counted in a gamma counter.

Solutions

1. Tris/HCl buffer (0.05M, pH 7.2) containing 0.1% BSA was prepared by dissolving (0.606gm) of Tris (hydroxymethyl) amino methane and (0.1gm) of BSA in (95 ml) distilled water (D.W). Then the pH was adjusted to pH 7.2 using HCl (0.2 M). The volume was completed to 100ml with D.W.

2. PEG 6000 (10%) was prepared by dissolving (10gm) of PEG 6000 in 100ml Tris/HCl buffer (0.05M, pH 7.2) containing 0.1% BSA.

Calculations

1. The bound fraction (B) represents the counted radioactivity in each tube, expressed in c.p.m i.e (^{125}I-anti hCG antibody/hCG) complex.

2. Total activity (T) represents the counted radioactivity in the tubes containing ^{125}I-anti hCG antibody only.

3. The (B/T)% ratio for each tubes was counted as follows:

4. $(B/T)\% = \dfrac{\text{Sample mean counts (B)}}{\text{Total activity mean counts (T)}} * 100$

5. The percent of binding value (B/T)% were plotted against the increasing amount of proteins.

- **The Effect of Different Concentrations of ^{125}I-anti hCG Antibody**

1. One hundred microliters containing the optimum protein amount (34.3µg) form the poled fractions that containing (685.7 µg.ml^{-1} protein) of malignant premenopausal breast tumor homogenate was added to increasing volumes (10, 15, 20, 25, 30, 35 and 40) µl; (0.0882-0.2058µg.ml^{-1}) of ^{125}I-anti hCG antibody

(1470µg.ml^{-1}) then the volume were made up to 250µl with Tris/HCl buffer (0.05M, pH 7.2) containing 0.1% BSA.

2. A set of tubes containing only the same increasing volumes of ^{125}I-anti hCG antibody (10, 15, 20, 25, 30, 35 and 40) µl were set a side until counting for total activity computation.

3. The tubes were incubated at 25°C for 60 min.

4. After incubation, the (^{125}I-anti hCG antibody/hCG) complex was estimated by following the steps 4, 5, 6 and 7 in the experiment of protein effect.

Solutions

Tris/HCl buffer (0.05M, pH 7.2) containing 0.1% BSA and PEG 6000 (10%) were prepared as described in the experiment of protein effect..

Calculations

1. The (B/T) percent values were determined according to the experiment of protein effect.

2. The percent of binding values (B/T)% were plotted against the concentration of ^{125}I-anti hCG antibody.

- The Effect of Different pH on the Binding

1. One hundred microliters containing the optimum protein amount (34.3µg) form the poled fractions that containing (685.7 µg.ml^{-1} protein) of malignant premenopausal breast tumor homogenate was added to 20µl (0.1176µg.ml^{-1}) of ^{125}I-anti hCG antibody, the mixtures volumes were made up to 250µl with Tris/HCl buffer (0.05M) containing 0.1% BSA of different pH (6.8, 7.0, 7.2, 7.4, 7.6, 7.8 and 8.0).

2. One additional tube containing only 20µl of the ^{125}I-anti hCG antibody were set aside until counting for total activity computation.

3. The tubes were incubated at 25°C for 60 min.

4. After incubation, the (^{125}I-anti hCG antibody/hCG) complex was estimated by following the steps 4, 5, 6 and 7 in the experiment of protein effect.

Solutions

Tris/HCl buffer (0.05M) containing 0.1% BSA and PEG 6000 (10%) in different PH were prepared as described in the experiment of protein effect.

Calculations

1. The values of (B/T)% were determined according to the experiment of protein effect.

2. The percent of binding value (B/T)% were plotted against their corresponding pH values.

- Time Course of the Binding

1. One hundred microliters containing the optimum protein amount (34.3µg) form the poled fractions that containing (685.7 µg.ml^{-1} protein) of malignant premenopausal breast tumor homogenate was added to 20µl of ^{125}I-anti hCG antibody (1470µg.ml^{-1}), the mixture volume was made up to 250µl with Tris/HCl buffer (0.05M, pH 7.2) containing 0.1% BSA.

2. One additional tube containing only 20µl of the ^{125}I-anti hCG antibody were set aside until counting for total activity computation.

3. All tubes were incubated at 25°C at different time interval (30, 60, 90, 120, 150 and 180) min.

4. After incubation, the (^{125}I-anti hCG antibody/hCG) complex was estimated by following the steps 4, 5, 6 and 7 in the experiment of protein effect.

5. To determine the time course of HCG binding to ^{125}I –anti hCG antibody at different temperatures. Steps 1, 2 and 3 in the same experiment were repeated at different temperatures (5, 37, 45°C).

Solutions

Tris/HCl buffer (0.05M, pH 7.2) containing 0.1% BSA and PEG 6000 (10%) were prepared as described in the experiment of protein effect.

Calculations

1. The (B/T) percent values were determined according to the experiment of protein effect.

2. The values of (B/T)% were plotted against the incubation time.

4.4. The Kinetic and The Thermodynamic Studies

- Determination of Affinity Constant (K_a) and the Maximal Binding Capacity (B_{max}) of hCG in Partially purification hCG from malignant premenopausal Breast Tissue with ^{125}I-anti hCG Antibody

1. One hundred microliters containing the optimum protein amount (34.3µg) form the poled fractions that containing (685.7

μg.ml⁻¹ protein) of malignant premenopausal breast tumor homogenate was incubated with increasing volumes (6, 8, 10, 12, 14, 16, 18 and 20 μl) of ^{125}I-anti hCG antibody (0.035-0.1176 μg.ml⁻¹). The final volumes were made up to 250 μl with Tris/HCl buffer (0.05M, pH 7.2, 0.1% BSA).

2. Two additional tubes containing only increasing volumes (6, 8, 10, 12, 14, 16, 18 and 20 μl) of ^{125}I-anti hCG antibody were set aside until counting for total activity computation.

3. The tubes were incubated for 60 min at 45°C.

4. After incubation, the steps 4, 5, 6 and 7 of the experiment of protein effect. were repeated.

5. The previous steps were performed at different temperature (5, 25 and 37°C), the time of incubation needed to get the equilibrium state were (60 min at 5 °C, 90 min at 25°C and 90 min at 37°C).

Solutions

Tris/HCl buffer (0.05M, pH 7.2, 0.1% BSA) prepared as mentioned in the experiment of protein effect.

Calculations

1. The values B/F ratio were determined where:-

B: Is the bound radioactivity (mean of counts c.p.m), which represents the (^{125}I-anti hCG antibody/hCG) complex.

F: Is the free radioactivity (mean of the counts c.p.m), which represented the non-bound ^{125}I-anti hCG antibody.

T: Is the total radioactivity mean of the counts.

F = Total counts (T) − Bound radioactivity (B)

2. The concentration of the (^{125}I-anti hCG antibody/hCG) complex in (ug/ml) that formed after time (t) was calculated from the following equation:-

$$B(mg/ml) = \frac{B(c.p.m)}{T(c.p.m)} \times \text{Concentration of } ^{125}\text{I-anti hCG antibody in the incubation medium (µg/ml)}$$

3. The affinity constant and the maximal binding capacity were determined according to Scatchard equation [212,259]: -

$$\frac{B}{F} = \frac{1}{K_d}(B_{max} - B)$$

$$K_a = \frac{1}{K_d}$$

Where:

Ka = Affinity constant

Kd = Dissociation constant

B$_{max}$ = Maximal binding capacity

4. (µg/ml), gives a linear relationship. The values of the affinity constant of the binding (Ka) at each temperature can be calculated from the slop of the straight line, while the value of the total concentration of hCG (B$_{max}$) in breast tumor tissue was calculated from the intercept with the x-axis.

- **Thermodynamic Studies of the Binding of hCG in Premenopausal Patients with Breast Tumor to ^{125}I-anti hCG Antibody**

According to the steps of the two experiments explained in time course of the binding and determination of kinetic parameters the thermodynamic parameters were calculated.

Calculations

1. The thermodynamic parameters of standard state (ΔH^o, ΔG^o, ΔS^o) were obtained from Van't Hoff plot, the values of the

natural logarithm of equilibrium constant (affinity constant Ka) obtained at different temperature were plotted against the reciprocal values of absolute temperatures in kelvin (1/T) was calculated according to the following equation: -

$$\ln Ka = \frac{\Delta S^o}{R} - \frac{\Delta H^o}{RT}$$

Where :-

ΔH^o : The enthalpy change of the standard state.

ΔS^o : The entropy change of the standard state.

R: The gas constant (8.31441 J.K^{-1} mol^{-1})

ΔH^o value obtaied from the linear relationship of the plot. The change in Gibbs free energy of the standard state (ΔG^o) was obtained from the following equation :-

$$\Delta G^o = -RT \ln Ka$$

While the standard state entropy change was obtained from

$$\Delta S^o = \frac{\Delta H^o - \Delta G^o}{T}$$

2. The thermodynamic parameters of the transition state were obtained from Arrhenius plot of $\ln k_{+1}$ values against 1/T values, that gives a linear relationship according to the following equation:-

$$\ln K_{+1} = \ln A - \left(\frac{Ea}{RT}\right)$$

Where :-

A : Arrhenius constant.

Ea: Apparent energy of activation.

T: Absolute temperature in kelvin.

The value of E_a of the binding reaction can be determined from the slop of the straight line.

The enthalpy of transition state (ΔH^*) was obtained from:-

$$\Delta H^* = E_a - RT$$

The free energy change of the transition state (ΔG^*) was calculated by using the following equation:-

$$\Delta G^* = -RT \ln K_{+1} + RT \ln \left(\frac{KT}{h}\right)$$

Where:-
K: Boltzmann constant = 1.38×10^{-23} J.deg^{-1}
h: Plank's constant = 0.662×10^{-33} J.S^{-1}

The change in entropy of the trasition state (ΔS^*) was calculated from the following equation:-

$$\Delta S^* = \frac{\Delta H^* - \Delta G^*}{T}$$

Result and Discussion

Partially purification of hCG by Gel Filtration Technique

To partially purified hCG from malignant breast tumor, gel filtration technique was preformed by using Sephadex G 100[223,251-253]. By this technique hCG was separated from aggregates and other protein having differ molecular weight, as shown in figure (4-1, A). This experiment revealed the presence of two different eluted components eluted with different elution volume corresponding to their different molecular weights. All fractions that have been collected, 1 ml for each fraction, were tested for hCG presence by assay method, then the fractions that have a higher binding activity were collected, pooled and determined there total protein. It was found that fractions containing hCG were begin within the void volume (V_o), fraction number 10, until the fraction number 16. The isolation of hCG from malignant permenopausal breast tumors group on Sephadex G-100 showed 4.6 folds of purification; as illustrated in table (4-1). The same isolation procedure was preformed for standard hCG as a comparison with hCG from malignant breast tumor. From this experiment one peak was obtained and as mention above all collected fractions, 1ml for each fraction, were tested for hCG presence by assay method. It was found that fractions containing hCG also begin with void volume fraction until the fraction number 16, as shown in figure (4-1,B). From these findings it seems that hCG which obtained from malignant breast tumor homogenate is in the same range of molecular weight of the standard hCG.

Table (4-1): Partial isolation of hCG by gel filtration. (All other details are explained in the text).

HCG Source	Total protein mg. ml^{-1}	Specifically bound of ^{125}I-anti hCG antibody to	Specifically bound ^{125}I-anti hCG antibody/mg protein	Purification fold

		hCG		
Crude extract	0.6	46.2	0.077	1.00
Gel filtration on sehpadex G-100	0.0343	12.1	0.353	4.6

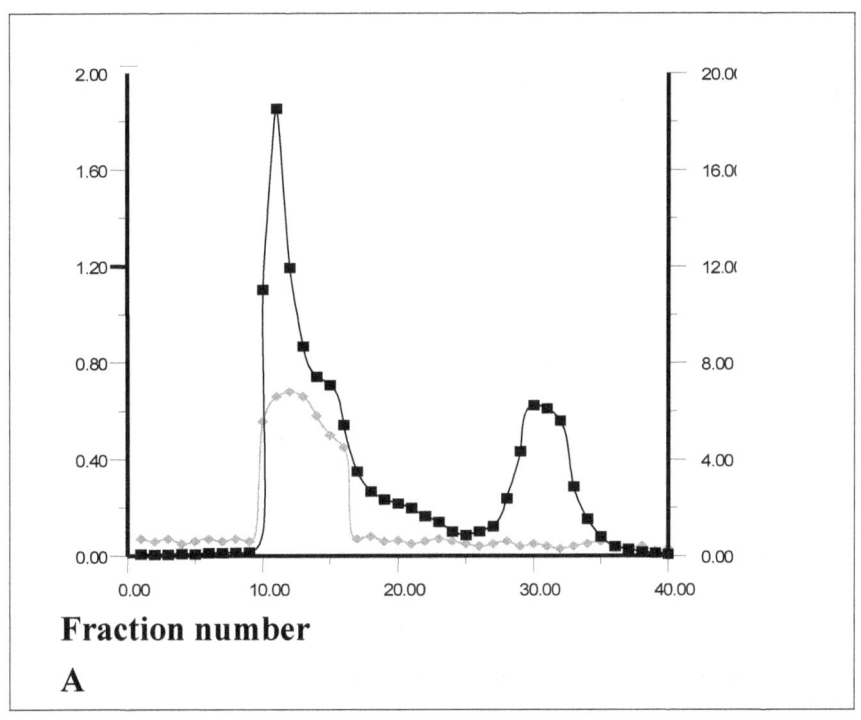

Fraction number

A

Figure (4-1): The elution profile of hCG from, A: Pre-malignant breast tumor.

The Choice of the Appropriate Conditions for the Binding of the Partially Purified hCG from malignant premenopausal Breast Tissue to ^{125}I-anti hCG Antibody

The Effect of Different Protein Concentration

The effect of increasing amount of partially purified hCG to a fixed amount of ^{125}I-anti hCG antibody to produce (^{125}I-anti hCG antibody/hCG) complex was shown in figure (4-2). At the beginning, the formation of small complex was occurred; then by increasing of the added hCG, large complex was preformed until reach maximum binding (8.4%). Excess addition of hCG make large complex less probable; causing solublization of the complex. The amount of partially purified hCG (34.3 µg) that needed for reaching maximum binding is much lower than the crude homogenate used in the previous experiment of protein effect in chapter two. According to the results, (34.3 µg) was used in all the subsequent experiments since it gives maximum value of binding.

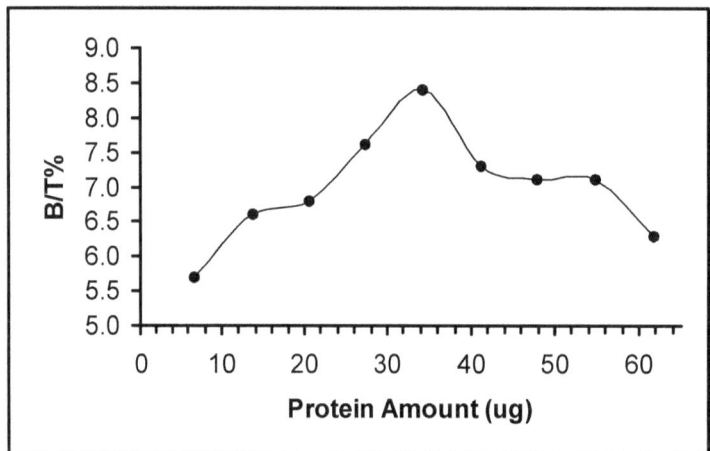

Figure (4-2): Influence of protein amount on the binding of ^{125}I-anti hCG antibody with partially purified hCG from malignant premenopausal breast tumor. (All other details are explained in the text).

The Effect of Different Concentrations of ^{125}I-anti hCG antibody

Effect of ^{125}I-anti hCG antibody concentration on the (^{125}I-anti hCG antibody/hCG) complex formation was investigated by incubated fixed amount of partially purified hCG with increasing concentration ^{125}I-anti hCG antibody for 60 mint at 25°C. Figure (4-3) is representative this binding and revealed that percent of binding increased by the amount of ^{125}I-anti hCG antibody added until it reaches the maximum binding (8.4%) at (0.1176 µg.ml^{-1}). This concentration is nearby the concentration needed for reaching the maximum binding with crude homogenate. From these observation it found the tracer concentration that needed for reaching the maximum binding is almost constant while the difference was in the amount of added protein. This is due to that crude homogenate is containing number of proteins; hCG is one of its contents and it seems that its concentration in the crude homogenate is very low as comparison for the needed protein from the crude and partially purified hCG. ^{125}I-anti hCG antibody (0.1176 µg.ml^{-1}) was used in all subsequent experiments since it gives maximum value of binding.

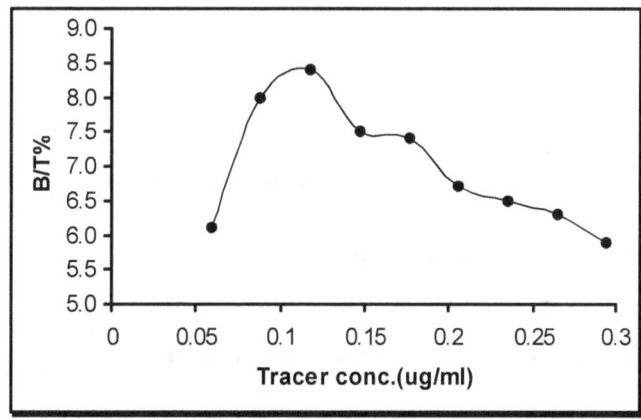

Figure (4-3): Influence of ^{125}I-anti hCG antibody concentration on its binding with partially purified hCG from malignant

premenopausal breast tumor. (All other details are explained in the text).

The Effect of Different pH on the Binding

The influences of pH on the binding of partially purified hCG with ^{125}I-anti hCG antibody is stated in figure (4-4). In this figure, optimum pH, giving maximum binding, was found to be 7.2 with decreasing in the binding at pH higher or lower than the optimum one. This result indicated that the induction of protonation-deprotonation process occurring within the ionizable groups of the amino acids present in the binding domain of hCG did not effected or changed by the isolation hCG from the crude homogenate media[224]. According to the results obtained, the pH of the buffer used in all subsequent experiments were adjusted to pH 7.2.

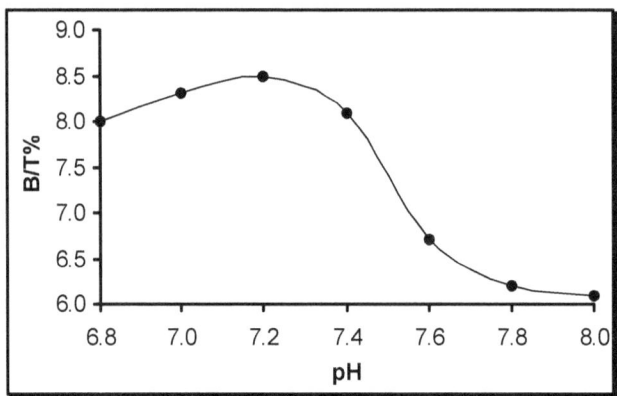

Figure (4-4): pH effect on the binding of ^{125}I-anti hCG antibody with partially purified hCG from malignant premenopausal breast tumor. (All other details are explained in the text).

Time Course of the Binding

The binding of partially purified hCG and ^{125}I-anti hCG antibody was carried out at different incubation time in four temperatures (5,25,37,45), as shown in figure (4-5). In this figure the present of binding shows conformance relation with elevation of temperature showing maximum binding (12.1%) at 45°C in 60 mint. These results are in an agreement with the previous experiment for malignant premenopausal breast tumor crude homogenate in chapter two. The binding in the two experiments shows it maximum binding at 45°C with shorter time for the partially purified hCG binding.

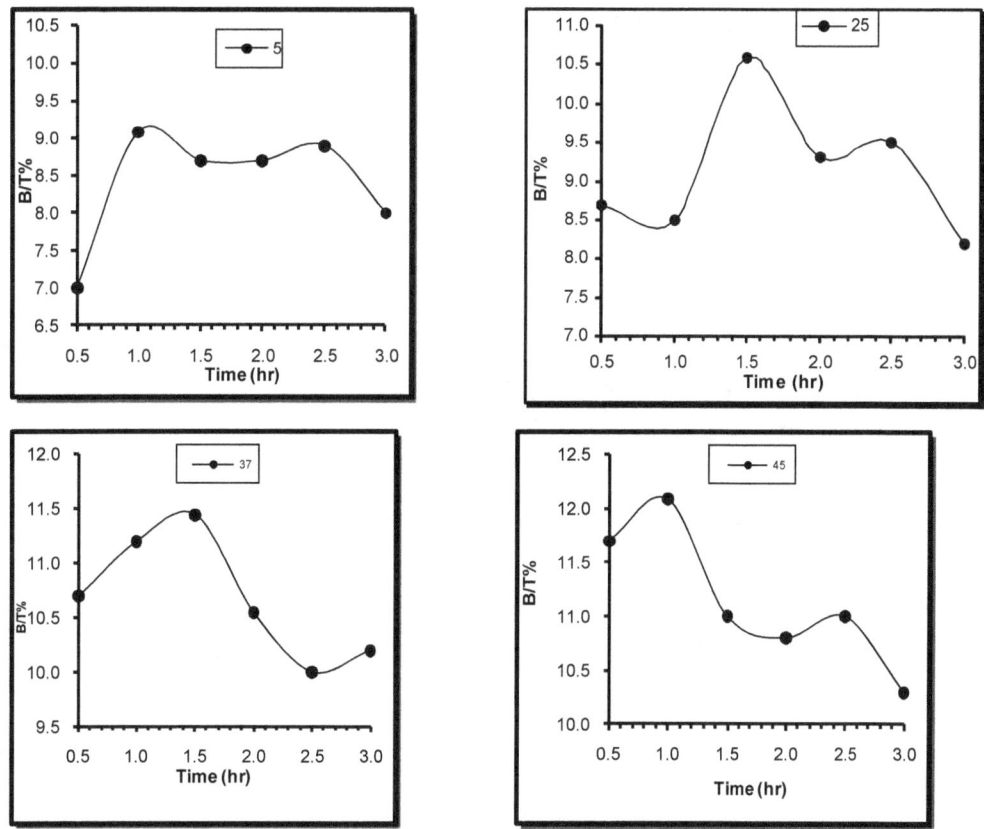

Figure (4-5): Time-Course of ^{125}I-anti hCG antibody binding to partially isolated hCG from malignant premenopausal breast tumor. (All other details are explained in the text).

Determination of Kinetic Parameters of ^{125}I-anti hCG antibody Binding with partially Purified hCG from malignant breast tumor

The time course of ^{125}I-anti hCG antibody binding with hCG from pre-menopausal malignant breast tumor was carried out to describe kinetic parameters of the binding. The simplest proposed model representing the binding of ^{125}I-anti hCG antibody with hCG could be expressed by the following equation:

$$^{125}\text{I-anti hCG Antibody} + \text{hCG} \underset{k_{-1}}{\overset{k_{+1}}{\rightleftharpoons}} [^{125}\text{I-anti hCG Antibody/hCG}]$$

Where-:

K_{+1}: is the rate of the association of ^{125}I-anti hCG antibody with hCG.

K_{-1}: is the rate of the reverse reaction of the dissociation of the complex formed under the same condition.

At equilibrium:

$$K_a = \frac{[^{125}\text{I-anti hCG Antibody/hCG}]}{[^{125}\text{I-anti hCG Antibody}][\text{hCG}]} \quad \ldots\ldots\ldots(2)$$

$$K_d = \frac{[^{125}\text{I-anti hCG Antibody}][\text{hCG}]}{[^{125}\text{I-anti hCG Antibody/hCG}]} \quad \ldots\ldots\ldots(3)$$

Thus:

$$K_a = \frac{1}{K_d} = \frac{K_{+1}}{K_{-1}} \quad \ldots\ldots\ldots\ldots\ldots\ldots\ldots\ldots\ldots(4)$$

Where-:

Ka: is the equilibrium constant of the association (affinity constant.

K_d: is the equilibrium constant of the dissociation ^{125}I-anti hCGantibody/hCG) complex.

The values of Ka and maximal binding capacity (B_{max}) were calculated from Scatchard plots at five different temperatures as shown in figure (4-6) and table (4-2).

As in table (4-2), the K_a and B_{max} values are temperatures depended. The K_a increased with increased temperature, in the following order 5>25>37>45 °C, to reach it highest value at 45°C (7.756 µg^{-1}.ml), while the concentration of partially purified hCG was determined to be (0.0143 µg^{-1}.ml) at 45 °C.

The value of dissociation constant (K_d) was calculated by using equation (4), which indicate that K_d values are decrease with increasing temperature to reach it lowest value (0.1289 µg^{-1}.ml) at 45 °C.

Table (4-2): The kinetic parameter of ^{125}I-anti hCG antibody binding to partially purified hCG in pre-menopausal malignant breast tumor homogenate.(All other details are explained in the text).

Temp °C	Binding Capacity B_{max}(µg.ml^{-1})	K_a (µg^{-1}.ml)	K_d (µg.ml^{-1})
5	0.0121	6.4672	0.1546
25	0.0122	6.872	0.1455
37	*0.0131*	7.066	0.1415
45	0.0143	7.756	0.1289

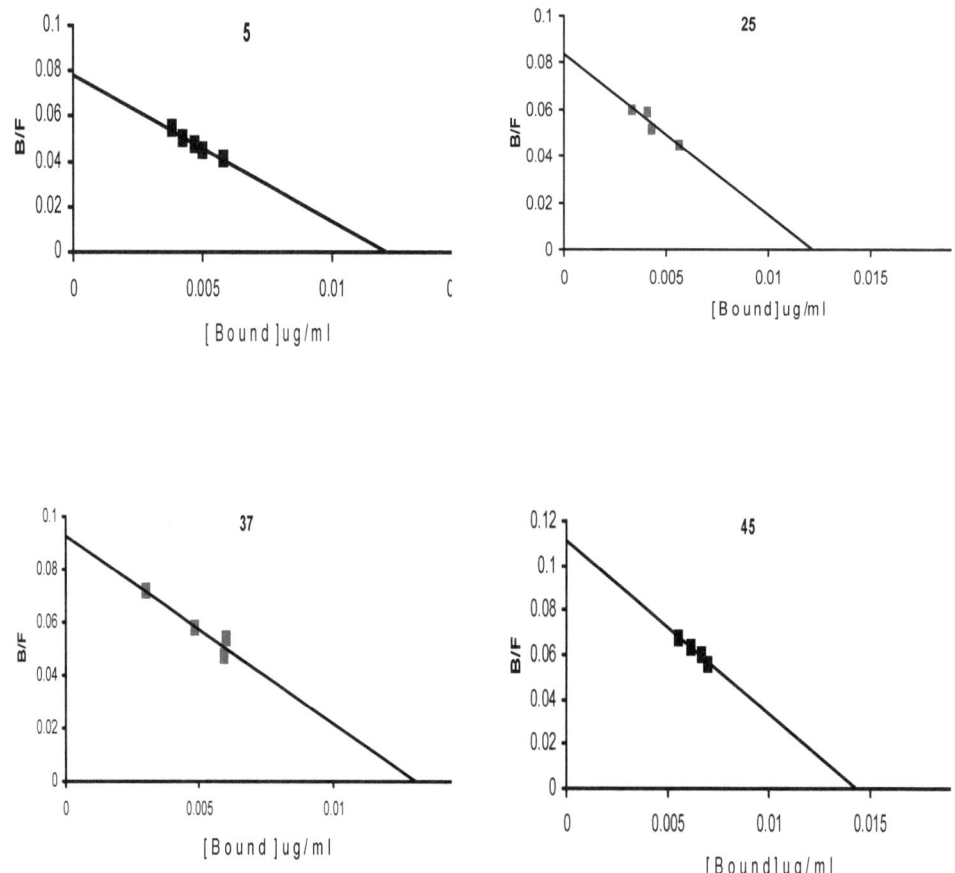

Figure (4-6): Scatchard plot of ^{125}I-anti hCG antibody binding to the partially purified hCG in malignant breast tumor at four different temperatures. (All other details are explained in the text).

Time course data in figure(4-5) were used to determine the reaction order of hCG binding to its specifically ^{125}I-anti hCG antibody. Because the binding is small and the most labeled antibody remains free and only small fraction binds even at equilibrium, i.e, $[Ab]_t \gg [AbAg]_e$

Thus:

$$[Ab]_t \gg \frac{[AbAg]_t [AbAg]_e}{[Ag]_t}$$

$[AbAg]_e$: is the concentration of (^{125}I-anti hCG/hCG)complex formed at equilibrium.

$[AbAg]_t$: is the concentration of (^{125}I-anti hCG/hCG) complex after time (t).

$[Ab]_t$: is the total concentration of ^{125}I-anti hCG antibody in µg. ml^{-1}.

$[Ag]_t$: is the total concentration of hCG in µg. ml^{-1}.

So the following equation could be used in order to fit the pseudo- first order kinetics[260]-:

$$Ln \frac{[AbAg]_e}{[AbAg]_e - [AbAg]_t} = K_{+1} t \frac{[Ab]_t [Ag]_t}{[AbAg]_e} \quad \dots\dots\dots(5)$$

Figure (4-7) shows the plot of $\ln \frac{[AbAg]_e}{[AbAg]_e - [AbAg]_t}$ against time in malignant breast tumor, which gives a straight line with a slope, equal to the observed value of first rat constant K_{obs} in min^{-1}. The rate constant (K_{+1}) in µg^{-1}.ml was calculated at four different temperatures by using the following equation[261]:

$$K_{obs} = K_{+1} \frac{[^{125}\text{I-anti hCG Antibody}]_t [hCG]_t}{[^{125}\text{I-anti hCG Antibody/hCG}]_e} \quad \dots\dots(6)$$

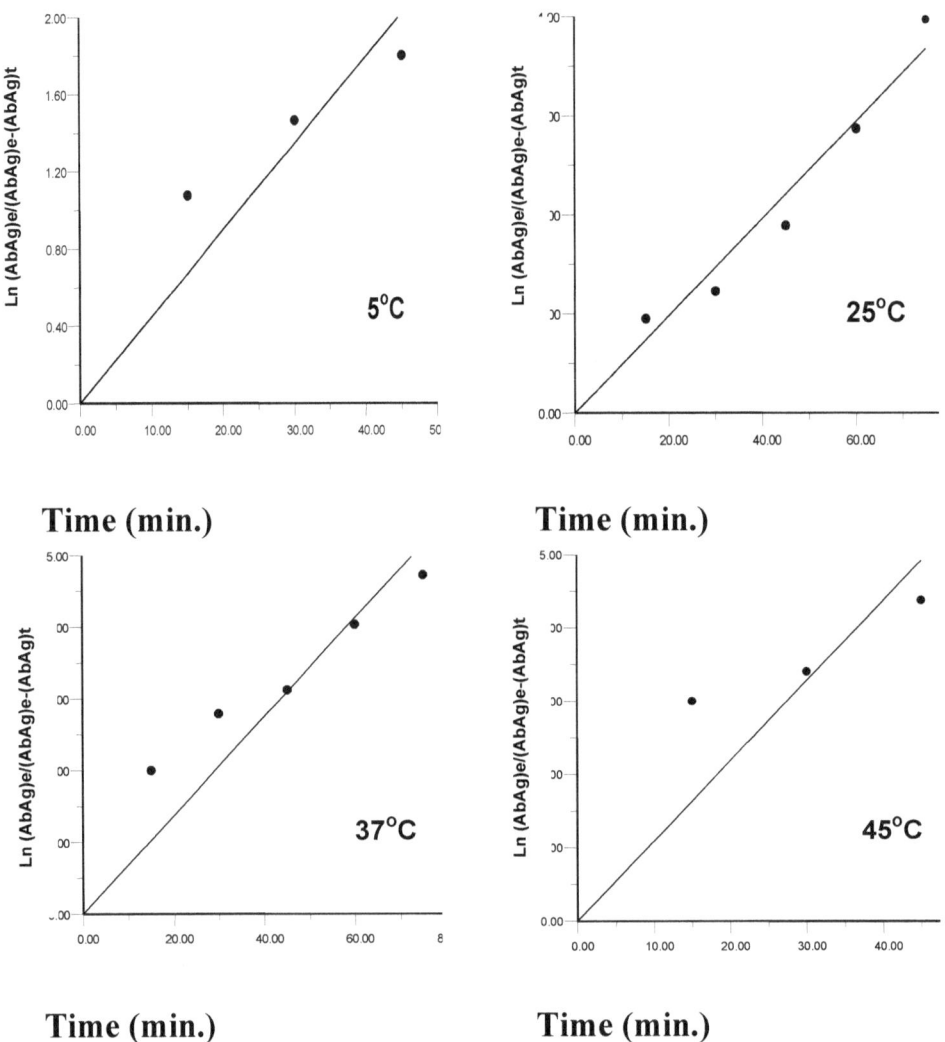

Figure (4-7): Kinetics of ^{125}I-anti hCG antibody binding to patrially purified hCG in malignant breast tumor at four different temperatures. (All other details are explained in the text).

The values of K_{-1} at four different temperatures were calculated by using equation (4). Half-life time of association (t ½)$_{ass.}$, Which represented the time needed for the formation of half amount of the complex at equilibrium was determined from the concentration of the complex at equilibrium and the time-course curve. The half-life time of dissociation (t ½)$_{diss}$, was calculated from the following relation:

$$(t_{1/2})_{diss.} = \frac{\ln 2}{k_{-1}} = \frac{0.693}{k_{-1}}$$

Table (4-3) summarized the values of K_{obs}, K_{+1}, K_{-1}, $(t_{1/2})_{ass.}$ and $(t_{1/2})_{diss.}$ at four different temperature. These values indicate that highest rate for the association reaction K_{+1} occurs at 45°C and the dependence of association and dissociation reaction rate on temperature.

Table (4-3): The effect of temperature on the kinetic parameter of ^{125}I-anti hCG antibody binding to partially purified hCG in pre-menopausal malignant breast tumor homogenate. (All other details are explained in the text).

Temp °C	k_{obs} x 10^{-2} (min^{-1})	k_{+1} x 10^{-1} µg^{-1}.ml.min^{-1}	k_{-1} x 10^{-2} min^{-1}	$(t_{1/2})_{as}$ (hr)	$(t_{1/2})_{dis}$ (hr)
5	4.48	3.37	5.21	3.69	13.3
25	4.89	4.26	6.20	2.5	11.1
37	6.87	5.97	8.45	1.66	8.2
45	10.92	8.66	11.1	1.08	6.2

The Thermodynamic Studies of ^{125}I-anti hCG antibody to the Partially Purified hCG in malignant breast tumor

The dependence of the equilibrium binding constant (affinity constant) for the binding of ^{125}I-anti hCG antibody with partially purified hCG on temperature (Van`t Hoff blote) was clarified in figure (4-8). Also table (4-4) clarify the values of thermodynamic parameters of Standard State of partially isolated hCG in malignant breast tumor.

The results indicate that ΔH°, in general, had very small value; indicating a favorable interaction between ^{125}I-anti hCG antibody and partially purified hCG. Also, the positive sign a certain that the reaction was nearly endothermic.

The negative values of ΔG° reflects the stability of the complex hence, the high affinity of the reactants. So this system is characterized by the sole contribution of ΔS° to the stability of the complex formed, while ΔH° has little or no effect[262]. The positive values of ΔS° suggest that the reaction spontaneity be entropically driven Entropy was the driven force for the occurrence of the binding. This indicates that the hydrophobic interactions played an important role are stabilizing the complex[263].

These include the non-covalent interaction which are fundamentally electrostatic in nature such as charge-charge, charge-dipole, dipole-dipole, charge-induced dipole, dipole-induced dipole interactions and hydrogen bonds . The sum of these types of interactions can yield some stabilization to the folded structure of the complex[264].

So the negative values of ΔG° showed that the overall reaction energetically favorable in the direction of complex formation.

Table (4-4): Thermodynamic parameters at standard state of ^{125}I-anti hCG antibody to the partially purified hCG in pre-malignant breast tumor homogenate.(All other details are explained in the text).

Temp°C	ΔH^o KJ.mol^{-1}	ΔG^o KJ.mol^{-1}	ΔS^o J.mol^{-1}.K^{-1}
5	2.952	-4.314	26.14
25	2.952	-4.775	24.93
37	2.952	-5.039	25.78
45	2.952	-5.415	26.31

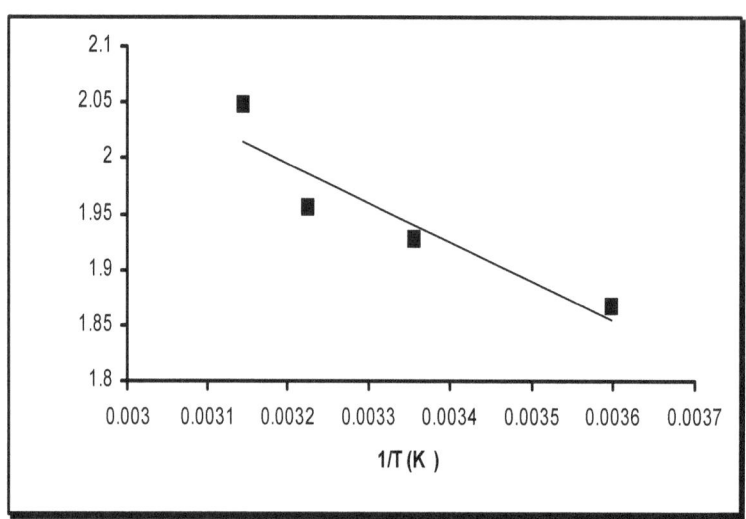

Figure (4-8): Van`t Hoff lpot for the binding of ^{125}I-anti hCG antibody to the partially purified hCG in pre-malignant breast tumor homogenate. (All other details are explained in the text).

Thermodynamic Parameters of Transition State

According to the transition state theory, the interaction of hCG with it ^{125}I-anti hCG antibody lead to the formation of an activated complex (transition states) then the formation of the final product:-

$$^{125}\text{I-anti hCG} + \text{hCG} \longrightarrow [^{125}\text{I-anti hCG/hCG}] \longrightarrow [^{125}\text{I-anti hCG/hCG}]$$

State(A) An Activated Complex Final Product
Transition State State(B)

The thermodynamic parameters of the transition state) ΔH^*, $\Delta G^*, \Delta S^*$ and Ea) could be determined from Arrhenius equation and the kinetic constant.

Figure (4-9) shows the Arrhenius plot of $\ln K_{+1}$ against $1/T$ values. Table (4-5) shows the values of thermodynamic parameters of transition state (Ea, ΔH^*, ΔG^*, ΔS^*). The value of Ea that determined from Arrhenius plot represents the apparent energy activation of the binding reaction and the required energy to overcome the energy barrier of the transition state for the formation of (^{125}I-anti hCG antibody/hCG) complex.. The positive value of ΔH^* shows that content of activated complex is more than that of purified species[265]. The high positive value of ΔG^* indicated that the formation of activated comples was a non spontaneous process and required a lot of energy(equal to Ea) to overcome the transition energy barrier and giving the final product. Also, the ΔG^* positive values is mainly attributed to the decrease in entropy of the transition state ($\Delta S^* < 0$). The high negative ΔS^* revealed that activated complex had more ordered structure than the reactant species($\Delta S^* < 0$).

The values of the thermodynamic parameters of the binding reaction, gave an overall idea about the nature of forces that regulate the formation of complex. It is proposed that the formation of a protein-ligand complex, occurs in two steps. The first stabilization of the complex by hydrophobic interactions and the second is the stabilization by short range interaction, such as electrostatic interaction, hydrogen bonding and vander wall's interactions[265]. Hydrophobic interactions contribute to the complex stability via high positive entropy change ($\Delta S^* > 0$), while electrostatic interactions, hydrogen bonding and vander waal's interactions contribute to the stability of the complex via negative entropy change($\Delta S^* < 0$)[265,266]. The thermodynamic data from this study indicate that the binding of ^{125}I- anti hCG antibody with hCG are entropically driven and in agreement with the concept that hydrophobic interaction play an important role in ^{125}I- anti hCG/hCG) interactions.

Table (4-5): Thermodynamic parameters at transition state of ^{125}I-anti hCG antibody to the partially purified hCG in pre-malignant breast tumor homogenate.(All other details are explained in the text).

Temp° C	Ea KJ.mol^{-1}	ΔH* KJ.mol^{-1}	ΔG* KJ.mol^{-1}	ΔS* J.mol^{-1}.K^{-1}
5	16.279	13.968	67.950	-194.18
25	16.279	13.802	75.095	-205.68
37	16.279	13.702	77.351	-205.32
45	16.279	13.635	78.433	-203.76

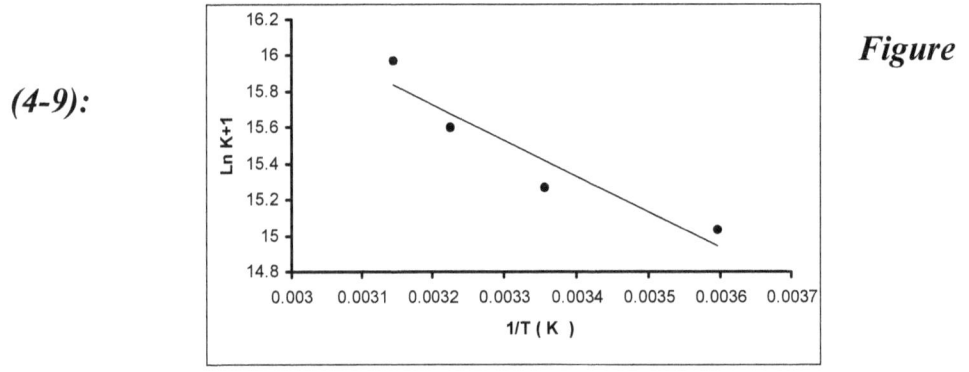

Figure (4-9): Arrhenius plot for the binding of ^{125}I-anti hCG antibody to the partially purified hCG in pre-malignant breast tumor homogenate.(All other details are explained

ChapterFive   Spectral studies on standard hCG ,Partially Purified hCG and I-anti hCG Antibody/ hCG complexes

Abstract

Gel filtration technique was used to separate free ^{125}I-anti hCG antibody from ^{125}I-anti hCG antibody/hCG complexes on Sephadex G-150 by using standard, benign and malignant breast tissue homogenates as hCG source. Also standard hCG was used as a comparison with partially purified hCG from malignant breast tumor homogenate that obtained by using gel filtration technique on sephadex G-100.

Ultraviolet spectroscopic studies were carried out to characterize the standard, benign and malignant complexes, standard and partially purified hCG. Factors affecting the absorption properties of them have been studied, such as pH, solvent polarity and pH titration. These factors show variable effect on the λ_{max} of each group.

Introduction

The protein is a complex molecules that several methods, such as ultraviolet spectra, have been developed to determined it structure[267,268]. *In this spectrum, even limited information about protein molecular is obtained, it can absorb radiation over a wide range of spectra giving a wide range of studies*[269].

Although several immunochemical, biochemical and biophysical studies was carried out to characterize protein-protein interaction, which is operative at almost every level of cell function, UV spectral method remain one of the most important spectral methods in immunology because it provides a sensitive, quantitative methodology for the study of antibody structure and specific ligand binding[270].

Many studies have been taken, under the supervising of Dr.Almudufer, to investigate many and different kind of proteins and steroids by studying there purified structure, their interaction and binding, studying the effect of different factor that might affect their absorbance [271-274]. **These studies reached to a conclusion that UV studies are useful for quantitative determination of complex formation and for studying these molecules which have a characteristics spectrum for each kind. hCG and its receptor of the proteins, which had, have their share of studying by UV spectra**[42,239,275-278]. **In this chapter hCG studying have been take as a complementary study for the previous studies.**

Materials and Methods

5.1. Materials

- **Chemicals**

The same chemicals and reagents mentioned in chapter two were used in the experiments of this chapter, beside of the standard hCG (immunotech, France), ^{125}I-anti hCG antibody(immunotech, France), Sephadex G-150, Sephadex G-100 and Blue dextran (2000) from Pharmacia fine chemicals (Sweden).

- **Instruments**

All instruments mentioned in chapter two and four were also used in the experiments of this chapter.

- **Patients**

The same benign and malignant premenopausal patient's tissues mentioned in chapter two were used in the following experiments.

5.2. Methods

- **Gel Filtration Technique for Separation of Free and Bound ^{125}I -anti hCG Antibody**

For Preparation of the Column, the dimensions (0.8x22 cm) were chosen according to the equation in the experiment of partially purification of hCG by gel filtration technique in chapter four. The Sephadex G-150 was used to separate free and bound ^{125}I-anti hCG antibody and eluent buffer and reagents was prepared as mentioned in the experiment of partially purification of hCG by gel filtration technique in chapter four. The void volume was determined and found to be 7 ml.

5.3. Separation Procedure of (^{125}I-anti hCG Antibody/hCG) Complex

- hCG from Benign (Fibrocyst) Breast Tumor Homogenate and its ^{125}I -anti hCG Antibody

1. Benign breast tumor homogenate, 140µl, containing (1.400µg. ml^{-1}) was incubated with 6µl of ^{125}I-anti hCG antibody (8.82µg. ml^{-1}) and complete the reaction to a final volume of 500µl with Tris/HCl buffer 0.05M pH 7.2. The tubes were incubated for 150 min at 4°C.

2. At the end of incubation, the mixture was applied to the surface of a Sephadex G-150 (0.8x22 cm) equilibrated with Tris/HCl buffer 0.05M, pH 7.2 containing 5mM EDTA and (0.02%) NaN$_3$. Elution was carried out using the same buffer to separate hCG bound to ^{125}I-anti hCG antibody from unbound (Free) ^{125}I-anti hCG antibody with a flow rate (12ml/hr.), and fraction volumes of 1 ml were collected.

3. Gamma counter counted the radioactivity of each fraction.

Solutions

1. Tris/HCl buffer (0.05M) at pH 7.2 was prepared by dissolving (0.606gm) of Tris (hydroxymethyl) amino methane, then the pH was adjusted to pH 7.2 using HCl (0.2M). The volume was completed to 100ml with D.W.

2. The eluent Buffer, Tris/HCl buffer (0.05M, pH 7.2) containing 5mM EDTA and (0.02%) NaN$_3$ was prepared as in the experiment of partially purification of hCG by gel filtration technique in chapter four.

Calculations

1. Radioactivity (c.p.m) of each eluted fraction was plotted against the fraction number.

2. The percent radioactivity was calculated by dividing the sum of the radioactivity of the fractions under each peak by the sum of radioactivity of all peaks appeared in the profile:

3. $\text{Percent radioactivity of each peak} = \dfrac{\text{Radioactivity per peak (c.p.m)}}{\text{Sum of radioactivity of all peaks (c.p.m.)}} \times 100$

- **hCG from Premenopausal Malignant (IDC) Breast Tumors Homogenate and Its ^{125}I-anti hCG Antibody**

1. Premenopausal malignant breast tumors, 66.6μl, containing (1.200μg.ml^{-1} protein) was incubated with 50μl of ^{125}I-anti hCG antibody (8.82μg.ml^{-1}) in a final volume 500μl with Tris/HCl buffer 0.05M, pH 7.2. The tubes were then incubated for 120 min at 45°C.

2. Steps 2 and 3 in experiment of hCG complex separation from benign tumor were repeated.

Solutions

1. Tris/HCl buffer (0.05M) at pH 7.2 was prepared as described in section (5.3.1)

2. The eluent buffer, Tris/HCl buffer (0.05M, pH 7.2) containing 5mM EDTA and (0.02%) NaN$_3$ was prepared as in the experiment of partially purification of hCG by gel filtration technique in chapter four.

Calculation

The same calculation that mentioned in experiment of hCG complex separation from benign tumor were repeated. was used to calculate the radioactivity; the absorbence and the percent of radioactivity of each eluted fraction.

- **Standard hCG and Its ^{125}I-anti hCG Antibody Reagents**

1. Standard hCG, 12.5μl, containing (1218.75 μg. ml^{-1} protein) was incubated with 50μl of ^{125}I-anti hCG antibody (73.5μg.ml^{-1}) in a final volume 500μl with Tris/HCl buffer 0.05M, pH 7.2. The tubes were then incubated for 60 min at 25°C.

2. Steps 2, and 3 in experiment of hCG complex separation from benign tumor were repeated.

Solutions

1. Tris/HCl buffer (0.05M) at pH 7.2 was prepared as described in experiment of hCG complex separation from benign tumor were repeated.

2. The eluent buffer, Tris/HCl buffer (0.05M, pH 7.2) containing 5mM EDTA and (0.02%) NaN$_3$ was prepared as in the experiment of partially purification of hCG by gel filtration technique in chapter four.

Calculation

The same calculation that mentioned in experiment of hCG complex separation from benign tumor were repeated. was used to calculate the radioactivity; the absorbence and the percent of radioactivity of each eluted fraction.

- ^{125}I-anti hCG Antibody Reagent

1. Fifty microliters of ^{125}I-anti hCG antibody (73.5μg.ml^{-1}) was completed to 500μl with Tris/HCl buffer 0.05M, pH 7.2, then this volume was injected to the column as mentioned in step2, then steps 3 were repeated.

2. Gamma counter counted the radioactivity of each fraction.

Solutions

1. Tris/HCl buffer (0.05M) at pH 7.2 was prepared as described in experiment of hCG complex separation from benign tumor were repeated.

2. The eluent buffer, Tris/HCl buffer (0.05M, pH 7.2) containing 5mM EDTA and (0.02%) NaN_3 was prepared as in the experiment of partially purification of hCG by gel filtration technique in chapter four.

Calculation

The same calculation that mentioned in experiment of hCG complex separation from benign tumor were repeated. was used to calculate the radioactivity; the absorbence and the percent of radioactivity of each eluted fraction.

5.4. Partially purification of hCG by Gel Filtration Technique

- *Preparation of the Column, Gel and Determination of Void Volume*

The Sephadex G-100 was used to isolate hCG and was prepared as mentioned in the experiment of partially purification of hCG by gel filtration technique in chapter four, the void volume was determined and found to be 10ml.

- Eluent Buffer Preparation

Buffers and reagents mentioned in the experiment of partially purification of hCG by gel filtration technique in chapter four. are used in this chapter. Other additional solutions are indicators in each experiment.

- **Samples Addition**

The same standard hCG and malignant premenopausal patient's tissues sample and procedure mentioned in the experiment sample and standard addition in chapter four were used in this experiment.

Calculations

The same Calculations mentioned in the experiment sample addition in chapter four were used in this experiment.

5.5. The UV Spectrum

- **The UV Spectrum of (^{125}I-Anti hCG Antibody / hCG) Complex of Benign Breast Tumor**

The gel filtration profile in experiment of hCG complex separation from benign tumor gave two peaks. The fractions under first peak were pooled and the absorption spectrum was scanned in UV Region (200-350nm) against the appropriate blank in the reference beam.

- **The UV Spectrum of (^{125}I-Anti hCG Antibody / hCG) Complex of Malignant Breast Tumor**

The gel filtration profile in experiment of hCG complex separation from malignant tumor gave two peaks. The fractions under first peak were pooled and the absorption spectrum was scanned in UV Region (200-350nm) against the appropriate blank in the reference beam.

- **The UV Spectrum of (^{125}I-Anti hCG Antibody / hCG) Standard hCG Complex**

The gel filtration profile in experiment of hCG complex separation from standard tumor gave two peaks. The fractions under first peak were pooled and the absorption spectrum was scanned in UV Region (200-350nm) against the appropriate blank in the reference beam.

- **The UV Spectrum of ^{125}I-Anti hCG Antibody**

Twenty microliter of ^{125}I-Anti hCG Antibody was completed to 1ml with eluant buffer and the absorption spectrum was scanned in UV Region (200-350nm) against the appropriate blank in the reference beam.

- **The UV Spectrum of Partially purified hCG from malignant premenopausal Breast tumor tissue**

The gel filtration profile in experiment of sample addition in chapter five gave two peaks. The fractions that gave higher radioactivity under the first peak were pooled and the absorption spectrum was scanned in UV region (200-350nm) against the appropriate blank in the reference beam.

- **The UV Spectrum of the Standard HCG**

The gel filtration profile in experiment of hCG complex separation from benign tumor gave one peak. The fractions under the peak were pooled and the absorption spectrum was scanned in

UV region (200-350nm) against the appropriate blank in the reference beam.

5.6. Factors Affecting the Absorption Properties of Benign, Malignant, Standard (^{125}I-Anti hCG Antibody/hCG) Complex, standard and partially purified hCG.

- pH Effect

Two hundred microliters of pooled fractions under the first peak of each group, which represents the (^{125}I-anti hCG antibody/hCG) complex was completed to 1 ml with different buffers at different pH values (4, 7.2, 11) then each of which was placed in 1cm cuvette in the sample beam and the buffer at the adjusted pH in the reference beam the absorption spectrum was measured in the area of (200-350nm).

Solutions

1. Tris-(hydroxymethyl)–aminomethan–hydrochloride(Tris-hydrochloride) buffer (0.05M) at pH 4 containing 5mM EDTA and (0.02%) NaN_3 was prepared by dissolving (0.788gm) of Tris hydrochloride, (0.08gm) of EDTA and Sodium azid 0.02% (weight: volume), then the pH was adjusted to pH 4 using NaOH (0.2M). The volume was completed to (100ml) with D.W.

2. Tris/HCl buffer (0.05M, pH 7.2) containing 5mM EDTA and (0.02%) NaN_3 was prepared as in the experiment of partially purification of hCG by gel filtration technique in chapter four.

3. Tris/HCl buffer (0.05M, pH 11) containing 5mM EDTA and (0.02%) NaN_3 was prepared as in the experiment of partially purification of hCG by gel filtration technique in chapter four.

- **Effect of Solvents Polarity**

1. Two hundred microliters of pooled fractions under the first peak of each group, which represents the (^{125}I-anti hCG antibody/hCG) complex was completed to 1ml with The eluent buffer in the presence of 20% ethanol.

2. The mixture was placed in the sample beam using 1cm cuvette against 20% ethanol prepared in the same buffer in the reference beam. The absorption spectrum was measured in the area of (200-350nm).

3. The experiment was repeated in the presence of 20% Glycerol, Ethylene glycol and Sucrose prepared in the same buffer.

Solutions

1. The eluent buffer, Tris/HCl buffer (0.05M, pH 7.2) containing 5mM EDTA and (0.02%) NaN_3 was prepared as in the experiment of partially purification of hCG by gel filtration technique in chapter four., then 20ml of Ethanol, Glycerol, Ethylene glycol, and 20gm of Sucrose was dissolved in it.

2. Tris/HCl buffer (0.05M, pH 11) containing 5mM EDTA and (0.02%) NaN_3 was prepared as in the experiment of partially purification of hCG by gel filtration technique in chapter four.

- **Effect of Urea, KCl and (Urea, KCl) Mixture**

Two hundred microliters of pooled fractions under the first peak of each group, which represents the (^{125}I-anti hCG antibody/hCG) complex were pipetted in a set of three tubes. The volume was completed to 1ml with The eluent buffer contains (0.03M KCl, 8M urea and mixture 1:1 of both 0.03M KCl and 8M Urea) respectively, then each sample was placed in 1cm cuvette in the sample beam and the buffer at the same pH in the reference beam. The absorption spectrum was measured in the area of (200-350nm).

Solutions

1. Eight molar of urea was prepared by dissolving 24.02g of urea in 50 ml of the eluent Buffer, 0.03M KCl was prepared by dissolving 0.2237g of the salt in 50ml of the eluent Buffer.

2. The eluent buffer Tris/HCl buffer (0.05M, pH 7.2) containing 5mM EDTA and (0.02%) NaN_3 was prepared as in the experiment of partially purification of hCG by gel filtration technique in chapter four.

- *Spectrophotometric pH Titration of the Complex*

Two hundred microliters of pooled fractions under the first peak of each group, which represents the (^{125}I-anti hCG antibody/hCG) complex was completed to 1ml with different eluent buffers pH values at ranging from 8 to 11. The maximum absorbency of each sample was measured at 295nm; the absorbency of λ_{max} at each pH value was plotted versus the corresponding pH. Other Two hundred microliters of pooled fractions under the first peak of each group was completed to 1ml with different buffer pH values at ranging from 4 to 8. The maximum absorbency of each sample was measured at 211nm. The absorbency of λ_{max} at each pH value was plotted against the corresponding pH.

Solutions

The eluent buffers (0.05M) at different pH were prepared as described in the experiment of hCG complex separation from benign tumor.

Results and Discussion

Gel Filtration Technique for Separation of Free and Bound ^{125}I-anti hCG Antibody

Gel filtration technique was used to separate (^{125}I-anti hCG Antibody/HCG) complexes from unbound ^{125}I-anti hCG Antibody for benign, malignant breast tumor and standard hCG, as shown in figures (5-1,A-C).

Filtration technique was preformed on G-150, yielding two peaks. First peak represents (^{125}I-anti hCG Antibody/HCG) complex, while the second peak represents the unbound (free) ^{125}I-anti hCG Antibody.

Separation process was depending upon the difference in the molecular weight of the compounds, so the first peak was presenting the complex and the second was presenting the unbound ^{125}I-anti hCG Antibody, due to that the complex has greater molecular weight than the unbound ^{125}I-anti hCG Antibody.

To insure this observation, unbound ^{125}I-anti hCG Antibody was applied alone to the same column as a comparison, as shown in figure (5-1,D). This experiment appeared only one peak in the same position of the second peak of the figures (5-1,A-C) which represent the unbound ^{125}I-anti hCG Antibody.

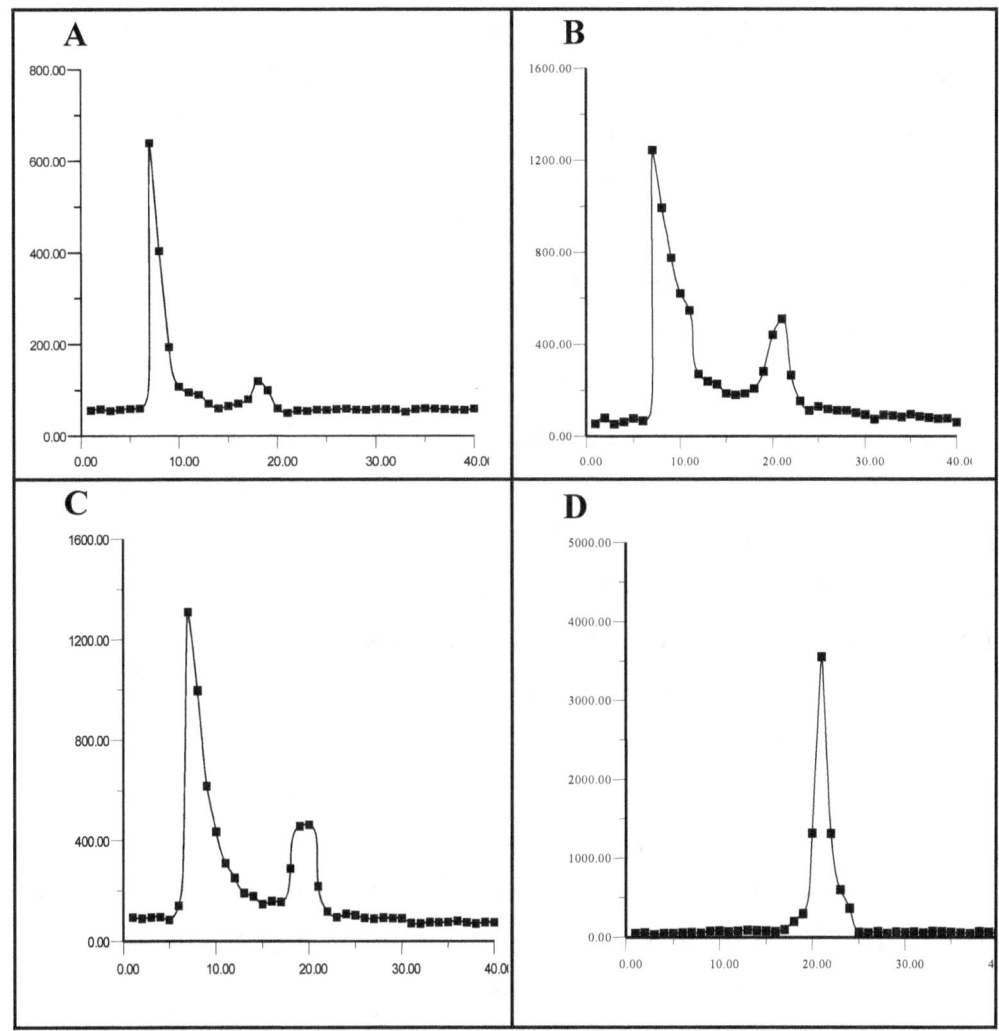

Figure (5-1): The elution profile of hCG from, A: benign breast tumor, B: pre-malignant breast tumor, C: standard hCG and the elution of ^{125}I-anti hCG Antibody (D). (All other detailes are explained in the text).

The UV spectra of (^{125}I-anti hCG Antibody/HCG) complexes, standard, partially purified hCG and ^{125}I-anti hCG Antibody

The UV spectra of (^{125}I-anti hCG Antibody/HCG) complexes at pH 7.2 shows one maximum wavelength at 222nm, 220nm and 220nm for standard hCG, benign and malignant breast tumor respectively, as shown in figure (5-2,A-C). This wavelength is assigned to the peptide bonds of the complex molecule[279], as shown in table (5-1).

The UV spectra of ^{125}I-anti hCG Antibody at pH 7.2 shows one maximum wavelength at 227nm, as in table (5-1), which could be a signed to the side chain chromophore of the amino acid residues[280], as in figure (5-2,D).

Figure (5-2,E and F) shows the UV spectra of standard and partially purified hCG at pH 7.2. It consisted of one maximum wavelength at 279nm for standard hCG, which shows a slit difference from the wavelength of partially purified hCG at 271nm. These wavelengths are assigned to the side chain chromophore of amino acids residues[280].

Table (5-1): The λ_{max} values of (^{125}I-anti hCG Antibody/HCG) complexes, ^{125}I-anti hCG Antibody, standard hCG and partially purified hCG. (All other details are explained in the text).

Fractions	λ_{max}
Standard (^{125}I-Anti hCG Antibody/ hCG) Complex	222 nm
Benign (^{125}I-Anti hCG Antibody/ hCG) Complex	220 nm
Malignant (^{125}I-Anti hCG Antibody/ hCG) Complex	220 nm
^{125}I-Anti hCG Antibody	277 nm

Standard hCG	279 nm
Partially purified hCG	271 nm

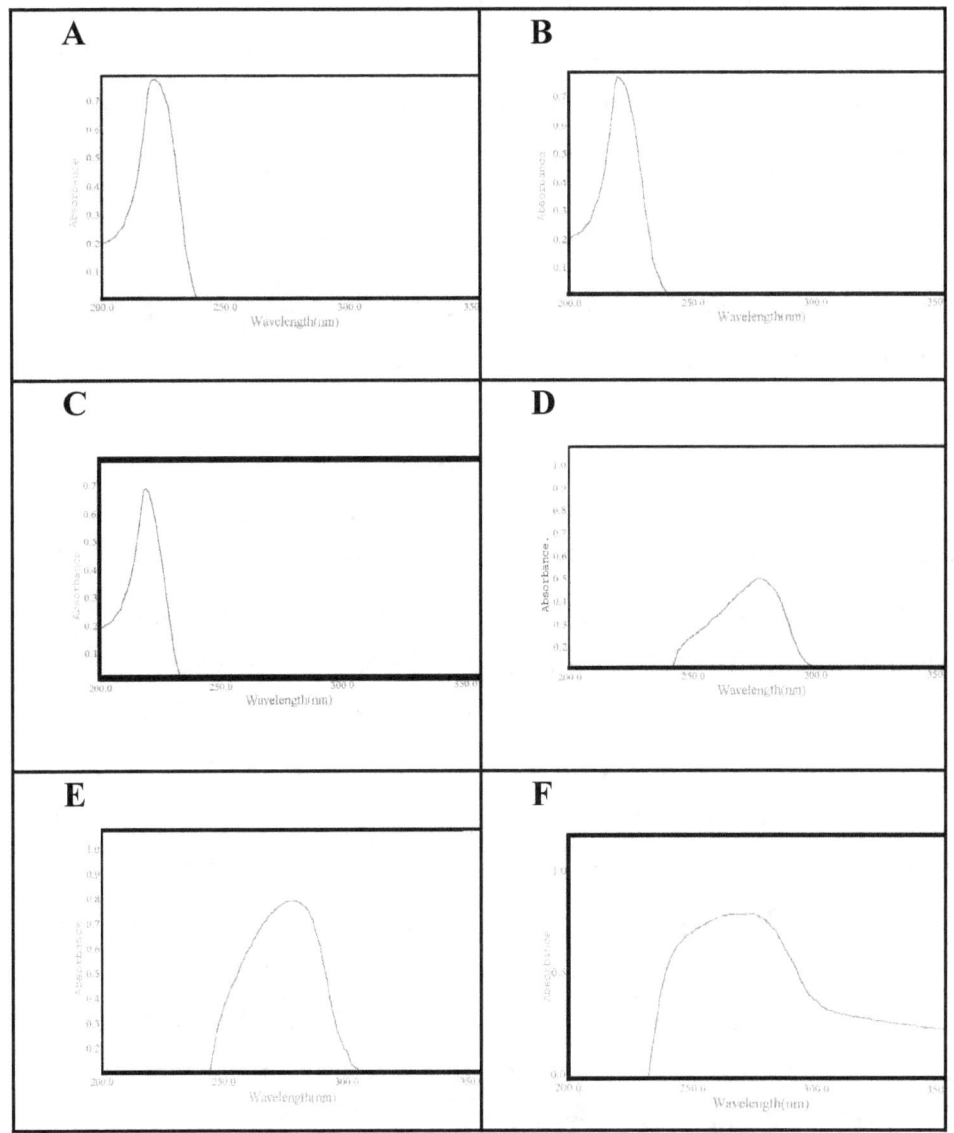

Figure (5-2): The UV spectrum at pH 7.2 for A: Standard complex, B: Benign complex, C: Malignant complex, D: ^{125}I-Anti hCG Antibody, E: Standard hCG and F: Partially purified hCG.

Factors Affecting the Absorption Properties of Standard, Benign, Malignant, (^{125}I-Anti hCG Antibody/hCG) Complex, standard and partially purified hCG.

Large number of environmental factors produces detectable change in λ_{max} and absorbency of protein molecules. Such factors are pH, polarity of the solvent[280].

pH Effect

The ionization state of ionizable chromophores in the protein molecular is determine by the pH of the solvents[280]. So, the λ_{max} values of UV spectrum for the three hCG complexes, standard and partially purified hCG were determined at three different pH (4, 7.2, 11), as shown in table (5-2). In acidic region, pH 4, one maximum wavelength was obtained at 226nm, 226nm, 225nm for standard hCG, benign and malignant breast tumor respectively. With increasing of pH value of the complexes medium from 4 to 7.2, the λ_{max} were slightly shift to a shorter wavelengths for the three complexes 222nm, 220nm and 220nm respectively, while, no λ_{max} was observed for the three complexes at pH 11. Standard and partially purified hCG show two maximum wavelength at 236nm, 279nm and 227nm, 271nm respectively, while at pH 7.2 only one maximum wavelength was obtained at 279nm and 271nm respectively. The disappearance of the λ_{max} of the polypeptide bound could be due to a conformation change in the protein[274]. When the pH value was increased from 7.2 to 11, two maximum wavelengths were obtained again at 240nm, 279nm and 220nm, 274nm for standard and partially purified hCG respectively. The spectrum shifts of protein produced by pH cannot be simply attributed to the inductive effect at vicinal charges, such spectral changes must therefore be attributed mainly to rearranged of secondary and tertiary structure, although the possibility of field effects due to unusually close conjunction of charges aromatic groups is not excluded[281].

Table (5-2): The effect of different pH on λ_{max} values of (^{125}I-anti hCG Antibody/HCG) complexes, standard hCG and partially purified hCG. (All other details are explained in the text).

Fractions	pH	λ_{max}
Standard (^{125}I-Anti hCG Antibody/ hCG) Complex	4	226nm
	7.2	222nm
	11	-
Benign (^{125}I-Anti hCG Antibody/ hCG) Complex	4	226nm
	7.2	220nm
	11	-
Malignant (^{125}I-Anti hCG Antibody/ hCG) Complex	4	225nm
	7.2	220
	11	-
Standard hCG	4	236nm 279nm
	7.2	279nm
	11	240nm 279nm
Partially purified hCG	4	227nm 271nm
	7.2	271nm

	11	220nm 274nm

Effect of Solvents Polarity

Significant solvent effects can be induced by use of a mixture of water and a substance of a reduced polarity such as ethanol, ethylene glycol, glycerol and sucrose[280]. Several spectra changes were obtained in the presence of these perturbants, like the alteration of λ_{max} positions and intensities of protein spectrum and the appearance of new chromophores on the surface of the protein[281,282]. When no effect on the maximum absorbency is observed, this indicated no interaction or any change that happened between the solvent and the molecules. When one band was observed, may be attributed to the amino acids buried in the internal region of the protein and surrounded by non-polar amino acids[280]. A blue shift in the wavelength in some solvents may be attributed to n $\rightarrow \pi^*$ transition and also may be attributed to hydrogen bonding of the solvent to the amino acid. The polar chromophores show a red shift if their hydrogen bonding to solvent molecules increases in the excited state, but a blue shift if their hydrogen bonding to solvent molecules decreases in the excited state[284,285]. Solvent may produce a blue shift by hydrogen bonding to the oxygen atom in the amino acids and withdrawing electrons from benzene chromophore[289]. The λ_{max} values under the effect of 20% of different solvents at pH 7.2 are shown in table (5-3). For (^{125}I-Anti hCG Antibody/ hCG) Complex of standard hCG, the λ_{max} for the amide groups of polypeptide bond showed longer wavelength than the λ_{max} original in the presence of 20% ethanol (225nm), ethylene glycol (232nm), glycerol (235nm) and sucrose (231nm). For benign and malignant breast tumor complexes, Ethanol has showed no effect on the λ_{max} values of two complexes, as comparing these values to the λ_{max} in the table (5-1). While 20% ethylene glycol, glycerol, and sucrose caused in red shift in the λ_{max} of both benign and malignant complexes. For standard hCG, it seems that 20 % ethanol have no effect on the position of the λ_{max}. While for partially purified hCG there was a slight red shift toward longer wavelength as comparing these values to the λ_{max} in the table (5-1). In 20% ethylene glycol, stander and partially isolated hCG, was a significant red shift in the λ_{max} at 291nm and 317nm respectively with appearance of a new λ_{max} at 229nm for the amide groups of the polypeptide bond of partially purified hCG. When 20% glycerol was used, a slight blue shift in the λ_{max} at 277nm with appearance of a new λ_{max} at 236nm observed for standard hCG. While for partially purified hCG, the original λ_{max} was disappeared with appearance of a new λ_{max} at 233nm. Even 20% sucrose has no effect on the original λ_{max} of standard hCG but it caused in the presence of a new λ_{max} at 236nm. The original λ_{max} of partially purified hCG showed a slight blue shift at 269nm and the appearance of a new λ_{max} at 219nm.

Table (5-3): The effect of 20% of ethanol, ethylene glycol, glycerol and sucrose on λ_{max} values of (^{125}I-anti hCG Antibody/hCG) complexes, standard and partially purified hCG. (All other details are explained in the text).

Fractions	λ_{max}

	20% Ethanol	20% Ethylene glycol	20% Glycerol	20% Sucrose
Standard (^{125}I-Anti hCG Antibody/ hCG) Complex	225nm	232nm	235nm	231nm
Benign (^{125}I-Anti hCG Antibody/ hCG) Complex	220nm	229nm	233nm	227nm
Malignant (^{125}I-Anti hCG Antibody/ hCG) Complex	220nm	227nm	233nm	228nm
Standard hCG	279nm	291nm	277nm 236nm	279nm 236nm
Partially purified hCG	247nm	317nm 229nm	233nm	269nm 219nm

Effect of Urea, KCl and (Urea, KCl) Mixture

In the presence of urea (8M), as shown in table (5-4), The λ_{max} of the polypeptide bond of hCG complexes showed significant red shift (15-23) for both standard and benign complex, while a slight red shift in the λ_{max} at 224nm for malignant complex was observed. These changes were concomitant with the appearance of a new λ_{max} at 279nm, 278nm and 278nm for the three complexes respectively.

A slight degrease in λ_{max} of standard hCG was obtained at 278nm, while a slight red shift than the original λ_{max} at 273nm and a new λ_{max} at 229nm was obtained for polypeptide bond of the partially purified hCG.

The blue shift could be due to that the protein is unfolded in the high concentration of urea, the chromophores buried in the interior are transformed into the solvent. This transfer produces a blue shift in the absorption of these chromophores, giving rise to a moderate decrease in the absorption at this wavelength. Also this transfer leads to a new λ_{max} to be appear.

Adding 0.03M KCl caused red shift for the λ_{max} of benign complex at 229nm with appearance of a new wavelength for the three complexes at 279nm, 278nm and 278nm respectively. When 0.03M KCl was added, no alteration in the position of the λ_{max} of standard hCG was detected. For partially purified hCG a slight red shift was observed for the λ_{max} at 272nm with a new λ_{max} at 230nm.

A slight blue shift was observed for the λ_{max} of both standard and malignant complexes at 219nm and 219nm respectively. While. Such a blue or a red shift can arise by introducing positive (K^+) or negative (Cl^-) charges near the chromophore (the amide group) which might interact directly with the π-electron system of the amide group.

When 8M urea was mixed with 0.03M KCl, there was no alteration in the λ_{max} position of the standard hCG, while there was a slight red shift (1nm) for the original wavelength at 272nm, like the same shift when 8M urea was used, with the appearance of a new wavelength at 233nm. For the three hCG complexes, a slight red shift was observed for standard and benign breast tumor complexes, while blue shift for malignant breast tumor complex was observed. It seems that the shift in the case of the mixture may be due to the effect of the 8M urea and 0.03M KCl. As seen, the changes in absorption near 230nm were larger than those near 278nm, which have been noted by Glazer who noted that solvent perturbation or denaturation the protein.

Table (5-4): The effect of 8M Urea, 0.03M KCl and mixture 8M Urea and 0.03M KCl on λ_{max} values of (^{125}I-anti hCG Antibody/HCG) complexes, standard and partially purified hCG. (All other details are explained in the text).

Fractions	λ_{max}		
	8M Urea	0.03 M KCl	Mixture 8M Urea and 0.03M KCl
Standard (^{125}I-Anti hCG Antibody/ hCG) Complex	237nm 279nm	219nm 279nm	224nm
Benign (^{125}I-Anti hCG Antibody/ hCG) Complex	243nm 278nm	229nm 278nm	229nm
Malignant (^{125}I-Anti hCG Antibody/ hCG) Complex	224nm 278nm	219nm 278nm	219nm
Standard hCG	278nm	279nm	279nm
Partially purified hCG	229nm 272nm	230nm 272nm	233nm 272nm

Spectrophotometric pH Titration of the Complex

Spectrophotometric pH Titration is the following of the change in absorbance of the chromophore with increasing pH. The study of protein structure require the determination of pka values for proton dissociation from ionizable amino acide chains, because these values give an indication of the location of the amino acid in the protein. This can often be done spectrophotometrically becouse dissociation often changes the spectrum of one of the chromophores, the observation of tyrosine dissociation was performed by measuring the absorption at 295nm (λ_{max} for the ionized form of tyrosine), and the observation of histidine dissociation was carried out by measuring the absorption at 211nm[283].

Figure (5-3) shows the pH titration curve of standard hCG and partially purified hCG for histidine and tyrosine respectively. Figure (5-3,A) shows that pka values for tyrosine are (3) for standard hCG and (2.8) for partially purified hCG. From the same curve it seems that about (40%) histidine residues of standard hCG are internal and a large arise in the absorbance at very high pH was observed. While, for the partially purified hCG about (35%) histidine residues are internal. The pka value from figure (5-3,B) appears to be about (40%) for tyrosyl residues are internal for standard hCG and (28%) tyrosine residue are internal for purified hCG, indicating the high content of histidine compared to the low content of tyrosine in the standard hCG and the partially purified hCG[288]

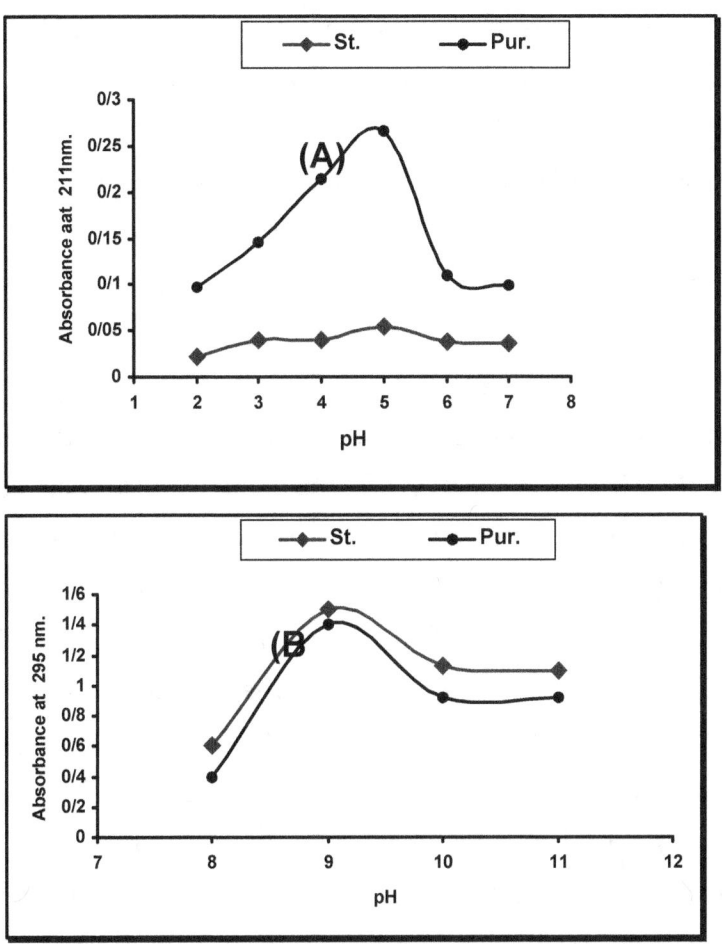

Figure (5-3): Spectrophotometric pH titration of standard hCG and partially purified hCG: (A) for histidine, (B) for tyrosine.

(St): Standard hCG, (Pur): partially purified hCG. (All other details are explained in the text).

Conclusions

1. Serum hCG level showed a slight different between the normal and breast tumors women patients. Therefore it appears that usefulness of β-hCG as a marker for the evaluation of breast tumors is evidently limited even with using highly specific and sensitive method (IRMA).

2. The development protocol for the assay (RIA) of hCG is capable to characterize these receptor and the procedure is suitable for the assessment of hCG receptor in benign and malignant breast tumor.

3. The binding of hCG receptor to ^{125}I-hCG is found to be highly influence by hCG receptor concentration pH, time and temperature with different values of these parameter for each group which indicate the presence of different forms of hCG receptors in each group.

4. The developed protocol for the assay (IRMA) for hCG determination is capable of characterize this hormone and the procedure is suitable for the assessment of hCG in benign and malignant breast tumor.

5. The binding of hCG is found to be highly influence by hCG concentration ,pH, time and temperature with different value of these parameters for each group which indicate that hCG has different properties in each group.

6. Partially purification of hCG from premenopausal breast tumor homogenate showed the presence of one form of this hormone and its binding reaction is time and temperature dependent. The reaction was found to be pseudo first order kinetics at (5,25,37,45).

7. Thermodynamic studies on the association of ^{125}I-anti hCG with partially purified hCG in malignant breast tumor

estimate that the binding reaction occurs spontaneously $\Delta G^o<0$ and entropically driven since $\Delta S^o>0$.

8. Spectroscopic studies indicate a detectable difference in λ_{max} value between standard hCG and hCG from malignant breast tumor and another detectable difference in the λ_{max} value between standard hCG complex and the complexes of hCG from benign and malignant breast tumors. This finding makes the UV spectra a good method for distinguish between normal and breast tumor patients.

Future work

According to the results obtained in this thesis, the flogging work are suggested for the future;

1. Investigation the presence of βhCG and hCG receptor in more cases and more type of breast tumor tissues.

2. Purification of hCG from different tumor types and analysis of its structure by using different method, such as UV spectra, mass spectrometry, amino acid sequence analysis and X ray.

3. Purification of hCG receptor from different breast tumor kinds, characterized its binding reaction condition and its spectroscopic analysis.

4. Cell culture from different women breast tumor type to study the effect of hCG on cell growth as a complement study for the effect of hCG on breast tumor progression.

5. Isolation of DNA from different breasts tumor kinds to study the presence of hCG receptor by using several techniques, such as Northen Blotting, Western blotting and ligand blotting.

References

1. Pierce LG, Parsons TF. *Ann Rev Biochem* 1981; 50: 465
2. Lapthorn AJ, Harris DC, Littlejohn A, Lustbader JW, Canfield RE, Machin KJ,
   Morgann FJ, Isancs NW. *Nature* 1994; 369: 455.
3. Birken S, Maydelman Y, Gawinowicz MA. *Methods* 2000; 21: 3.
4. http://journals endocrinology.org/jme/022/0185/0220185.pdf.
5. http://web.indstate.edu/thcme/mwking/hormone-table.html
6. Greenspan FS, Garduer DG. *"Basic and clinical Endocrinology"*, 6th ed. 2001.
   Mc Graw Hill Compaies. USA. Chapter 16, pp577.
7. Donaldson ES, VanNagell JR, Pursell S, Gay EC, Meeker WR, Kashmiri R,
   Voorde J. *Cancer* 1980; 45: 948.
8. Thotakura NR, Weintraub BD, Bahl OP. *Mol Cell Endocrinol* 1990; 70: 263.
9. Shao K, Balasubramanian SV, Pope CM, Bahl OP. *Mol Cell Endocrinol* 1998;
   146:39.
10. O'connerJF, Birken S, Lustbader JW. *Endocr Rev* 1994; 15:650.

11. Lapthorn AJ, Janes RW, Isaacs NW, Wallace BA. *Nature Structureal Biology* 1995;2: 266.

12. Iles RK, chard T. *J Mol Endocrinol* 1993; 10: 217.

13. Berger P, Bidart JM, Delves PS, Dirnhofers. *Endocrinology* 1996; 125: 33.

14. Shao K, Purohil S, bahl OP. *Mol Cell Endocrinol* 1996; 122: 173.

15. Wu H, Lustbader JW, Liu Y, Canfield RE, Hendrickson WA. *Structure* 1994; 2: 545.

16. Lobel AJ, Harris DC, littejohn A, Lustbader JW. *Endocrinology* 1999; 10: 261.

17. http://www.hcglab.com/index.html.

18. Beriken S, Kovalevskaya G, O'Connor J. *Mol Cell Endocrinol* 1996; 125: 121.

19. Cole LA. *Clin Chem* 1997; 43: 2233.

20. Birken S, Cawinowicz MA, Kardana A, Cole LA. *Endocrinology* 1991; 129: 1551.

21. O'Connor JF, Birken S, Lustbader JW, Krichevsky A, Chen Y, canfield RE. *Endocr Rev* 1994; 15: 650.

22. Alfthan H, Stenman UH. *Mol cell Endocrinol* 1996; 125:107.

23. Wide L, Lee JY, Rasmussen C. *J Clin Endocrinol Metab* 1994; 78: 1419.

24. Hoermann R, Berger P, Spoettl G, Gillesberger F, Kardana A, Cole LA,

Mann K. *Clin Chem* 1994; 40: 2306.

25. Kardana A, Cole LA. *J Clin Endocrinol Metab* 1994; 79: 761.

26. Rotmensch S, Liberati M, Kardana A, Copel JA, Ben-Rafael Z, Cle LA. *Am J*

   Ostet Gynecol 1996; 174: 609.

27. Kardana A, Elliott MM, Gawinowicz MA, Birken S, Cole LA. *Endocrinology*

   1991; 129: 154.

28. KovalevsKaya G, BirKen S, Kakuma T, Schlatterer J, O'Connor JF. *Clin Chem*

   1999; 45: 68.

29. Kohorn EI, Cole L. *Int J Gynecol Cancer* 2000; 10: 330.

30. Alfthan H, Haglund C, Dabek J, Stenman UH. *Clin Chem* 1992;38: 1981.

31. Birken S, Gawinowiz MA, Maydelman Y, Milgrom T. *J Endocrinol* 2001;

   171: 131.

32. Cole LA, Briken S. *Mol Endocrinol* 1988; 2: 825.

33. Udagawa A, Okamoto T, Nomura S, Matsuo K, Suzuki H, Mizutani S. *Mol*

   Cell Endocrinol 1998; 139:171.

34. Schroeder HR, Halter CM. *Clinical Chemistry* 1983; 29: 667.

35. Birken S, Armstrong EG, Kolks MA, Cole LA, Agosto GM, Krichevsky A,

   Vaitukaitis JL, Canfieled RE. *Endocrinology* 1988; 123: 572.

36. De Medciros SF, Amato F, Matthews CD, Norman RJ. *Obstet Gynecol* 1992;

   80: 223.

37. Yoshimura M, Nishimura R, Murotani A, Miyamoto Y, Nakagawa T, Hasegawa K, Koizumi T, Shii K, Baba S, Tsubota N. *Cancer* 1994; 73: 2745.

38. Carter PG, Iles RK, Neven P, Ind TE, Shepherd JH, Chard T. *Gynecol Oncol* 1994; 55: 271.

39. Carter PG, Iles RK, Neven P, Ind TE, Shepherd JH, Chard T. *Br J Cancer* 1995; 71: 350.

40. NishimuraR, Koizumi T, Morisue K, Yamanaka N, Lalwani R, Yoshimmra M, Nakagawa T, Shii K, Hasegawa K, Baba S. *Cancer Res* 1995; 55:1479.

41. Odell WD, Griffin J, Sawitzke A. *Trends Endocrinol Metab* 1990; 1: 418.

42. Methal AA. "*Biochemical Characterization of Human Gynecologic Cancer*" Ph.D. 2002. Thesis supervised by Al-Mudhaffar SA and Al-Ani At, College of Scoence, Baghdad University.

43. Elliott M, Kardana A, Lustbader J, Cole L. *Endocrine* 1997; 7: 15.

44. Reimer T, Koczan D, Muller H, Friese K, Krause A, Thiesen HJ, Gerber B. *J Mol Endocrinol* 2000; 24: 33.

45. Hu XC, Chow LW. *Anticancer Res* 2001; 21: 421.

46. Bilchik AJ, Nora DT, Saha S, Turner R, Wiese D, Kuo C, Ye X, Morton DL, Hoon DS. *Arch Surg* 2002; 137: 1377.

47. Posillico EG, Handwerger S, Tyrey L. *Biol Reprod* 1985; 32: 1101.

48. Charies worth MC, McCormicK DJ, Bergert ER, Vutyavanich T, Hojo H,
Ryan RJ. *Endocrinology* 1990; 127: 2977.

49. Chen F, Wang Y, Puett D. *Mol Endocrinol* 1992; 6:914.

50. Dimhofer S, Lechner O, Madersbacher S, Klieber R, deLeeuw R, Wich G,
Berger P. *J Endocrinol* 1994; 140: 145.

51. Shindelman JE, Ortmeyer AE, Wada HG, Stockdale F, Sussman HH. *Int J*
Cancer 1980; 25: 599.

52. SeppalaM, Wahlstrom T. *Int J Cancer* 1980; 26: 267.

53. Grossmann M, Trautmann ME, Poertl S, Hoermann R, Berger P Arrold R,
Mann K. *Eur J Clin Invest* 1994; 24: 131.

54. Akar AH, Gervasi G, Blacker C, Wehmann RE, Blithe DL, Nisula BC. *J*
Endocrinol 1990;125: 477.

55. Rodgers M, Mitchell R, Lambert A, Peers N, Robertson WR. *Clin*
Endocrinol 1992; 37: 558.

56. Daiter E, Braunstein GD, Snvder PJ, Coutifaris C, Mastroianni L, Pavlou
SN, Strauss JF. *J Clin Endocrinol Metab* 1994; 78: 1293.

57. O'Connor J, Birken S, Lustbader JW, Krichevsky A, Chen Y, Canfield RE.
Endocr Rev 1994; 15: 650.

58. Birken S, Maydelman Y, Gawinowicz MA, Pound A, Liu Y, Hartree SA.
Endocrinology 1996; 137: 1402.

59. Birken S, Chem Y, Gawinowicz MA, Agosto GM, Canfield RE, Hartree AS.
Endocrinology 1993; 133: 985.

60. Policastro PF, Danicls-McQueen S, Carle G, Boime I. *J Biol Chem* 1986;
261: 5907.

61. Bo M, Boime I. *J Biol Chem* 1992; 267: 3179.

62. Talmadge K, Vamvako poulso NC, Fidden JC. *Nature* 1984; 307: 37.

63. Albanese C, Cdin IM, Crowley WF. *Recent Prog Horm Res* (Abstra) 1996;
51: 23.

64. Ringler GE, Strauss JF. *Endocr Rev* 1990; 11: 105.

65. Sawai K, Matsuzaki N, Kameda T. *J Clin Endocrinol Metab* 1995; 80: 1449.

66. Merz WE. *Eur J Endocrinol* 1996; 135: 269.

67. Nakanishi T, Kohroki J, Suzuki S, Ishizaki J, Hiromori Y, Takasuge S, Itoh N,
Watanabe Y, Utoguchi N, Tanaka K. *J Clin Endocrinol Metab* 2002; 87: 2830.

68. Licht P, Cao H, Lei ZM, Rao ChV, Marz WE. *Endocrinology* 1993; 133: 3014.

69. Rao ChV, *Early Pregnancy Biol Med* 1997;3:1.

70. Kisielewska J, Flint APF, Ziecik AJ. *J Endocrinol* 1996; 148: 175.

71. Zygmunt M, Herr F, Keller-Schoenwetter S, Kunzi-Rapp K, Munstedt K, Rao ChV, Lang U, Preisser KT. *J Clin Endocrinol Metab* 2002; 87:5290.

72. Mann RJ, Keri RA, Nilson JH. *Recent Prog Horm Res* 2003; 58: 343.

73. Lei ZM, Toth P, Rao ChV, Pridham D. *J Clin Endocrinol Metab* 1993; 77: 863.

74. Lei ZM, Rao ChV. Endocrinology of trophoblast tissue. In Beckr K, Rabar R. *"Principle and Practice of Endocrinology"* 2002, 3ed ed. Philadelphia:Lippincott Williams and Wilkins.

75. Nagy AM, Glinoer D, Picelli G,Delogne-Desnoeck J, Fleury B, Courte C, Kanfman JM, Robyn C, Menris S. *J Endocrinol* 1994; 140: 513.

76. Johnson MR, Abbas A, Nicolaides KH. *J Endocrinol* 1994; 143: 309.

77. Toth P, Lukace H, Hiatt ES, Reiol KH, Iyer V,Rao ChV. *Brain Res* 1994;654:181.

78. Rao ChV. *J Bellevue Obstet Gynecol Soc* 1999; 15: 26.

79. Toth P, Li X, Rao ChV. *J Clin Endocrinol Metab* 1994; 79:307.

80. Rao ChV. *Seminars in Reproductive Medicine* 2001; 19: 7.

81. Zygmunt M, Hahn D, Munstedt K, Bischof P, Lang U. *Placenta* 1998; 19:587.

82. Eta E, Ambrus G, Rao ChV. *J Clin Endocrinol Metab* 1994; 79: 1582.

83. Toth P, Li X, Lei ZM, Rao ChV. *J Clin Endocrinol Metab* 1996; 81: 1283.

84. McGregor WG, Kuhn RW, Jaffe RB. *Science* 1983; 220: 306.

85. Rao ChV, Lie X, toth P. *J Clin Endocrinol Metab* 1993; 77: 1706.

86. Al- Hadder AA, Lei ZM, Rao ChV. *Biol Reprod* 1997; 56: 1071.

87. Naughton MA, Merrill DA, McManus LM, Fink LM, Berman E, White MJ,

Martinez-Hernandez A. *Cancer Res* 1975; 35:1887

88. Yorde DE, Hussa OR, Garancis JC, Pattillo RP. *Lab Invest* 1979; 40: 391..

89. Shi QJ, Lei ZM, Rao ChV, Lin J. *Endocrinol* 1993; 132: 1387.

90. Zhang W, Lei ZM, Rao ChV,. *Life Sci* 1999; 65:2083.

91. Lie ZM, Rao ChV, *Seminars in Reproductive Medicine* 2001; 19: 103.

92. Brannstein GD, Kamdar V, Roso J, Swaminathan N,Wade ME. *J Clin Endocrinol Metab* 1979; 49: 917.

93. Csaba G, Kovacs P. *Acta Protozool* 2000; 39: 191.

94. Sussman HH, Weintraub BD, Rosen SW. *Cancer* 1974; 33:820.

95. Tseng A, Horning SJ, Freiha FS, Resser KJ, Hannigan JF, Torti MF. *Cancer* 1985; 56: 2534.

96. Ugrinska A, Bombardieri E, Stokket MP, Crippa F, Pauwels EK. *QJ Nucl Med (Abstract)* 2002, 46: 88.

97. Murata T, Ihara S, Nakayama T, Nakagwa S, Higashiguchi T, Imai T, Nakano H. *Pathology International* 1999; 49: 816.

98. Konishi I, Koshiyama M, Mandai M. *Gynecol Oncol* 1997; 65: 273.

99. Tao YX, Bao S, Ackermann DM. *Biol Reprod* 1997; 56: 67.

100. Lojun S, Bao S, Lei ZM, Rao ChV. *Biol Reprod* 1997; 57: 1202.

101. Butler SA, Laidler P, Porter JR. *J Mol Endocrinol* 1999; 22: 185.

102. Vaitukaitis J, Braunstein G, Ross G. *Am J Obstet Gynecol* 1972; 133: 751.

103. Cole LA, Shahabi, Rinne KM. *Prenat Diagn* 1999; 15: 24.

104. James T, Kurtzman, Wilson H, Rao ChV. *Seminars in Reproductive Medicine*
    2001; 19: 63.

105. Birken S, Krichevsky A, O`Connor J, Schlatter J, Cole L, Kardana A, Canfield
    R. *Endocrine* 1999; 10: 137.

106. Birkenfeld S, Noiman G, Krispin M, Schwarz S, Zakut H. *Eur J surg Oncol*
    1989; 15: 103.

107. Hoerman R, Gerbes AL, Spoettl G, Jungest D, Mann K. *Cancer Res* 1992; 52:
    1520.

108. Grossmann M, Hoermann R, Gocze PM, OH M, Berger P, Mann K. *Eur J Clin*
    Inv 1995; 25: 867.

109. Gailani S, CHU TM, Nussbaun A, Ostrander M, Christoff N. *Cancer* 1976; 38:
    1684.

110. Bidart JM, Bellet D. *Endocrinol Metab* 1993; 4: 285.

111. Acevedo HF, Hartsock RJ. *Cancer* 1996; 68: 2388.

112. Lamerz R, Stoetzer OJ, Mezger J, Brandt A, Darso WM, Wilmanns W.
    Anticancer Res 1999; 19: 2421.

113. Alfthan H, Haglund C, Roberts P, Stenman UH. Cancer Res 1992; 52: 4628.

114. Schulter EM, Mulder C, Van-Kamp GJ, Kereman SP. *Anticancer Res* 1997;

17:1255.

115. Acevedc HF, Krichevsky A, Campbell-Acevdo EA, Galyon JC, Buffo MJ,
   Hartsock RJ. *Cancer* 1992; 69: 1818.

116. Bhalang K, Kafrawy AH, Miles DA. *Cancer* 1999; 85: 757.

117. McManus L, Naughton M, Martinez HA. *Cancer Res* 1976; 36: 3476.

118. Sheth NA, Adil MA, Nadkarni JJ, Rajpal RM, Sheth AR. *Gynecol Oncol* 1980;
   11: 321.

119. Schwarz RU, Petzoldt B, Waldschmidt. *Anticancer Res* 1997; 17: 1255.

120. Ind T, lles RK, Shepherd J, Chard T. *Br Obstet Gynaecol* 1998; 104: 1024.

121. Mohabeer J, Bukly CH, Fox H. *Gynecol Oncol* 1983; 16: 78.

122. Ruddon RW, B`yran AH, Hanson CA, Perini F. *J Biol Chem* 1981; 256: 5189.

123. Collins RJ, Wong LC. *Gynecol Oncol* 1989; 33; 99-107.

124. De-Bruijn HW, Ten-Horr KA, Krans M, Vander-Zee AC. *Br J Cancer* 1997;
   75: 1217.

125. Kumar S, Talwar GP, Biswas DK. *J Natl Cancer Inst* 1992; 84: 42.

126. Triozzi PL, Stevens VC, Aldrich W, Powell J, Toded CW, Newman MJ. *Oncol*
   Rep 1999; 6: 7.

127. Moulton HM, Yoshihara PH, Mason DH, Iveren PL, Triozzi PL. *Clin Cancer*
   Res 2002; 8: 2044.

128. http://aventis-pasteur.co.th/whatnew/index.phtml?page=15.

129. http://members.aol.com/Cabacteria/trial.html.

130. Marcillic I, Troalen F, Bidart JM, Ghillami P, Ribrag V, Escudier B,
   Malassagne B. *Cancer Res* 1992; 52: 3901.

131. Syrigos KN, Fyssas I, Konstandoulakeis MM, Harvington KJ, Popadpouls. *Gut*
   2001; 42: 88.

132. IIes RK, Jenkins BJ, Blandy JP, Chard T. *Br J Uurol* 1989; 64: 241.

133. Regelson W. *Cancer* 1995; 76:1299.

134. Acevedo HF, Hartsockk RJ. *Cancer* 1996; 78: 2388.

135. Gebauer G, Muller-Ruchholtz W. *Anticancer Res* 1997; 17: 2731.

136. IIes RK, Chard T. *J Mol Endocrinol* 1989; 2: 107.

137. Acevodo HF, Tong JY, Hartsoch RJ. *Cancer* 1995; 76; 1467.

138. Acevodo HF, Krichevsky A, Campbell-Acevedo EA. *Cancer* 1992, 69:
   1829.

139. Acevodo HF, Krichevsky A, Campbell-Acevedo EA. *Cancer* 1992, 69:
   1818.

140. Gillott DJ, Iies RK, Chard T. *Br J Cancer* 1996; 73: 323.

141. Bridges PJ, Wright DJ, Buford WI, Ahmad N, Hernandez-Fonseca H,
   McCormick ML, Schrick FN, Dailey RA, Lewis PE, Inskeep EK. *J Anim Sci*
   2000; 78: 2942.

142. Russo IH, Russo J. *Eur J Cancer Prev* 1993; 3: 101.

143. Russo J, Russo IH. *Cancer Epidemiol Biomarkers Prev* 1994; 3: 353.

144. Russo J, Russo IH. *Cancer Lett* 1995; 90: 81.

145. Russo IH, Russo J. *J Natl Inst Environm Heal Sci* 1996; 104: 938.

146. Russo J, Hu YF, Tang X, Russo IH. *J Natl Cancer Ins Monogr* 2000; 27: 17.

147. Russo J, Hu YF, Sliva IDCG, Russo IH. *Microse Res Tech* 2001; 52: 204.

148. http://wwwioncolink.com/templates/types/section.cfm?c=3&s=s

149. Russo J, Lynch H, Russo IH. *Breast J* 2001; 7: 278.

150. Silva IDCG, Hu YF, Russo IH, Ao X, Salicioni AM, Yang X, Russo J. *Int J*

  Oncol 2000; 16: 231.

151. Russo IH, Russo J. *Radiation Res* 2001; 155:151.

152. Casicato DA, Lowitz BB. *"Manual of Clinical Oncology"* 2000, 4th ed.

  Chapter10, pp 218-237.Philadelphia. Awolters Klnwer Company.

153. Rao ChV. *Obstet Gynecol* 2000; 96: 783.

154. Russo IH, Koszalka M, Russo J. *Br J Cancer* 1991; 64: 481.

155. Russo IH, Russo J. *J Cell Biochem Suppl* 2000; 34: 1.

156. Gebauer G, Fehm T, Beck EP, Berkolz A, Licht P, Jager W. *Breast Res Treat*

  2003; 77: 125.

157. Russo IH, Koszalka M, Russo J. *Carcinogenesis* 1990; 11: 1849.

158. Russo IH, Koszalka M, Gimotty PA, Russo J. *Br J Cancer* 1990; 62: 243.

159. Russo IH, Russo J. *J Mammary Gland Biol Neoplasia* 1998; 3; 49.

160. Russo J, Russo IH. *Endocr Related Cancer* 1997; 4; 7.

161. Mgbonyebi OP, Tahin Q, Russo J, Russo IH. *Proc Am Assoc Cancer Res* 1996;
37: 1564.

162. Tahin Q, Mgbonyebi OP, Russo J, Russo IH. *Proc Am Assoc Cancer Res* 1996;
37; 1622.

163. Russo IH, Ao X, Tahin S, Russo J. *Proc Am Assoc Cancer Res* 1996; 37: 1066.

164. Lie ZM, Rao ChV. Protective Role of Human Chorionic Gonadotrophin and
Lutinizing Hormone Aginst Breast Cancer.. In: Barnea ER, Springer. 2001:
209.

165. Lojun S, Bao S, Lei ZM, Rao ChV. *Biol Reprod* 1997; 57: 1202.

166. Jiang X, Russo IH, Russo J. *Int J Oncol* 2002; 20: 77.

167. Srivastava P, Russo J, Russo IH. *Carcinogensis* 1997; 18: 1799.

168. Srivastava P, Russo J, Russo IH. *Anticancer Res* 1998; 18: 4003.

169. Huynh H. *Int J Oncol* 1998; 13: 571.

170. Klinge CM, Rao ChV. The Cell Membrane Receptors, In: Sciarrg J.
"Gynecology and Obstetrics", 2000. Harper and Row, Philadelphia, Volume 5,
chapter 5.

171. Loosfelt H, Misrahi M, Atger M, Salesse R, Thi MTVH-L, Jolivet A,

Guiochon-Mantel A, Sar S, Jallal B, Garnier J, Milgrom E. *Science* 1989; 245: 525.

172. Rao ChV, Lei ZM. *ARTA* 1990; 1: 241.
173. Cheng KW, Nathwani PS, Leung PCK. *Endocrinol* 2000; 141: 2340.
174. Ziecik AJ, Stanch PD, Tilton JE. *Endocrinol* 1986; 119: 1159.
175. Hu YL, Lei ZM, Roa ChV. *Endocrinol* 1996; 137:3897.
176. Hu YL, Lei ZM, Roa ChV. *Life Sci* 1998; 63: 2157.
177. Hu YL, Lei ZM, Huang ZH, Rao ChV. *Breast J* 1999; 5: 186.
178. Rao ChV. *J Physiol Pharmacol* 1996; 47: 41.
179. Rao ChV. *J Bell Obstet Gynecol Soci* 1999; 15: 26.
180. Konishi I, Koshiyama M, Mandai M. *Gynecol Oncol* 1997; 65: 273.
181. Sun T, Lei ZM, Rao ChV. *Mol Cell Endocrinol* 1997; 131:97.
182. Licht P, Russu V, Wildt L. *Semin Reprod Med* 2001; 19: 37.
183. Mizrachi D, Shemesh M. *Mol Cell Endocrinol* 1999; 157: 191.
184. Lin J, Lojun S, Lie ZM, Wu WX, Peirer SC, Rao ChV. *Mol Cell Endocrinol* 1995; 111: 13.
185. Taoy X, Heit M, Lei ZM, Rao ChV. *Am J Obstet Gynecol* 1998; 179: 1026.
186. Bird J, Li X, Lie ZM, SanFilippo J, Yussman MA, Rao ChV. *J Clin Endocrinol Metab* 1998; 83: 1776.

187. Pabon JE, Li X, Lei ZM, Sanfilippo J, Yussman MA, Rao ChV. *J Clin Endocrinol Metab* 1996; 81: 2397.

188. Thompson DA, Othman MI, Lei ZM. *Life Sci* 1998; 63: 1057.

189. Hill JB, Alsip NL, Rao ChV, Asher EF. *Am J Obstet Gynecol* 1997; 176; 150.

190. Eblen A, Bao S, Lei ZM, Nakajuma ST, Rao ChV. *J Clin Endocrinol Metab* 2001; 86: 2643.

191. Tao YX, Lei ZM Rao ChV. *Lfi Sci* 1997; 60: 1297.

192. Meduri G, Charnaux N, Loosfelt H, Jolivet A, Spyratos F, Milgrom E. *Cancer Res* 1997; 56; 857.

193. Jiang X, Russo IH, Russo J. *Int J Oncol* 2002; 20:735.

194. Vaitukaiti JL, Brasunstein Gd, Rossd GT. *Ann J Obstet Gynecol* 1972; 113;751.

195. Sheth NA, Suraiya NJ, Sheth AR, Ranadive KJ, Jussawalla DJ. *Cancer* 1977; 39: 1693.

196. Castro A, Busshbaum P, Nadji M, Voigt W, Tabei S, Morales A. *Acta Endocrinol* 1980; 94: 511.

197. Jorge C, Monteiro MP, Ferguson KM, McKinna A. *Cancer* 1984;53: 957.

198. Loganath A, Peh Kl, Gunasegram R, Thiagaraj D, Cheah E, Ratnam SS. *Pathology* 1988; 22: 275.

199. Yorde DE, Hussa RO, Garancis JC, Pattillo RA. *Lab Invest* 1979; 40: 391.

200. Wachner R, Wittekind C, Vonkleist S. *Eur J Cancer Clin Oncol* 1984; 20: 679.

201. Wahren B, Lidbrink E, WallGren A, Eneroth P, Zajicek J. *Cancer* 1978; 42: 1870.

202. Hoon DS, Sarantou T, Doi F, Chi DD, Kuo C, Conrad Ajj, Schmid P, Turner R, Guiliano A. *Int J Cancer* 1996; 69: 369.

203. Taback B, Chan AD, Kuo CT, Bostick PJ, Wang HJ, Giuliano AE, Hoon DS. *Cancer Res* 2001; 61: 8845.

204. Giovangrandi Y, Parfait B, Asheuer M, Olivi M, Lidereau R, Vidaud M, Bioche I. *Cancer Lett* 2001; 168: 93.

205. Span PN, thomas CM, Heuvel JJ, Bosch RR, Schalken JA, Locht L, Mensink EJ, Sweep CG. *J Endocrinol* 2002; 172: 489.

206. Pond-Tor S, Rhodes RG, Dahlberg PE, Leith JT, McMichael J, Dahlberg AE. *Breast Cancer Res Treat* 2002; 72: 45.

207. Brzezinski A, Schenker JG. *Gynecol Endocrinol* 1997; 11: 357.

208. Thompson SA, Johnson MP, Brook SC. *The Prpstate* 1982; 3: 45.

209. Chamberlain J, Jargarinec N, Ofner P. *Biochem J* 1966; 99: 610.

210. Janson JC, Ryden L*"Protein Purification"* 1998. John Wiley and Sons. Inc. Part I, pp3-13.

211. Lawry OH, Rosebrough NJ, Farr AL, Randell RJ. *F Biol Chem* 1951; 193:265.

212. McArthur JW. *Prog Gynecol* 1963; 4: 146.

213. Sheth NA, Suraiya JN, Ranadive KJ, Sheth AR. *Br J Cancer* 1974;30: 566.

214. Tormey DC, Waalkes PT, Simon RM. *Cancer 1977*; 39: 2391.

215. Bradlow HL, Schwartz MK, Fleisher M, Nisselbaum JS, Boyan R, O`Connor J, Fukushima DK. *J Ciln Endocrinol Metab* 1979; 49: 778.

216. Martinez L, Castilla JA, Blanco N, Peran F, Herruzo A. *Int J Gynecol Obstet* 1995; 48: 187.

217. Butzow R, Huhtaniemi I, Clayton R, Wahlstrom T, Andersson LC, Seppala M. *Int J Cancer* 1987; 39: 498.

218. Tormey DC, Waalkes TP, Ahmann D, Gehrke CW, Zumwatt RW, Snyder J, Hansen H. *Cancer* 1975; 35:1095.

219. Walker RA. *J Clin Path* 1978;31:245.

220. Hu XC, Loo WT, Clow LW. *J Surg Oncol* 2003; 82:228.

221. Troccoli R, Battistelli S, Marcheggiani F, Sessa M, Fronduti A. *Int J Biol Markers* 1990;5:133.

222. Agnantis NJ, Patra, Khaldi L, Filis S. *Eur J Gynecol Oncol* 1992;13:461.

223. Mori KF. *Endocrinol* 1970;86:97.

224. Haro LS, Talaments FG. *Mol Cell Endocrinol* 1985; 43:199.

225. Devlin TM. "*Text Book of Biochemistry with Clinical Correlation*" 1986. 2nd.ed; John Wiley and Sons:p:66,125.

226. Lanja EO "*Molecular Characterization of Luteinnizing Hormone in Some of*

Ovarian Tumor" 2002. M.Sc. thesis supervised by Al-Mudhaffar SA, College

of Science, Baghdad Univ.

227. Burtis CA, Ashwood ER. " *Tietz Text Book of Clinical Chemistry*" 1994. 2nd

ed.; W B Saunders Company; P: 148.

228. Segaloff D, Ascoli M. *Endocrine Rev* 1993; 14: 324.

229. Uilenroek JTJ, Linden RV. *Acta Endocrinol* 1983; 103: 413.

230. Toth P, Lukacs H, Gimes G, Seestyen A, Pasztor N, Paulin F, Rao ChV.

Reproduct Biol 2001; 1: 5.

231. Purification, Properties HB, Composition S. *J Biol Chem* 1986; 261: 9450.

232. Minegishi T, Kusuda S, Dufau ML. *J Biol Chem* 1987; 262: 17138.

233. Ziecik AJ, Jedlinska M, Rzucidlo SJ. *Acta Endocrinol* 1992; 127: 185.

234. Reshef E, Lei ZM, Rao ChV, Pridham DD, Chegini N, Luorsky SL. *J Clin*

Endocrin Meta 1990; 70: 421.

235. Fralish G, Dattilo B, Puett D. *Mol Endocrinol* 2003; 17: 1192.

236. Gromoll J, Wistue J, Terwort N, Godmann Muller T, Simoni M. *Biol Reprod*

2003; 69:75.

237. Galet C, Min L, Narayanan R, Kishi M, Weigel NL, Ascoli M. *Mol Endocrinol*

2003; 17:411.

238. Dufau LM, Charreau EH, Catt KJ. *J Biol Chem* 1973; 248:6973.

239. Pandian MP, Bahl OP. *Arch Bioch Biph* 1973; 182:420.

240. Bruch RC, thotakura NR, Bahl OP. *J Biol Chem* 1986;261: 5.

241. Roitt I, Brostoff J, Male D. *"Immunology"* 1998. 5th ed., London, Mosby

Philadelphia st. Louis.

242. Dad liker WB, Satussure VA. *Immunochemistry* 1970; 7: 799.

243. Bellisario R, Bahl OP. *J Biol Chem* 1975; 25:3837.

244. Stepien A, Derecka K, Gawronska B, Bodek G, Zwierzchowski L, Shemesh M,

Ziecik AJ. *J Physiol Pharmacol* 2000; 51: 917.

245. Dufau ML, Ryan DW, Baukal AJ, Catt KJ. *J Biol Chem* 1975; 250: 4822.

246. Kusuda S, Dufau ML. *J Biol Chem* 1986; 261: 16161.

247. Selvi PT, Ashish B, Murthy GS. *Cur Sci* 2002; 82: 1442.

248. Fiddes JC, Coodman HM. *Nature* 1979; 281: 351.

249. Boorstein WR, Vamarkopoulos NC, Fiddes JC. *Nature* 1982; 300: 419.

250. Nagy AM, Jauniaux E, Jurkovic D, Meuris S. *J Endocrinol* 1994; 142: 511.

251. Ashitaka Y, Tokura Y, Tane M, Mochizuki M, Tojo S. *Endocrinol* 1970;

86:233.

252. Donini S, D`Alessio I, Donini P. *Acta Endocrinol* 1975; 69: 749.

253. Morgan FJ, Canfield RE. *Endocrinol* 1971; 88: 1045.

254. Birken S, Chen Y, Gawinowiz MA, Lustbader JW, Pollak S, Agosto G, Buck

R, O`Connor J. *Endocrinol* 1993; 133: 1390.

255. Kaplan GN, Maffezzoli RD, Chrambach A. *J Clin Endocr* 1972; 43: 370.

256. Birken S, Kovalevskaya G, O`Connor J. *Mol Cell Endocrinol* 1996; 125: 121.

257. Moss J, Ross PS, Agosto G, Birken S, Canfield RE, Vaughan M. *Endocrinol* 1978; 102: 415.

258. Scopes RK. "*Protien Purification principles and Practice*" 1982. New York, Springer Verlag; pp162,197.

259. Scatchard G. Ann NY *Acad Sci* 1949; 51: 660.

260. Seely DH, Wang WY, Salhanick HA. *Biochem Biophy Acta* 1980; 632:535.

261. Segel IH. "*Biochemical Calculation*" 1979. 3rd ed., John Willey and Sons,Inc. pp311.

262. Nemeth G, Scheraga HA. *J Phys Chem* 1962; 66: 1773.

263. Waelbroeck M, Van-Obberghen E, De-Meytes P. *J Biol Chem* 1979; 254: 7736.

264. Williams CA, Chase MW. "*Methods in Immunology and Immuno Chemistry*" 1971. 5th ed., New York: Academic Press, Vol. III, Chapter 13.

265. Laport DC, Wierman EM, Storm DI. *Biochem* 1980; 19: 3814.

266. Blumenthal DK, Stull JT. *Biochem* 1982; 21: 2386.

267. Scheraga HA. "*Protein Structure*" 1961. New York, Academic Press, chap 6: pp175-287.

268. Mathews, Ch K, Holde KE. *"Bio Chemistry"* 1990. California, the
   Benjamin/Cummings Publishing Co.

269. Leach SJ. *"Physical Principles and Techniques of protein Chemistry"* 1969.
   New York, Academic Press, Part A, Chap.3, pp. 102-170.

270. http://www.biophysics.org/btol/img/C.Royer.pdf

271. Pri HS. *"Biochamical Study on Prolactin and some Tumor Marker in*
   Breast Tumor" 2000. Ph.D. thesis supervised by Al-Mudhaffar SA, College of
   Science, Baghdad University.

272. Maysoon KH. " *Molecular Characterization of Testosterone Receptor in*
   Mammary Tissues Effected by Tumors" 2001. M.Sc. thesis supervised by Al-
   Mudhaffar SA, College of Science, Baghdad University.

273. Wasan AA. *"Biochamical Study on Alfa Feto Protein and some Tumor*
   Marker in Gastric Cancer" 2000. Ph.D. thesis supervised by Al-Mudhaffar
   SA, College of Science, Al-Monstansiriyah University.

274. Salwa HN. *"Biochemical Characterization of CA 15-3 in Sera and Tissues of*
   Breast Tumors" Ph.D. thesis supervised by Al-Mudhaffar SA, College of
   Science, Al-Monstansiriyah University.

275. Retanabanangkoon K, Keutmann HT, Kitzmann K, Ryan RJ. *J Biol Chem*
   1983; 258: 14527.

276. Bhowmick N, Huang J, Puett D, Isaacs NW, Laphthorn AJ. *Mol Endocrine* 1996;10: 1147.

277. Roche PC, Ryan RJ, MCCormick DL. *Endocrinol* 1992; 131: 268.

278. Hong S, Ji I, Ji T. *Mol Endocrinol* 1999; 13: 1285.

279. Delvin TM. "*Text Book of Biochemistry with Clinical Correlation*" 1986. 2nd.ed; John Wiley and Sons:p:66.

280. Freifrlder D. "*Physical Biochemistry, Application to Biochemistry Molecular Biology*" 1982. 2nd ed.; San Francisco: Freeman WH and Company. Chapter 14, pp494-591.

281. San Y, Bovey DA. J AM *Chem Soc* 1960; 235: 2818.

282. Brealy GJ, Kkaska M. J *Am Chem Soc* 1950; 77: 4462.

283. Leach SJ, Scharaga HA. *J Biol Chem* 1960; 235: 2827.

284. Leach SJ, "*Physical Principles and Techniques of Protein Chemistry*" 1969. New York, Acadimic Press, Part A, Chapter 3, pp 101-170.

285. Bayliss NS, McRae EG. *J Phys Chem* 1954; 58:1002.

286. Laskowski MM, Leach SD, Scheraga HA. *J Am Chem Soc* 1960; 5:71.

287. Herskowits TT, Laskowski MJ. *Biol Chem* 1962; 2481.

www.ingramcontent.com/pod-product-compliance
Lightning Source LLC
Chambersburg PA
CBHW062346220526
45472CB00008B/1719